Bone Substitute Biomaterials

Related titles:

Novel biomaterials for bone regeneration
(ISBN 978-0-85709-804-7)

Non-metallic biomaterials for tooth repair and replacement
(ISBN 978-0-85709-244-1)

Bioactive glasses
(ISBN 978-1-84569-768-6)

Woodhead Publishing Series in Biomaterials: Number 78

Bone Substitute Biomaterials

Edited by

Kajal Mallick

AMSTERDAM • BOSTON • CAMBRIDGE • HEIDELBERG • LONDON
NEW YORK • OXFORD • PARIS • SAN DIEGO
SAN FRANCISCO • SINGAPORE • SYDNEY • TOKYO
Woodhead Publishing is an imprint of Elsevier

Woodhead Publishing Limited is an imprint of Elsevier
80 High Street, Sawston, Cambridge, CB22 3HJ, UK
225 Wyman Street, Waltham, MA 02451, USA
Langford Lane, Kidlington, OX5 1GB, UK

British Library Cataloguing in Publication Data
A catalogue record for this book is available from the British Library.

Library of Congress Control Number: 2014939192

ISBN 978-0-85709-497-1 (print)
ISBN 978-0-85709-903-7 (online)

For information on all Woodhead Publishing publications
visit our website at http://store.elsevier.com/

Typeset by Newgen Knowledge Works Pvt Ltd, India

Printed and bound in the United Kingdom

Contents

Contributor contact details

(*=main contact)

Editor

K. K. Mallick
Warwick Manufacturing Group
School of Engineering
Gibbet Hill Road
University of Warwick
Coventry, CV4 7AL, UK

Email: k.k.mallick@warwick.ac.uk

Chapter 1

S. Sprio,* M. Sandri, M. Iafisco, A. Ruffini, S. Minardi and A. Tampieri
Institute of Science and Technology for Ceramics, ISTEC-CNR,
National Research Council
Via Granarolo 64
48018 Faenza, Italy

Email: simone.sprio@istec.cnr.it

Chapter 2

P. Bhattacharya and G. H. van Lenthe*
KU Leuven – University of Leuven
Department of Mechanical Engineering, Biomechanics Section
Celestijnenlaan 300C, bus 2419
3001 Leuven, Belgium

Email: Pinaki.Bhattacharya@kuleuven.be; Harry.vanLenthe@kuleuven.be

Chapter 3

E. B. Hunziker
Centre of Regenerative Medicine for Skele
Department of Clinical Research
University of Bern
Bern, Switzerland

Email: ernst.hunziker@dkf.unibe.ch

Chapter 4

T. J. Blokhuis
University Medical Center Utrecht
Department of Surgery/Traumatology
G04-228; P.O.Box 85500
3508 GA Utrecht, The Netherlands

Email: T.J.Blokhuis@umcutrecht.nl

Chapter 5

V. Guarino, M. G. Raucci, A. Ronca, V. Cirillo and L. Ambrosio*
Institute for Polymers, Composites and Biomaterials National Research Council
Mostra d'Oltremare, Pad.20

V.le Kennedy 54
80125, Naples, Italy

Email: vguarino@unina.it; ambrosio@unina.it

Chapter 6

K. K. Mallick and J. Winnett*
Warwick Manufacturing Group
Gibbet Hill Road
University of Warwick
Coventry, CV4 7AL, UK

Email: j.winnett@warwick.ac.uk

Chapter 7

A. Bansiddhi*
Department of Materials Engineering
Kasetsart University
50 Ngamwongwan Road
Chatuchak
Bangkok 10900, Thailand

Email: fengakb@ku.ac.th

D. C. Dunand
Department of Materials Science and Engineering
Northwestern University
Cook Hall
2200 Campus Drive
Evanston, Illinois 60208, USA

Email: dunand@northwestern.edu

Chapter 8

G. C. Wang
CIC biomaGUNE
Laboratorio de Biosuperficies 2
Paseo Miramón nº 182
Edificio empresarial C, 20009
San Sebastián-Donostia
Guipuzcoa, Spain

Email: lanceplus2@gmail.com

Z. F. Lu and H. Zreiqat*
Biomaterials and Tissue Engineering Research Unit
School of AMME
The University of Sydney
Sydney 2006, Australia

Email: zufu.lu@sydney.edu.au; hala.zreiqat@sydney.edu.au

Chapter 9

C. S. Bahney* and R. S. Marcucio
Department of Orthopaedic Surgery
Orthopaedic Trauma Institute
University of California, San Francisco (UCSF)
San Francisco General Hospital
Building 9, 3rd Floor
San Francisco, California, 94110, USA

Email: Chelsea.Bahney@UCSF.edu; Ralph.Marcucio@UCSF.edu

Chapter 10

J. Venkatesan and S. K. Kim*
Department of Marine-bio Convergence Science and Marine Bioprocess Research Center
Pukyong National University
Busan, 608-737, South Korea

Email: sknkim@pknu.ac.kr; venkatjchem@pknu.ac.kr

Chapter 11

L. Grøndahl* and C. S. Goonasekera
School of Chemistry and Molecular Biosciences
The University of Queensland
Cooper Rd.
Brisbane, QLD 4072, Australia

Email: l.grondahl@uq.edu.au; chandhi.goonasekera@uqconnect.edu.au

K. S. Jack
Centre for Microscopy and Microanalysis
The University of Queensland
Level 1 AIBN (Building 75)
Brisbane, Australia

Email: k.jack@uq.edu.au

Chapter 12

S. A. Clarke* and P. Walsh
School of Nursing and Midwifery and School of Chemistry and Chemical Engineering
Queen's University Belfast
Medical Biology Centre
97 Lisburn Rd
Belfast, BT9 7BL, UK

Email: s.a.clarke@qub.ac.uk; pamela.walsh@qub.ac.uk

Woodhead Publishing Series in Biomaterials

1 **Sterilisation of tissues using ionising radiations**
Edited by J. F. Kennedy, G. O. Phillips and P. A. Williams

2 **Surfaces and interfaces for biomaterials**
Edited by P. Vadgama

3 **Molecular interfacial phenomena of polymers and biopolymers**
Edited by C. Chen

4 **Biomaterials, artificial organs and tissue engineering**
Edited by L. Hench and J. Jones

5 **Medical modelling**
R. Bibb

6 **Artificial cells, cell engineering and therapy**
Edited by S. Prakash

7 **Biomedical polymers**
Edited by M. Jenkins

8 **Tissue engineering using ceramics and polymers**
Edited by A. R. Boccaccini and J. Gough

9 **Bioceramics and their clinical applications**
Edited by T. Kokubo

10 **Dental biomaterials**
Edited by R. V. Curtis and T. F. Watson

11 **Joint replacement technology**
Edited by P. A. Revell

12 **Natural-based polymers for biomedical applications**
Edited by R. L. Reiss et al

13 **Degradation rate of bioresorbable materials**
Edited by F. J. Buchanan

14 **Orthopaedic bone cements**
Edited by S. Deb

15 **Shape memory alloys for biomedical applications**
Edited by T. Yoneyama and S. Miyazaki

16 **Cellular response to biomaterials**
Edited by L. Di Silvio

17 **Biomaterials for treating skin loss**
Edited by D. P. Orgill and C. Blanco
18 **Biomaterials and tissue engineering in urology**
Edited by J. Denstedt and A. Atala
19 **Materials science for dentistry**
B. W. Darvell
20 **Bone repair biomaterials**
Edited by J. A. Planell, S. M. Best, D. Lacroix and A. Merolli
21 **Biomedical composites**
Edited by L. Ambrosio
22 **Drug–device combination products**
Edited by A. Lewis
23 **Biomaterials and regenerative medicine in ophthalmology**
Edited by T. V. Chirila
24 **Regenerative medicine and biomaterials for the repair of connective tissues**
Edited by C. Archer and J. Ralphs
25 **Metals for biomedical devices**
Edited by M. Ninomi
26 **Biointegration of medical implant materials: Science and design**
Edited by C. P. Sharma
27 **Biomaterials and devices for the circulatory system**
Edited by T. Gourlay and R. Black
28 **Surface modification of biomaterials: Methods analysis and applications**
Edited by R. Williams
29 **Biomaterials for artificial organs**
Edited by M. Lysaght and T. Webster
30 **Injectable biomaterials: Science and applications**
Edited by B. Vernon
31 **Biomedical hydrogels: Biochemistry, manufacture and medical applications**
Edited by S. Rimmer
32 **Preprosthetic and maxillofacial surgery: Biomaterials, bone grafting and tissue engineering**
Edited by J. Ferri and E. Hunziker
33 **Bioactive materials in medicine: Design and applications**
Edited by X. Zhao, J. M. Courtney and H. Qian
34 **Advanced wound repair therapies**
Edited by D. Farrar
35 **Electrospinning for tissue regeneration**
Edited by L. Bosworth and S. Downes

36 **Bioactive glasses: Materials, properties and applications**
Edited by H. O. Ylänen
37 **Coatings for biomedical applications**
Edited by M. Driver
38 **Progenitor and stem cell technologies and therapies**
Edited by A. Atala
39 **Biomaterials for spinal surgery**
Edited by L. Ambrosio and E. Tanner
40 **Minimized cardiopulmonary bypass techniques and technologies**
Edited by T. Gourlay and S. Gunaydin
41 **Wear of orthopaedic implants and artificial joints**
Edited by S. Affatato
42 **Biomaterials in plastic surgery: Breast implants**
Edited by W. Peters, H. Brandon, K. L. Jerina, C. Wolf and V. L. Young
43 **MEMS for biomedical applications**
Edited by S. Bhansali and A. Vasudev
44 **Durability and reliability of medical polymers**
Edited by M. Jenkins and A. Stamboulis
45 **Biosensors for medical applications**
Edited by S. Higson
46 **Sterilisation of biomaterials and medical devices**
Edited by S. Lerouge and A. Simmons
47 **The hip resurfacing handbook: A practical guide to the use and management of modern hip resurfacings**
Edited by K. De Smet, P. Campbell and C. Van Der Straeten
48 **Developments in tissue engineered and regenerative medicine products**
J. Basu and J. W. Ludlow
49 **Nanomedicine: Technologies and applications**
Edited by T. J. Webster
50 **Biocompatibility and performance of medical devices**
Edited by J-P. Boutrand
51 **Medical robotics: Minimally invasive surgery**
Edited by P. Gomes
52 **Implantable sensor systems for medical applications**
Edited by A. Inmann and D. Hodgins
53 **Non-metallic biomaterials for tooth repair and replacement**
Edited by P. Vallittu
54 **Joining and assembly of medical materials and devices**
Edited by Y. (Norman) Zhou and M. D. Breyen
55 **Diamond-based materials for biomedical applications**
Edited by R. Narayan
56 **Nanomaterials in tissue engineering: Fabrication and applications**
Edited by A. K. Gaharwar, S. Sant, M. J. Hancock and S. A. Hacking

57 **Biomimetic biomaterials: Structure and applications**
Edited by A. J. Ruys
58 **Standardisation in cell and tissue engineering: Methods and protocols**
Edited by V. Salih
59 **Inhaler devices: Fundamentals, design and drug delivery**
Edited by P. Prokopovich
60 **Bio-tribocorrosion in biomaterials and medical implants**
Edited by Y. Yan
61 **Microfluidic devices for biomedical applications**
Edited by X-J. James Li and Y. Zhou
62 **Decontamination in hospitals and healthcare**
Edited by J. T. Walker
63 **Biomedical imaging: Applications and advances**
Edited by P. Morris
64 **Characterization of biomaterials**
Edited by M. Jaffe, W. Hammond, P. Tolias and T. Arinzeh
65 **Biomaterials and medical tribology**
Edited by J. Paolo Davim
66 **Biomaterials for cancer therapeutics: Diagnosis, prevention and therapy**
Edited by K. Park
67 **New functional biomaterials for medicine and healthcare**
E.P. Ivanova, K.Bazaka and R. J. Crawford
68 **Porous silicon for biomedical applications**
Edited by H. A. Santos
69 **A practical approach to spinal trauma**
Edited by H. N. Bajaj and S. Katoch
70 **Rapid prototyping of biomaterials: Principles and applications**
Edited by R.Narayan
71 **Cardiac regeneration and repair Volume 1: Pathology and therapies**
Edited by R-K. Li and R. D. Weisel
72 **Cardiac regeneration and repair Volume 2: Biomaterials and tissue engineering**
Edited by R-K. Li and R. D. Weisel
73 **Semiconducting silicon nanowires for biomedical applications**
Edited by J.L. Coffer
74 **Silk biomaterials for tissue engineering and regenerative medicine**
Edited by S. Kundu
75 **Biomaterials for bone regeneration: Novel techniques and applications**
Edited by P. Dubruel and S. Van Vlierberghe
76 **Biomedical foams for tissue engineering applications**
Edited by P. Netti

77 **Precious metals for biomedical applications**
Edited by N. Baltzer and T. Copponnex
78 **Bone substitute biomaterials**
Edited by K. Mallick
79 **Regulatory affairs for biomaterials and medical devices**
Edited by S. Amato and R. Ezzell
80 **Joint replacement technology Second edition**
Edited by P. A. Revell
81 **Computational modelling of biomechanics and biotribology in the musculoskeletal system: Biomaterials and tissues**
Edited by Z. Jin
82 **Biophotonics for medical applications**
Edited by I. Meglinski
83 **Modelling degradation of bioresorbable polymeric medical devices**
Edited by J. Pan
84 **Perspectives in total hip arthroplasty: Advances in biomaterials and their tribological interactions**
S. Affatato
85 **Tissue engineering using ceramics and polymers Second edition**
Edited by A. R. Boccaccini and P. X. Ma

This book is dedicated to the editor, Dr Kajal Mallick, who sadly passed away before publication

Part I

Properties of bone substitute biomaterials in medicine

1

Bone substitutes based on biomineralization

S. SPRIO, M. SANDRI, S. PANSERI, M. IAFISCO, A. RUFFINI, S. MINARDI and A. TAMPIERI, Institute of Science and Technology for Ceramics, ISTEC-CNR, Italy

DOI: 10.1533/9780857099037.1.3

Abstract: The establishment of new and effective approaches to regenerative medicine requires a high interdisciplinary character between materials science, biology and surgery; this new paradigm is today fostering large advances in materials science and a more clear definition of the key aspects related to bone tissue regeneration. In this respect new bio-inspired approaches for the synthesis of biomaterials, such as *in vitro* biomineralization, enables the production of a new generation of smart, multifunctional and stimuli-responsive biomimetic materials for hard tissue regeneration. Among these advances, the implementation of new functionalities triggered by external magnetic signals may open the way to new therapeutic approaches to increase the regenerative potential of the scaffold, also reducing invasiveness during the implantation surgery.

Key words: biomimesis, bone regeneration, osteochondral regeneration, biomineralization, stimuli-responsive, magnetism.

1.1 Introduction

Nowadays, the development of novel materials and devices exhibiting smart behaviour and able to function as complex systems is an emerging and increasingly crucial necessity. For this reason, over the last decade, materials science has gradually evolved into a new interdisciplinary topic, which includes nanotechnology and, more recently, also the life sciences. Indeed, there is a growing consciousness of the limitations of conventional manufacturing methodologies in synthesizing highly complex devices, due to the lack of cost- and labour-effectiveness; moreover, worries about process reproducibility strongly increase with device complexity, so that the manufacture of such devices at reasonable costs and at adequate production scales for their widespread use is a serious concern.

In this respect, new approaches of fabrication are being explored, in particular by devoting special attention to the reproduction of complex structures following the examples found in nature. Such materials are of great interest due to their versatile applications in many application areas, spanning the biomedical through energetics to the environment, health and safety (EHS) fields. In particular, the new approaches of regenerative medicine require the development of materials displaying complex, non-homogeneous, physical–chemical features and structure, in compliance with the complex nature of human tissues and organs, and being able to exhibit smart behaviour, to follow the ever changing physiological environment of living organisms.

For a decade, one of the most amazing natural processes, biomineralization, was reproduced in the laboratory to develop materials and devices mimicking multifunctional anatomical regions. Biomineralization is a complex ensemble of concomitant phenomena, driving the development of complex biological structures, associating highly organized protein/carbohydrate matrices which function as templates for the nucleation and organization at the nanoscale of inorganic nanostructured phases (Berman *et al.*, 1993; Lowenstam and Weiner, 1989). The formation and organization of these structures occurs through information exchange at the molecular level between the organic and inorganic component. Such a process is able to induce chemical–physical and structural complexity not achievable by any conventional manufacturing process as well as outstanding properties including the capacity to intelligently respond to the environmental stimuli, also including the ability of self-renewal and self-regeneration upon limited damages.

The formation of human hard tissues is governed by self-assembly and organization of collagen molecules in a complex 3D structure, which acts as a template for simultaneous mineralization with nanocrystalline apatite. The heterogeneous nucleation, growth and specific orientation of the mineral nanograins are mediated by various chemical, physical, morphological and structural control mechanisms, controlled by the organic matrix at different size levels; these unique features influence cell behaviour and phenotype development and drive the formation/organization of living tissues.

This chapter illustrates the lab-based biomineralization processes, settled to synthesize hybrid hydroxyapatite (HA)/collagen scaffolds for regeneration of bone and osteochondral tissues. The reproduction of the natural phenomena yielding new bone formation allowed synthetic hybrid composites to be obtained, expressing high biomimesis and resulting in very good regenerative ability. Such devices were developed and brought to clinical application and the current research is now devoted to improving the behaviour of the existing materials to extend the field of application to regeneration of large, load-bearing, multifunctional anatomical regions. In this respect, the chapter will introduce new approaches to develop composite polymeric matrices able to mediate biomineralization processes, to achieve hybrid constructs with improved mechanical and elastic behaviour. Next, we will outline the new perspectives which may be opened up by new chemical/biochemical functionalization enabling the scaffolds to trigger specific cell phenotypes and/or to function as intelligent drug delivery systems capable of releasing bioactive molecules and drugs upon establishment of specific physiological conditions, such as inflammatory states. In conclusion, an overview on potential future trends in this field will explore the intriguing potential offered by novel superparamagnetic biohybrid devices. The recent development of a bioactive superparamagnetic HA opens new possibilities to produce scaffolds with remote controlling, avoiding any cytotoxic effect. The application of local weak magnetic fields may provide a new tool to assist and direct cell behaviour, thus increasing the osteogenic and angiogenic capacity of the new bone scaffolds and this may open new perspectives in regenerative medicine.

1.2 Key aspects driving the regeneration of hard connective tissues

The new concepts of regenerative medicine are intended to provide new innovative therapies able to restore the original functions of diseased tissues, with beneficial secondary effects of reducing healthcare costs, the duration of hospital stay as well as to significantly reduce the physical and psychological pain of patients. However, technological limitations still exist which weaken the establishment of regenerative therapies, since the regeneration of organized and multifunctional anatomical tissues (like osteocartilaginous regions) requires scaffolds able to show compositional and structural complexity. In fact, to trigger the correct cascade of biological events that lead to tissue regeneration, cells need to be exposed to an adequate array of chemical–physical, structural and morphological signals whose presentation follows precise spatial and temporal patterns. This requires the establishment of suitable strategies in designing scaffolds for bone or osteochondral tissue regeneration so as to reproduce such signals and provide cells with information compelling them to express specific phenotypes.

The structural organization of living tissues presents nano-sized elements hierarchically organized on different scale levels which allow them to exist and function with the minimum energy expenditure and an optimization of the available resources (Meyers, 2008). Native organisms form and evolve following a bottom-up scheme, able to self-adapt at all levels of hierarchy to the different modifications of the surrounding environment, partly due to its initial non-specificity (Fratzl and Weinkamer, 2007). Following this scheme, nano-sized natural elements spontaneously assemble to form structures with a high level of organization, up to the macroscopic scale, upon exchange of information at a molecular scale that regulate the formation and organization of mineral phases in contact with a protein-based matrix acting as promoter of heterogeneous nucleation.

In the case of the formation of hard connective tissues such as bone and osteochondral tissue, a collagen-based matrix, also containing non-collagenous macromolecular chains, forms upon extrusion by fibroblast cells and spontaneous self-assembly from nanofibrils to larger fibres. During this process, the heterogeneous nucleation of nanonuclei in the mineral phase takes place through activation of several control mechanisms, thus forming three-dimensional hybrid (organic–inorganic) composites showing superior physicochemical and texturing properties. In particular, chemical factors mediate the precipitation of ions naturally present in the extracellular environment (e.g., Ca^{2+}, PO_4^{3-}, Mg^{2+}, CO_3^{2-}, K^+, Na^+, SiO_4^{4-}), which concur to form nanonuclei of the mineral phase in correspondence of specific loci, situated at the gaps left by the staggering of tropocollagen molecules (Olszta *et al.*, 2007; Tampieri *et al.*, 2005, 2011). The mineral phase is a nanocrystalline HA ($Ca_{10}(PO_4)_6(OH)_2$), though the resulting Ca:P ratio is less than 1.67, that is, a calcium-deficient HA (LeGeros and LeGeros, 1984; Rey, 1998). The charge imbalance provoked by the calcium vacancies is partially restored by the incorporation of several foreign ions in the crystal sites of calcium (e.g., Mg^{2+}, K^+, Na^+, Sr^{2+},) and phosphate (e.g., CO_3^{2-}, SiO_4^{4-}). The lack of calcium and the presence of foreign ions concur to hamper the crystallization of the mineral phase, thus resulting in a poorly crystalline phase (in comparison with HA phases synthesized by conventional wet methods) with nano-sized crystals, characterized by a much higher specific surface and thus high solubility and bioavailability at physiological pH (Gomez-Morales *et al.*, 2013; LeGeros and LeGeros, 1984; Rey, 1998; Sprio *et al.*, 2008). Since bone is a living tissue that is continuously undergoing remodelling and repair, the small size and/or non-stoichiometry of the crystals presumably endows the mineral phase with the bioavailability influenced by the solubility constant at physiological pH, needed for bone resorption by osteoclasts (bone-resorbing cells).

Each of the aforementioned foreign ions carries out a specific function, each being extremely important for the biochemistry of new bone formation and bone remodelling (Bigi *et al.*, 1992; Celotti *et al.*, 2006; Landi *et al.*, 2004, 2005, 2006, 2008; LeGeros, 1991; Matsko *et al.*, 2011; Pietak *et al.*, 2007; Schwarz, 1973). In bone mineral carbonate ions mainly substitute phosphate (B-type position) and, to a smaller extent, hydroxyl (A-type position) (Boskey, 2006). The substitution of carbonate ions for phosphate ions is the major source of structural disorder in bone mineral and consequently increases its chemical reactivity (dissolution and thermal decomposition). In particular, B-type carbonation enhances the apatite solubility without altering the surface polar property and the affinity of the osteoblast cells. Further, CO_3^{2-} ions also exist in non-apatitic domains, mainly located in a hydrated layer surrounding apatite crystals; their proportion is higher in young bone, thus representing a reservoir of ions promoting remodelling processes. Magnesium is associated with the first stages of bone formation (about 5 mol%) and decreases with increasing calcification and with the ageing of the individual (Bigi *et al.*, 1992). The presence of magnesium increases the nucleation kinetics of HA on collagen fibres but retards its crystallization, giving rise to small grains (< 20 nm), thus increasing the bioavailability of the mineral phase; for this reason, the magnesium concentration is higher in cartilage and in young bone than in mature tissues. Silicon is also an essential element for healthy skeletal and cartilage growth and development in higher biological organisms (Matsko *et al.*, 2011), especially high in the metabolically active state of the cell; silicon is also known to bind to glycosaminoglycan macromolecules and has been shown to play a role in the formation of cross-links between collagen and proteoglycans (Pietak, 2007; Schwarz, 1973) thus resulting in the stabilization of bone matrix molecules and preventing their enzymatic degradation. Strontium is a trace element closely related to calcium, which accumulates in the skeleton, depending upon the skeletal site. Indeed Sr content is higher in cancellous than in cortical bone and 2–4 times higher in young bone than in mature bone tissue; moreover it is more abundant in femur diaphyses, lumbar vertebrae and iliac crests (Dahl *et al.*, 2001).

Besides the increased chemical instability, the bioactivity of the biological apatite is also governed by the unique size and shape of the HA nuclei. Indeed, the environment of the nucleation sites acts as a *spatial factor* confining the growth of nuclei and providing constraint in their shape. The crystal state of the substrate induces the nucleation and growth of the apatite nanocrystals along a specific crystallographic direction. Crystal growth is limited to very few unit cells of the apatite lattice, with values ranging from 30–50 nm in length, 15–30 nm in width and 2–10 nm in thickness (Eppell *et al.*, 2001; Jackson *et al.*, 1978; Lowenstam and Weiner, 1989; Robinson and Watson, 1952; Traub *et al.*, 1989; Weiner and Price, 1986).

The selectivity in the site of mineral phase nucleation and the physical confinement exerted by the organic template also result in a *structural factor*, inducing specific orientation of the apatite nanocrystals and exposure of crystal planes favouring topotactic adhesion of proteins such as osteocalcin. The apatite platelets develop along the long axis of collagen, so that the apatite crystals grow preferentially along the *c* axis of the hexagonal apatite lattice. Finally, *morphologic factors* act at the macroscopic scale, driving the neo-formed hybrid composite to take a complex architecture on different dimensional scales (morphogenesis), in strict relationship with the various phenomena described above. Due to their complex hierarchical organization, the 3D hybrid (organic–inorganic) composites thus developed can manifest complex functions and a specific interaction with cells, in turn driving their capacity of continuous remodelling (Sprio *et al.*, 2012).

1.3 Biomineralization processes to obtain collagen/hydroxyapatite composites as regenerative bone and osteochondral scaffolds

Biological systems store, process and transfer information at the molecular level. In this respect, the chemical state and the nanostructured morphological features presented by collagen act as cues driving the formation of bones as well as of the complex structure of multifunctional anatomical regions (e.g., the articular tissues, the periodontium). Hence, by exploiting the information inherent in the biopolymer structure, new bottom-up processes can be settled and directed by soft chemistry procedures simulating the conditions present in the biological environments.

The reproduction of the complex bone structure – made of elements characterized by a defined chemistry and with a morphology hierarchically organized down to the molecular scale – can be obtained by employing in-lab synthesis techniques where hybrid organic–inorganic composites are spontaneously built, driven by biological-like mechanisms. This can be achieved upon activation of the different control mechanisms inherent to the polymeric matrix leading to self-assembly/self-organization/mineralization with biologically competent mineral phases.

Since type I collagen is the main protein constituent of tendon and bone tissue (Lin *et al.*, 1999; Nimni, 1988) it can be used as a template for biomimetic mineralization process generating bone-like structures. In this respect, starting from aqueous solutions at human body temperature (i.e., 37°C) bone-like HA nanocrystals have been nucleated on self-assembling collagen fibres, exploiting the ability of the negatively charged carboxylate groups of collagen to bind the calcium ions of HA (Tampieri *et al.*, 2003).

Type I collagen, obtained after purification, elimination of telopeptides and virus inactivation (Tampieri *et al.*, 2003), can be processed as aqueous acidic suspension, at pH 3–3.5. In these conditions collagen is aggregated as dispersed nano-sized fibrils. The suspension can be mixed with a H_3PO_4 aqueous solution, and dropped with controlled kinetics into a basic $Ca(OH)_2$ suspension, stabilized at pH $\approx$ 12. Under such conditions two phenomena simultaneously occur: (i) a self-assembly process yielding the formation of collagen nanofibrils which progressively aggregate into ever thicker fibres, as pH increases; (ii) the formation of the apatite phase (induced by the presence of Ca^{2+} and PO_4^{3-} ions) which is the more stable calcium phosphate in such pH and temperature conditions (Elliott, 2002). After synthesis, the resulting material is a hydrogel which can be subjected to controlled freeze-drying that, by tailoring the kinetics of freezing/cooling and the freezing temperature, can result in a flexible tool to impart specific morphologies to the final construct. Additional modifications improving physical/mechanical strength and chemical stability against bioresorption can be carried out by using suitable cross-linking agents such as butanediol diglycidyl ether (BDDGE) and Genipin. During biomineralization the Ca^{2+} ions link to the COO^- functional groups present on collagen fibres, so that the formation of apatite results as a heterogeneous nucleation process where the subsequent growth of apatite nuclei is governed and confined by the crystallographic features and the morphologic nano-cues presented by the polymeric matrix (Sprio *et al.*, 2012).

In this respect, evidence of the heterogeneous nucleation of the HA phase comes from Fourier-transform infrared (FTIR) spectroscopy, transmission electron microscopy (TEM) investigations and flat film X-ray diffraction (XRD) analysis (Sprio *et al.*, 2012). In particular, such investigations point out that a chemical link exists between carboxyl groups exposed by the collagen matrix and the Ca^{2+} surface sites of the apatite lattice. The size of HA nuclei formed inside the collagen fibres and grown parallel to the fibres is limited to a few nanometres; this feature is accompanied by a preferential orientation of HA crystal domains along the (002) direction, as well as by a preferential growth of the HA nuclei along the direction of orientation of the typical collagen molecular axial spacing at 0.29 nm. The preferential orientation of HA crystals results in the exposure of the [100] face to the extracellular environment, which is supposed to promote the early binding *in vivo* of osteocalcin, a protein strongly involved in bone formation and remodelling, which has a strong and specific affinity for this type of crystal face (Mann, 2001). Despite the small crystalline size of the mineral phase obtained, XRD analysis shows that the [002] reflection is narrower and enhanced in intensity, as an effect of the preferential growth of the hexagonal apatite lattice along its *c* axis.

During the in-lab biomineralization process, the kinetics of fibril aggregation allows the mineral phase to nucleate on the collagen fibrils prior to their assembly, resulting in them being chemically linked and physically embedded into the polymeric matrix, which progressively organizes into a gel-like 3D network (Fig. 1.1). This process is extremely flexible (Landi *et al.*, 2005; Sprio *et al.*, 2008); it can be carried out in the presence of different foreign ions, present in traces in the natural mineral component of bones and with specific activity in the formation, stabilization and maturation of bone tissue. Among these ions, the most relevant are magnesium, silicon and strontium which can be introduced in the reaction vessel as $MgCl_2$ (Mg/Ca = 5–12 mol%), $Si(CH_3COO)_4$ (SiO_4/PO_4 = 5–8 mol%) and $Sr(NO_3)_2$ (Sr/Ca = 3–5 mol%).

The presence of magnesium also induces spontaneous carbonation of the mineral phase, as evidenced by FTIR, specifically in the B-position (i.e., phosphate site), with no evidence of carbonation in the A-position (i.e., hydroxyl site), probably prevented by the steric hindrance imposed

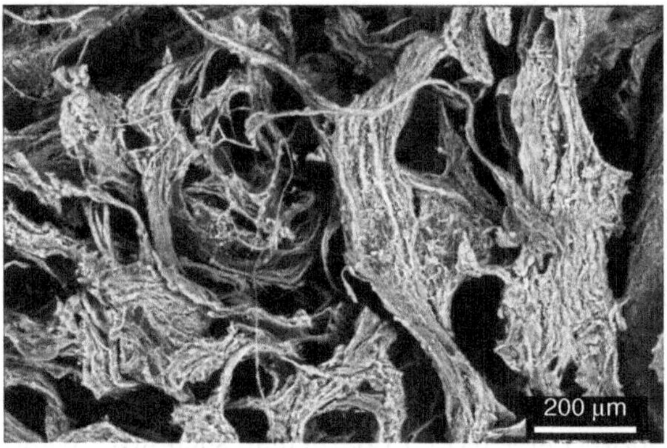

1.1 SEM micrograph of a self-assembled collagen matrix mineralized with nanonuclei of HA.

by the collagenous matrix that hampered access to the OH^- sites of the mineral phase. Chemical analysis by inductively coupled plasma-optical emission spectroscopy (ICP-OES) also confirms the carbonation occurred; in fact the increase of carbon in the apatite phase thus obtained is always accompanied by a reduction in the amount of phosphorus (Tampieri *et al.*, 2003).

The biomineralization process carried out in the conditions described above result in the heterogeneous nucleation of a biomimetic apatite phase onto the collagen fibrils, forming a 'true' bone-like composition:

$$(Ca, Mg, Sr)_{10-x/2}\,(PO_4)_{6-x-y}\,(CO_3)_x\,(SiO_4)_y\,(OH)_{2-2y}. \qquad [1.1]$$

The introduction of substituting ions during lab-based biomineralization not only affects composition but also crystal growth. In particular, besides the interaction with collagen, the presence of magnesium activates topotactic control mechanisms which drive specific preferential orientation but limit the growth of the mineral phase (Sprio *et al.*, 2012). The lower crystal order of the nuclei of biomimetic HA, compared to the stoichiometric phase (i.e. $Ca_{10}(PO_4)_6(OH)_2$), is evidenced by both XRD and FTIR analyses. Moreover the nuclei of HA with stoichiometric composition have an acicular shape, which turns into a globular shape in the case of multiple ion substitution. In this respect, Fourier-transform analysis of high-resolution transmission electron microscopy (HRTEM) images confirms the very short-range crystal order in the structure of the nucleated mineral phase, thus exhibiting features of a poorly crystalline calcium phosphate. Such features are accompanied by surface characteristics related to the presence of multiple substituting ions, in particular magnesium. In fact Mg-substituted HA is characterized by a reduced grain growth in comparison with the stoichiometric phase, thus resulting in a higher number of small nuclei that strongly increases the surface area of the mineral phase, and hence the density of biologically active surface sites; in this respect the surface Mg^{2+} sites are characterized by a higher affinity with water in comparison with Ca^{2+} (Bertinetti *et al.*, 2009). Cell-surface interaction includes specific chemical–physical linkages between cells and material, where cell adhesion and spreading pave the way to cell proliferation, finally providing a surface which is well covered by cells. Cell–material surface interaction and cell adhesion are complex processes involving the reorganization of cytoskeleton proteins further stimulated in this case by the ordered alignment of needle-like HA crystals along their *c* axis on collagen fibres. In this respect both of the phenomena described above result in increased bioactivity, since the water adsorption is preliminary to the linking of proteins favouring focal cell adhesion. This behaviour effectively explains the higher presence of magnesium in newly formed and immature human bone tissue (Bigi *et al.*, 1992). The *in vitro* cell behaviour of MG-63 human osteoblast-like cells reflect this finding well, resulting in a complete spreading of cells on the biomineralized collagen fibres so that they are nearly indistinguishable from the biomaterial (Fig. 1.2). The self-assembly and growth processes led to composites exhibiting a pseudo-plastic behaviour, similar to bone tissue, and mechanical properties close to the values found for trabecular bone at corresponding values of porosity (scaffold porosity = 60%: Young's modulus = 1.5–2 GPa, bending strength = 2–3 MPa; scaffold porosity = 40%: Young's modulus = 5–6 GPa, bending strength = 15 MPa) (Tampieri *et al.*, 2005).

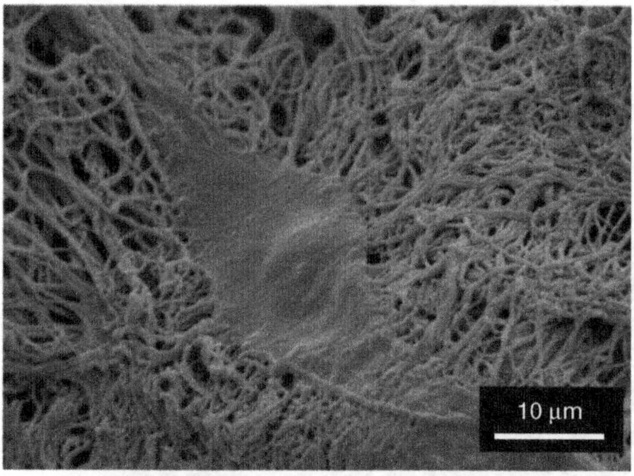

1.2 An MG-63 human osteoblast-like cell completely spread and integrated into hybrid HA/collagen composite.

1.3.1 Graded biohybrid devices for regeneration of multifunctional tissues

Biologically inspired mineralization processes represent a flexible tool to develop integrated scaffolds for regeneration of multifunctional anatomical regions. Indeed, complex tissues like osteochondral tissues and the periodontium can be mimicked by suitable modifications of features such as the extent of mineralization and morphology of the biopolymeric matrices. A promising strategy to develop osteochondral substitutes lies in the generation of heterogeneous scaffolds, integrating distinct layers mimicking the mineralized and non-mineralized tissues. Such an approach is based on the recognition of the different requirements to regenerate the cartilaginous and bony parts of an osteochondral defect. Due to their capacity to store and process information at a molecular level, lab-based biomineralization processes reproducing different histological areas in the osteochondral tissue were performed by simply varying the degree of mineralization and the 3D organization of collagen fibres (Tampieri *et al.*, 2008). Chemically and morphologically graded hybrid materials have been generated, built by stacking composite layers with decreasing degrees of mineralization, respectively mimicking the subchondral bone, the mineralized cartilage (tidemark) and the hyaline cartilage (Fig. 1.3), this latter formed by collagen fibres enriched with hyaluronic acid. In particular, in the cartilage-like layer the biologically inspired synthesis reproduces some cartilaginous environmental cues, such as the formation of a columnar-like structure converging towards the external surface where it forms horizontal flat ribbons (Fig. 1.4), resembling the morphology of the *lamina splendens*. This feature is induced by the propagation of a planar ice front (Schoof *et al.*, 2001) during the freeze-drying cycle; further, the addition of hyaluronic acid (rich

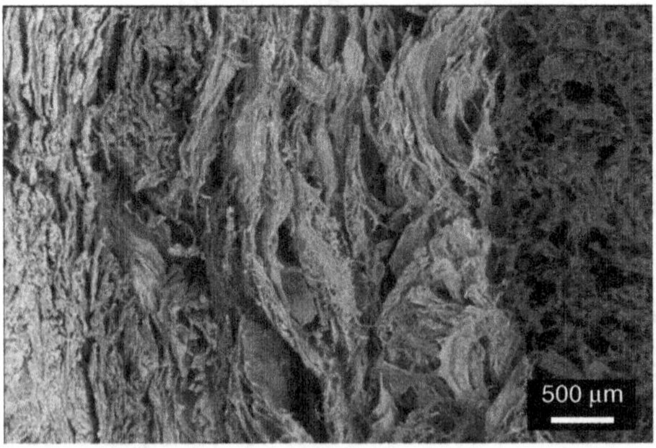

1.3 Microstructure of a three-layered osteochondral scaffold: from right to left the layers mimic subchondral bone, mineralized cartilage and hyaline cartilage.

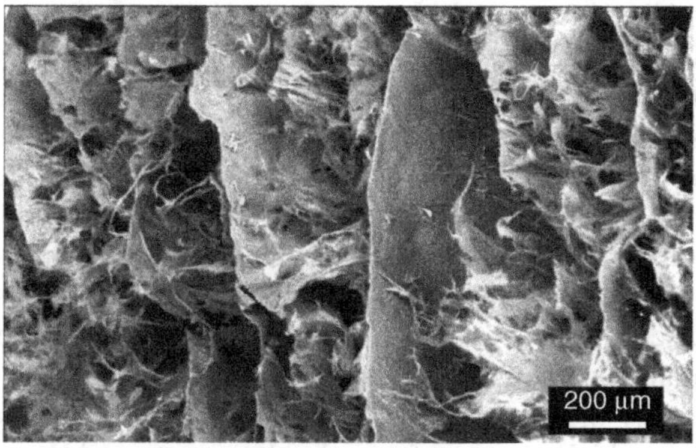

1.4 Detail of the cartilage-like layer in hybrid osteochondral scaffold.

in polar hydrogen atoms) in the upper cartilaginous layer creates bridges between the collagen fibres and improves the hydrophilic behaviour of the construct.

In vivo studies on sheep demonstrated that the chemical–physical and morphostructural characteristics of the biologically inspired graded composites functioned as biomimetic cues able to differentially support and direct the formation of different tissues (i.e., bone and cartilage) in the different histological layers (Kon *et al.*, 2010). In particular, also in a case of implantation without any seeded cells, the formation of newly hyaline-like tissue and rather orderly patterns of tissue repair was detected, with a strong proteoglycan staining and columnar rearrangement of chondrocytes, as well as an underlying, well-structured subchondral trabecular bone. These results demonstrated the high healing potential inherent in the biomimetic structure and chemistry of the scaffold; as occurs in native healthy tissues, bone marrow stem cells were directly recruited from the underlying subchondral bone and specific differentiation in bone or cartilage cells was driven by the features of the different layers of the graded scaffold (Tampieri *et al.*, 2011). The high elasticity of the scaffolds provided good shape memory that allowed easy deformation and adaptation to bone cavities by press-fit. The ease of handling and use of scaffolds is a very relevant requirement for surgeons and helps to greatly increase the level of reliability and confidence. Moreover, the bone-like features of the scaffolds allowed complete resorption and replacement with new healthy tissue, which was detected only six months after implantation (Kon *et al.*, 2010).

1.4 Composite biopolymeric matrices able to mediate biomineralization

Natural polymers such as cellulose, alginate and chitosan derive from biological structures (e.g., plants, algae, exoskeletons); hence they have characteristics which show promise for activating the same control mechanisms allowing bio-inspired assembly and mineralization (Mano *et al.*, 2007). Following this approach, new polymeric blends can be developed (Li *et al.*, 2005; Svensson *et al.*, 2005; Tangsadthakun *et al.*, 2006; Venkatesan and Kim, 2010; Wang *et al.*, 2009; Whitaker *et al.*, 2001) generating nanocomposite polymeric matrices able to mediate biomineralization.

The natural source of alginates makes this class of biopolymers attractive for safe and effective clinical applications. Indeed the plant-derived polymers avoid disease transmission while being able to provide micro-environments similar to those presented by the extracellular matrix (ECM). Alginates are a family of unbranched binary co-polymers extracted from marine brown algae which consist of strictly alternating residues of mannuronic and guluronic acid. In recent years there have been various reports on the generation of alginate-based, porous 3D scaffolds for potential application in tissue engineering. Those scaffolds are mostly created by freeze-drying of alginate gels to generate suitable porosity, often in combination with calcium phosphates and other additives (Bernhardt *et al.*, 2009; Dittrich *et al.*, 2007; Gelinsky *et al.*, 2007).

Chitosan is a biocompatible and bioerodible cationic polymer that has been used in the biomedical industry for wound healing and bone induction, drug and gene delivery, and in scaffolds for cell growth and cell encapsulation. Its pH-sensitive structure, which exhibits an open 3D porosity, allows the possibility of controlling the molecular interactions with collagen (Sionkowska *et al.*, 2004) and obtaining biomimetic bone scaffolds with improved mechanical properties (Li *et al.*, 2005; Wang *et al.*, 2009). The influence of blending polymers on the assembly of pre-nucleation mineral clusters on collagen fibres and subsequent apatite crystallization is still unclear; however, the amine group present in chitosan can bind Ca^{2+} ions (Venkatesan and Kim, 2010) so that biomineralization can occur on both the interpenetrated biopolymeric matrices.

Cellulose, a linear homopolymer of glucose $(C_6H_{10}O_5)_n$ with n ranging from 500 to 5000, is the most widespread polymeric material in nature. It is biocompatible and it easily integrates into a host tissue without chronic inflammation. The methods used to synthesize cellulose include the enzymatic/chemical/physical top-down approaches used to isolate it from plants (regenerated cellulose) and the bottom-up production from glucose by bacteria known as bacterial nanocellulose (BNC). Cellulose is easily machinable and is available in a wide range of forms and shapes. Its mechanical properties are close to those of native tissue. BNC represents a hydrogel that is produced by strains of the *Gluconacetobacter xylinus* bacterium. It consists of a supramolecular, three-dimensional network of nano-sized fibres (40–60 nm in diameter) with a high specific surface area (100–200 m^2/g). It features a high purity, a high degree of polymerization (up to 8000), a high crystallinity (up to 90%), a high and controllable water content (up to 99%) and superior mechanical properties.

The biosynthesis of cellulose is not only performed in plants, but also by bacteria, algae and fungi. Bacteria like *G. xylinus* are able to synthesize BNC at the interface between air and a culture medium. This extracellularly produced cellulose consists of an ultra-fine network of pure D-glucose. In contrast to plant cellulose, BNC has a higher water adsorption capacity and higher mechanical strength, in particular under wet conditions. Furthermore the material is mouldable *in situ* and easily sterilizable. BNC is of great potential for medical applications and has been used in tissue engineering of bone and cartilage (Czaja *et al.*, 2005; Klemm *et al.*, 2001; Mueller *et al.*, 2006; Svensson *et al.*, 2006; Zaborowska *et al.*, 2010). The solubility and biodegradation of BNC can be adjusted by the generation of aldehyde groups or incorporation of susceptible substances like glucosamine.

When compared with other natural biodegradable polymers, BNC displays many unique properties including high water uptake capacity, high crystallinity, an ultra-fine nanofibril network structure and importantly has remarkable mechanical properties despite the fact that it contains up to 99% water. The water-holding ability is the most probable reason why BNC implants do not elicit any foreign body reaction. Moreover, the nanostructure and morphological similarities with collagen make BNC attractive for imparting increased hydrophilicity and elasticity to bone and osteochondral scaffolds, as well as for cell immobilization and cell support (Chen *et al.*, 2009; Gatenholm and Klemm, 2010). The feature of increased elasticity can provide bone and osteochondral scaffolds with improved shape memory, which is suitable for minimally invasive, press-fit, surgical procedures for implantation into osteochondral defects.

In the development of biopolymeric blends, solution casting methods can be used, followed by a pH-controlled co-fibration process, to be performed in optimized conditions of concentration and pH. The control of pH, ionic strength and polymer concentration can trigger the activation of the specific functional groups involved in the chemical interaction among the polymeric components, also keeping available the specific functional groups able to bind Ca^{2+} ions. The control of the environmental conditions of synthesis enables the simultaneous mineralization of the two polymeric components during co-assembly/3D organization and the generation of fibrous composite scaffolds for bone regeneration (Fig. 1.5). The high flexibility of the process allows control of the characteristics of the nucleating apatite phase and an increase in its osteogenic ability by multiple ion substitutions, as obtained by biologically inspired mineralization of collagen matrices (see also Section 1.3). The structural and mechanical properties of the new composite scaffolds can be tailored by varying the volume concentration of the different polymeric components. Indeed, the characteristics of relatively higher stiffness

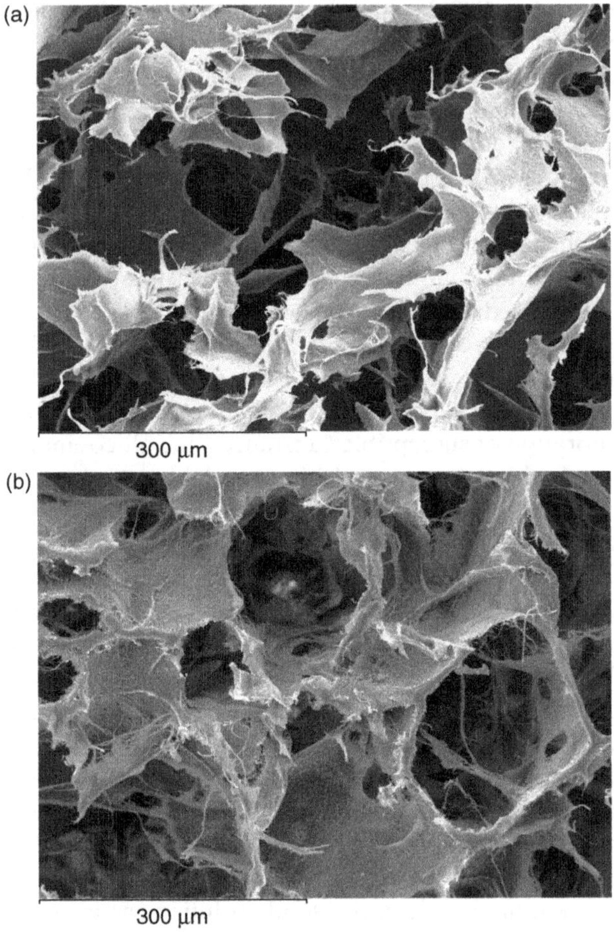

1.5 Microstructure of a blend of collagen–chitosan (a) non-mineralized, (b) mineralized with HA nuclei.

exhibited by cellulose and chitosan and the possibility of controlling their interactions with collagen at a molecular level can impart improved biomechanical performance to collagen-based mineralized scaffolds, so as to significantly extend the surgical applications to the regeneration of larger bone defects. Moreover, similarly to what can be performed in collagen-based hybrid devices, the morphological/structural features of the new scaffolds may be also controlled by chemical cross-linking and controlled freeze-drying, to impart defined fibre organization and pore morphology.

Depending on the chemical features of the various polymeric components and on the ratio of the different polymers comprising the blend, a suitable density of charged sites able to link Ca^{2+} ions can be obtained, thus defining the extent of mineral phase that can be heterogeneously nucleated. In consequence, as already obtained with collagen matrices, biodevices with different extents of mineralization can be generated and engineered to produce 3D graded composite constructs mimicking multifunctional anatomical regions, such as osteochondral tissues and the periodontium. In this respect, scaffolds endowed with increased mechanical properties and elasticity, that is, suitable for regeneration of larger osteochondral areas, can be obtained.

1.5 New intelligent bone scaffolds: functionalized devices able to respond to specific environmental conditions

In the last decade novel approaches for scaffold functionalization are becoming established to improve the performance of implanted biodevices towards increased biomimesis and capacity of regenerating/repairing diseased tissues. Functionalization can be the decoration of the scaffold surface with bioactive cues interacting with cells and the ECM to trigger tissue-specific phenotypes. In this respect the use of biomaterials with a high density of active surface sites is desirable, possibly presenting functional groups suitable for linking with various molecules.

Hybrid fibrous constructs obtained by bio-inspired self-assembly and mineralization processes can be considered as multipurpose platforms for surface functionalization, offering highly specific surface areas and a high density of active sites on the surface of both the polymeric matrix and the exposed nanograins of the mineral phase.

Bone tissue requires the promotion of angiogenesis, establishing a multi-branched vascular network to increase the delivery of oxygen and nutrients vital for bone formation and repair. In this respect lower oxygen tension is required to stimulate osteoblasts to secrete vascular endothelial growth factor (VEGF) inducing angiogenesis (Genetos and Lee, 2009).

Contrary to bone, reduced local oxygen concentration is known to play an important role in cartilage differentiation and subchondral bone formation. Particular attention has been drawn to the role of oxygen tension in the regeneration of articular cartilage (O'Driscoll *et al.*, 1997). Cartilage is avascular and contains chondrocytes that are responsible for the generation of the ECM that constitutes the architectural structure of the tissue. It was found that the levels of local oxygen tension may be strictly correlated to the activity of chondrocytes, which influence the quality of the cartilaginous tissues developed in different areas of articular regions (Hirao *et al.*, 2006; Wernike *et al.*, 2008; Zhou *et al.*, 2004) and, in consequence, its damaging effect upon tissue diseases such as osteoarthritis (Pool *et al.*, 1993) which may alter the normal oxygen tension gradient through the tissue.

1.6 SEM image of alginate beads.

The local tension of oxygen can be influenced by introducing suitable chemical moieties. In particular, quinone derivatives are functional groups renowned for their ability to sequester oxygen (Batchelor-McAuley *et al.*, 2011). They can be linked, for instance, to selected areas of multifunctional hybrid scaffolds mimicking different anatomical regions (e.g., osteochondral regions, see also Section 1.3), to provide different oxygen levels in compliance with the requirements for regeneration of bone and cartilage. An interesting and novel tool to improve scaffold functionalization is the linking of suitable multi-branched molecules (i.e., dendrimers) as unique nanoscale signalling units able to selectively modify charge, chemical and biochemical activity, topology and architecture of scaffold surfaces, as well as to significantly increase the number and density of the exposed sites on the biomaterial surface (Lloyd *et al.*, 2008; Tang *et al.*, 1996; Tomalia *et al.*, 2007).

Nanobeads have also attracted significant attention in the drug and gene delivery arenas. They can be developed in a variety of materials, including biopolymers such as chitosan, gelatine and alginate, by appropriate casting and cross-linking methods (Fig. 1.6). These nanoparticles can be engineered to host and entrap different biological substances such as nucleic acid-based materials and drugs and also for cell encapsulation (Agnihotri *et al.*, 2004; Haas *et al.*, 2005; Sung *et al.*, 2007; Yao *et al.*, 2008) and can easily be integrated into biomimetic scaffolds as smart release systems (Fig. 1.7). This feature can also be enhanced by incorporating suitable stimuli-responsive polymers.

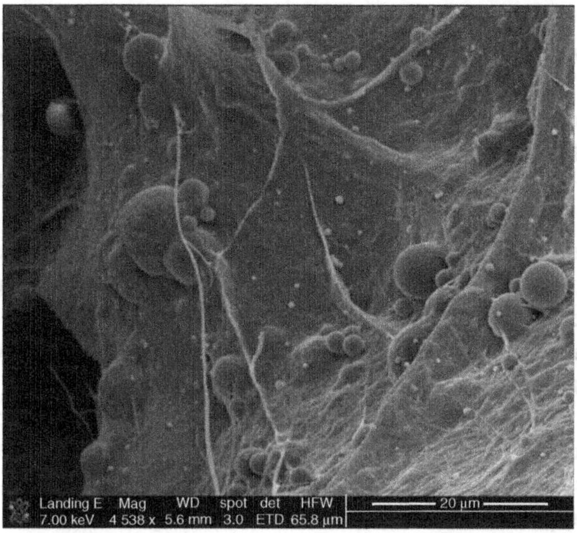

1.7 Alginate beads incorporated in a collagen-based hybrid bone scaffold.

Stimuli-responsive polymers are an emerging class of materials that exhibit changes in their properties upon even slight variations of environmental conditions, such as pH, temperature, electromagnetic fields and ionic strength as well as a variety of biochemical stimuli (Ward and Georgiou, 2011). Some of the relevant properties which can result affected by environmental stimuli are the structural organization of the hydrogel, the solubility and the hydrophilic/hydrophobic character. These features are very interesting for biomedical applications since the high sensitivity of such polymeric systems can allow precise responses to local inflammatory states or external stimuli, so as to be activated only when required (Schmaljohann, 2006).

Hydrogels which are pH-sensitive are based on macromolecular networks able to modify their volume when electrostatic repulsion of certain functional groups in the polymer network is activated/deactivated. In accordance with environmental conditions, pH-sensitive hydrogels can absorb or release various amounts of water while maintaining a distinct three-dimensional structure; hence, they can release different chemical/biochemical moieties in a controlled manner, to activate local therapies such as tissue healing and regeneration.

Among pH-sensitive biopolymers, chitosan is currently receiving attention for medical and pharmaceutical applications (Gupta *et al.*, 2002). Chitosan is a natural poly-cationic polymer that possesses useful properties such as biocompatibility, biodegradability and established physiochemical, structural and immunological properties, it is non-toxic, easily available and has gel-forming capability. When pH is below 6.5, chitosan in solution carries a high positive charge density, one charge per basic glucosamine unit, and swells in acidic pH. Polymer blends in the form of interpenetrating polymer networks can be developed (Chen *et al.*, 2008) to adjust some hydrogel properties such as hydrophilicity and/or to tailor the response of the hydrogel to the environmental stimuli. In this respect, chitosan can be blended with collagen to generate fibrous composite polymeric matrices able to mediate biomineralization processes (see Fig. 1.5), that can be directed to achieve new smart biohybrid bone and osteochondral scaffolds integrating pH-sensitive systems.

Thermo-responsive polymers and hydrogels exhibit volume phase transitions at certain critical temperatures, causing a sudden change in the solvation state, thus influencing hydrophobicity and solubility. Therefore, they can function as smart drug delivery systems, especially in the case of critical temperatures close to that of the human body (Schmaljohann, 2006).

In a different approach, thermolabile molecular ligands can be used as bridge molecules between the biomaterial and the chemical/biochemical moieties acting as surface functionalization (Erogbogbo *et al.*, 2011; Howes *et al.*, 2010). In such a system, bond breakage can be induced by providing local hyperthermia, and this can be activated through external control, with the aid of superparamagnetic media placed in the implant site and the application of suitable magnetic fields. Nowadays the hyperthermia effect is widely used, especially in anti-cancer therapies, to provoke localized cell apoptosis. Currently, iron oxide nanoparticles, particularly magnetite, are widely used as heating media; however long term cytotoxicity of magnetite poses several concerns for its use *in vivo* and is now becoming a matter of open discussion among scientists (Lewinski *et al.*, 2008; Singh *et al.*, 2010). In this respect, recently developed apatite phases with intrinsic superparamagnetic behaviour (Fe-HA) could pave a new way to solve this dilemma (Tampieri *et al.*, 2012). Fe-HA nanoparticles can be synthesized by ionic substitution of Ca^{2+} with Fe^{2+} and Fe^{3+} ions in specific ratios as well as in defined independent crystal sites of calcium. Fe-HA nanoparticles can be introduced and driven in the target tissue by magnetic guidance and activated by remote magnetic control once at the correct site. Due to its composition, Fe-HA is biocompatible and bioresorbable and it does not leave any harmful by-products in the body, hence it is a promising candidate to replace the magnetite nanoparticles, thus representing a breakthrough in the field of drug delivery systems and, more widely, in both the fields of regenerative medicine and of nanomedicine applied to therapeutics and diagnostics. The Fe-HA phase can be used as a base material to develop novel smart devices for bone regeneration; in this respect, recent investigation demonstrated the feasibility of heterogeneous nucleation of the Fe-HA phase on biopolymeric matrices, as illustrated in the next section.

1.6 Future trends in regenerative medicine: superparamagnetic hybrid bone scaffolds

In spite of the recent advances in the development of implants for hard tissue regeneration, some limitations in their application still exist. More specifically, due to the establishment of new lifestyles and the need to have an active life even in advanced age, musculo-skeletal diseases require healing therapies able to ensure full recovery of the body functionality and shorter rehabilitation times, which also impact on the healthcare costs. Some of the key factors related to these concerns can be identified in: (i) high chemical–physical and structural biomimesis of the host tissue; (ii) the ability to control and direct the formation of a suitable vascular network to sustain the reconstruction of organized and functional tissues; and (iii) the early stabilization of the scaffold soon after implantation.

With respect to these critical issues, biohybrid mineralized composites developed in the last decade can now offer high mimicry of multifunctional human tissues; the currently available data obtained from implantation in animal and human clearly prove that such biodevices can regenerate distinct adjacent tissues and be progressively integrated and resorbed, so that no traces of the implants are visible after a few months, as described in Section 1.3.

On the other hand, aside from the use of scaffolds with a high osteogenic/chondrogenic character, a fundamental requirement for the development of well-organized bone is the establishment of extensive angiogenesis, ensuring the complete morphological and biological maturation of the newly formed tissue, especially in the case of dimensionally critical defects. Indeed, angiogenesis is strongly involved in the initiation and promotion of tissue regeneration via the growth of new capillary blood vessels that allow continuous delivery of oxygen and nutrients, as well as the elimination of waste products. VEGF is the fundamental molecule promoting angiogenesis (Ylä-Herttuala *et al.*, 2005), however uncontrolled VEGF expression may lead to increased vascular leakiness and oedema, as well as the induction of vascular tumours (angiomas). (Banfi *et al.*, 2005, 2012; Ozawa *et al.*, 2004).

Such problems are often related to the presence of large bone defects, indeed vascularization of large 3D scaffolds often prove to be lacking or insufficient, limited to the peripheral areas of the implant.

Another key requirement to achieving fast and effective osteointegration is the early and stable fixation of the scaffold; the implant fixation represents a major clinical problem due to the difficulties in obtaining a stable interface between host bone and the scaffold. Currently, in the treatment of small osteochondral lesions, most surgeons do not use any fixation system, the stability of the scaffold being granted only by the fibrin clot and by the congruency between the prepared lesion site and the geometry of the scaffold. In the treatment of extensively damaged bone, fixation is achieved by extremely complicated and invasive procedures (external fixations, intramedullary nails, plates and screws), requiring continuous control and, very often, multiple surgical interventions with an obvious increase in health cost and potential complications.

With respect to these issues, the development of scaffolds with magnetic properties may represent a new approach revealing new options that cannot be adequately addressed by other methods currently in use. In particular, superparamagnetic devices can be remotely activated/deactivated by external magnetic fields; magnetic activation may open the way to novel regenerative scaffolds with multiple functions. For instance, magnetic stimulation can enhance cell activity to produce new healthy and organized tissues, as suggested by literature (Ba *et al.*, 2011; Dini and Abbro, 2005; Huang *et al.*, 2006); hence, new personalized approaches could be established by development of new therapies designed to favour tissue regeneration in specific clinical cases. Another possible application is the use of weak local magnetic fields as signals to recall and drive specific biochemical cues with magnetic marking (Plate I, see the colour section between pages 186 and 187). Indeed, mimicking endogenous growth factor production and delivery is still to be solved with conventional scaffolds (Laschke *et al.*, 2006; Patel *et al.*, 2008; Schieker *et al.*, 2006). The idea of utilizing magnetic scaffolds for additional control of angiogenesis *in vivo* was proposed (Bock *et al.*, 2010). Such scaffolds can be manipulated *in situ* by appropriate application of magnetic stimuli to attract or deliver angiogenic and other bioactive factors, in turn linked to magnetic nanocarriers, in order to provide long-lasting assistance to tissue neo-formation and organization.

Magnetic forces can also aid improvements in the fixation of scaffolds at the implant site; a superparamagnetic implant can be stabilized by external magnetic bandages; this can help to minimize the use of external fixing media and possibly reduce the hospitalization time and pain in the patient.

In view of the establishment of these new intriguing applications, the selection of suitable superparamagnetic media to be incorporated in the bone scaffolds is nowadays limited to iron oxides (see also previous section); however the recent development of a novel, bioactive, apatite phase with intrinsic superparamagnetic properties (Tampieri *et al.*, 2012) suggests that biomimetic scaffolds with intrinsic superparamagnetic properties may be produced, potentially with no risks or harmful effects due to incorporation of harmful nanoparticles in the human tissues. Recently, upon suitable modification of the lab-based biomineralization process, defined amounts of di- and trivalent iron ions were incorporated in the apatite lattice during heterogeneous nucleation on self-assembling collagen fibres, giving rise to hybrid biomimetic composites mineralized with superparamagnetic Fe-HA nuclei (data still not published). Further, the flexibility of the synthesis process allowed tailoring of the extent of mineralization of the superparamagnetic hybrid composites. Consequently 3D hybrid constructs with features mimicking different mineralized and non-mineralized tissues could be synthesized and merged into morphologically/compositionally and magnetically graded composites able to mimic multifunctional anatomical regions (see also Section 1.3). The newly developed biodevices associate high chemical–physical and structural mimesis with the ability of external activation that can prospectively enhance and expand their potential use in regenerative medicine as well as in nanotherapeutics/nanodiagnostics. All these advances can be combined with the use of composite biopolymeric blends, as described in Section 1.4; in this respect, the mineralization process can be adapted to the new matrices, so that the properties of the resulting hybrid construct may be endowed with different and improved properties, in turn opening the way to new and still unexplored applications.

1.7 Conclusions

The development of hybrid systems based on biologically inspired processes of self-assembly/organization and mineralization is a new bottom-up approach that promises to provide new biomimetic solutions for regenerative medicine. Basically, the success of scaffolds capable of mimicking in detail complex, multifunctional anatomical regions showed that chemical–physical and morphostructural biomimesis (patents: WO2006092718, WO2007045953, EP1447104) are key features that can uniquely favour a real and efficient regeneration of organized and healthy tissues. In this respect, the huge complexity of biological tissues is transferred to synthetically obtained constructs thanks to the flexibility in the process that enables the controlled nucleation of mineral phases with a wide spectrum of composition and properties on a variety of polymeric blends, as well as the possibility of acting on fibre and pore orientation by easy lab-based processes, to tailor specific properties such as stiffness, elasticity and hydrophilicity. By tailoring the properties of the final construct, biodevices that can be adapted to different personalized therapies can be produced. In this respect, due to the very high socioeconomic impact of degenerative diseases involving hard human tissues, the possibility of providing regeneration in an ever increasing number of clinical cases is a very relevant challenge and would be a significant breakthrough in medicine.

In a wider perspective, further advances in the investigation of natural phenomena and their translation into lab-based, biologically inspired processes may extend their application beyond the biomedical field. For instance, preliminary data on the performances of biomineralized fibrous composites as media for capturing nanoparticles from defined air flows and with reduced pressure drop (data not published), should encourage further research towards the application of biologically inspired processes into filtering devices for a wide span of applications. The possibility of use of self-assembling devices for many different applications may result in a breakthrough in the current manufacturing concepts and generate, in forthcoming decades, new, highly flexible, economically viable and environmentally friendly fabrication technologies for mass production of complex smart devices with designed, tailored functionalities.

1.8 Acknowledgements

The research leading to these results has received funding from the European Union's Seventh Framework Programme ([FP7/2007–2013]) under grant agreement n° 246373, OPHIS.

The authors also acknowledge Dr E. Tasciotti of TMHRI (The Methodist Hospital Research Institute, Houston, TX, USA) for the fruitful collaboration and discussion.

1.9 References

Agnihotri, S.A., Mallikarjuna, N.N. and Aminabhavi, T.M. (2004) Recent advances on chitosan-based micro- and nanoparticles in drug delivery, *J Con Rel*, **100**, 5–28.

Ba, X., Hadjiargyrou, M., DiMasi, E., Meng, Y., Simon, M., Tan, Z. and Rafailovich. M-H. (2011) The role of moderate static magnetic fields on biomineralization of osteoblasts on sulfonated polystyrene films, *Biomaterials*, **32**(31), 7831–7838.

Banfi, A., von Degenfeld, G. and Blau, H.M. (2005) Critical role of microenvironmental factors in angiogenesis, *Curr Atheroscler Rep*, **7**, 227–234.

Banfi, A., von Degenfeld, G., Gianni-Barrera, R., Reginato, S., Merchant, M.J., McDonald, D.M. and Blau, H.M. (2012) Therapeutic angiogenesis due to balanced single-vector delivery of VEGF and PDGF-BB, *FASEB J*, **26**(6), 2486–2497.

Batchelor-McAuley, C., Dimov, I.B., Aldous, L. and Compton, R.G. (2011) The electrochemistry of quinizarin revealed through its mediated reduction of oxygen, *PNAS*, **108**(50), 19891–19895.

Berman, A., Hanson, J., Leiserowitz, L., Koetzle, T.F., Weiner, S. and Addadi, L. (1993) Biological control of crystal texture: a widespread strategy for adapting crystal properties to function, *Science*, **259**(5096), 776–779.

Bernhardt, A., Despang, F., Lode, A., Demmler, A., Hanke, T. and Gelinsky, M. (2009) Proliferation and osteogenic differentiation of human bone marrow stromal cells on alginate/hydroxyapatite scaffolds with anisotropic pore structure, *J Tissue Eng Reg Med*, **3**, 54–62.

Bertinetti, L., Drouet, C., Combes, C., Rey, C., Tampieri, A., Coluccia, S. and Martra, G. (2009) Surface characteristics of nanocrystalline apatites: effect of Mg surface enrichment on morphology, surface hydration species, and cationic environments, *Langmuir*, **25**(10), 5647–5654.

Bigi, A., Foresti, E., Gregorini, R., Ripamonti, A., Roveri, N. and Shah, J.S. (1992) The role of magnesium on the structure of biological apatites, *Calc Tissue Int*, **50**(5), 439–444.

Bock, N., Riminucci, A., Dionigi, C., Russo, A., Tampieri, A., Landi, E., Goranov, V.A., Marcacci, M. and Dediu, A. (2010) A novel route in bone tissue engineering: magnetic biomimetic scaffolds, *Acta Biomater*, **6**, 786–796.

Boskey, A.L. (2006), 'Mineralization, structure and function of bone', in Seibel, M.J., Robins, S.P. and Bilezikian, J.P. (eds) *Dynamics of Bone and Cartilage Metabolism*, 201–212, Academic Press.

Celotti, G., Tampieri, A., Sprio, S., Landi, E., Bertinetti, L., Martra, G. and Ducati, C. (2006) Crystallinity in apatites: how can a truly disordered fraction be distinguished from nanosize crystalline domains?, *J Mater Sci: Mater Med*, **17**(11), 1079–1087.

Chen, S., Mingzhu, L., Shuping, J. and Bin, W. (2008) Preparation of ionic-crosslinked chitosan-based gel beads and effect of reaction conditions on drug release behaviors, *Int J Pharm*, **349**(1–2), 180–187.

Chen, Y.M., Xi, T.F., Zheng, Y.D., Guo, T.T., Hou, J.Q., Wan, Y.Z. and Gao, C. (2009) In vitro cytotoxicity of bacterial cellulose scaffolds used for tissue-engineered bone, *J Bioact Compat Polym*, **24**, 137–145.

Czaja, W., Krystynowicz, A., Bielecki, S. and Brown Jr, R.M. (2006) Microbial cellulose – the natural power to heal wounds, *Biomaterials*, **27**, 145–151.

Dahl, S.G., Allain, P., Marie, P.J., Mauras, Y., Boivin, G., Ammann, P., Tsouderos, Y., Delmas, P.D. and Christiansen, C. (2001) Incorporation and distribution of strontium in bone, *Bone*, **28**(4), 446–453.

Dini, L. and Abbro, L. (2005) Bioeffects of moderate-intensity static magnetic fields on cell cultures, *Micron*, **36**(3), 195–217.

Dittrich, R., Tomandl, G., Despang, F., Bernhardt, A., Hanke, T., Pompe, W. and Gelinsky, M. (2007) Scaffolds for hard tissue engineering by ionotropic gelation of alginate – influence of selected preparation parameters, *J Am Ceram Soc*, **90**, 1703–1708.

Elliott, J.C. (2002) Calcium phosphate biominerals, *Rev Min Geochem*, **48**(1), 427–453.

Eppell, S.J., Tong, W., Katz, J.L., Kuhn, L. and Glimcher, M.J. (2001) Shape and size of isolated bone mineralites measured using atomic force microscopy, *J Orthop Res*, **19**, 1027–1034.

Erogbogbo, F., Yong, K.T., Roy, I., Hu, R., Law, W.C., Zhao, W., Ding, H., Wu, F., Kumar, R., Swihart, M.T. and Prasad, P.N. (2011) *In vivo* targeted cancer imaging, sentinel lymph node mapping and multi-channel imaging with biocompatible silicon nanocrystals, *ACS Nano*, **5**(1), 413–423.

Fratzl, P. and Weinkamer, R. (2007) Nature's hierarchical materials, *Progr Mater Sci*, **52**, 1263–1334.

Gatenholm, P. and Klemm, D. (2010) Bacterial nanocellulose as a renewable material for biomedical applications, *MRS Bull*, **35**, 208–213.

Gelinsky, M., Eckert, M. and Despang, F. (2007) Biphasic, but monolithic scaffolds for the therapy of osteochondral defects, *Int J Mater Res*, **98**, 749–755.

Genetos, D.C. and Lee, C.M. (2009) HIF-1 alpha regulates hypoxia-induced EP1 expression in osteoblastic cells, *J Cell Biochem*, **107**(2), 233–239.

Gómez-Morales, J., Iafisco, M., Delgado-López, J.M., Sarda, S. and Drouet, C. (2013) Progress on the preparation of nanocrystalline apatites and surface characterization: overview of fundamental and applied aspects, *Prog Cryst Growth Char Mat*, **59**(1), 1–46.

Gupta, P., Vermani, K. and Garg, S. (2002) Hydrogels: from controlled release to pH-responsive drug delivery, *Drug Discovery Today*, **7**(10), 569–579.

Haas, J., Ravi Kumar, M.N.V., Borchard, G., Bakowsky, U. and Lehr, C-M. (2005) Preparation and characterization of chitosan and trimethyl-chitosan-modified poly-(ε-caprolactone) nanoparticles as DNA carriers, *AAPS Pharm Sci Tech*, **6**, E22–E30.

Hirao, M., Tamai, N., Tsumaki, N., Yoshikawa, H. and Myoui, A. (2006) Oxygen tension regulates chondrocyte differentiation and function during endochondral ossification, *J Biol Chem*, **281**, 31079–31092.

Howes, P., Green, M., Levitt, J., Suhling, K. and Hughes, M. (2010) Phospholipid encapsulated semiconducting polymer nanoparticles: their use in cell imaging and protein attachment, *J Am Chem Soc*, **132**, 3989–3996.

Huang, H.M., Lee, S.Y., Yao, W.C., Lin, C.T. and Yeh, C.Y. (2006) Static magnetic fields up-regulate osteoblast maturity by affecting local differentiation factors, *Clin Orthop Relat Res*, **447**, 201–208.

Jackson, S.A., Cartwright, A.G. and Lewis, D. (1978) The morphology of bone-mineral crystals, *Calcif Tissue Res*, **25**, 217–222.

Klemm, D., Schuhmann, D., Udhardt, U. and Marschet, S. (2001) Bacterial synthesized cellulose – artificial blood vessels for microsurgery, *Progr Pol Sci*, **26**, 1561–1603.

Kon, E., Delcogliano, M., Filardo, G., Fini, M., Giavaresi, G., Francioli, S., Martin, I., Pressato, D., Arcangeli, E., Quarto, R., Sandri, M. and Marcacci, M. (2010) Orderly osteochondral regeneration in a sheep model using a novel nano-composite multilayered biomaterial, *J Orthop Res*, **28**, 116–124.

Landi, E., Tampieri, A., Celotti, G., Vichi, L. and Sandri, M. (2004) Influence of synthesis and sintering parameters on the characteristics of carbonate apatite, *Biomaterials*, **25**(10), 1763–1770.

Landi, E., Tampieri, A., Celotti, G., Langenati, R., Sandri, M. and Sprio, S. (2005) Nucleation of biomimetic apatite in synthetic body fluids: dense and porous scaffold development, *Biomaterials*, **26**(16), 2835–2845.

Landi, E., Tampieri, A., Mattioli-Belmonte, M. and Celotti, G. (2006) Biomimetic Mg- and Mg,CO_3-substituted hydroxyapatites: synthesis characterization and *in vitro* behavior, *J Eur Ceram Soc*, **26**(13), 2593–2601.

Landi, E., Sprio, S., Sandri, M., Celotti, G. and Tampieri, A. (2008) Development of Sr and CO_3 co-substituted hydroxyapatites for biomedical applications, *Acta Biomater*, **4**(3), 656–663.

Laschke, M.W., Harder, Y., Amon, M., Martin, I., Farhadi, J., Ring, A., Torio-Padron, N., Schramm, R., Rücker, M., Junker, D., Häufel, J.M., Carvalho, C., Heberer, M., Germann, G., Vollmar, B. and Menger, M.D. (2006) Angiogenesis in tissue engineering: breathing life into constructed tissue substitutes, *Tissue Eng*, **12**, 2093–2104.

LeGeros, R.Z. (1991), Calcium phosphates in oral biology and medicine, *Monographs in Oral Science*, **15**, 1–201.

LeGeros, R.Z. and LeGeros, J.P. (1984), 'Phosphate minerals in human tissues', in Nriagu, J.O. and Moore, P.B., *Phosphate Minerals*, Springer-Verlag, New York.

Lewinski, N., Colvin, V. and Drezek, R. (2008) Cytotoxicity of nanoparticles, *Small*, **4**, 26–49.

Li, X., Feng, Q., Liu, X., Dong, W. and Cui, F. (2005) Collagen-based scaffolds reinforced by chitosan fibres for bone tissue engineering, *Polym Int*, **54**, 1034–1040.

Lin, H., Clegg, D.O. and Lal, R. (1999) Imaging real-time proteolysis of single collagen I molecules with an atomic force microscope, *Biochem*, **38**, 9956–9963.

Lloyd, A.W., Oliver, G.W.J., Standen, G., Santin, M. and Meikle, S.T. (2008) Biomaterial with functionalised surfaces. Patent: WO2008068531.

Lowenstam, H.A. and Weiner, S. (1989), *On Biomineralization*, Oxford University Press, New York.

Mann, S. (2001), *Bio-mineralization: Principles and Concepts in Bioinorganic Materials Chemistry*, Oxford University Press.

Mano, J.F., Silva, G.A., Azevedo, H.S., Malafaya, P.B., Sousa, R.A., Silva, S.S., Boesel, L.F., Oliveira, J.M., Santos, T.C., Marques, A.P., Neves, N.M. and Reis, R.L. (2007) Natural origin biodegradable systems in tissue engineering and regenerative medicine: present status and some moving trends, *J R Soc Interf*, **4**, 999–1030.

Matsko, N.B., Žnidaršič, N., Letofsky-Papst, I., Dittrich, M., Grogger, W., Štrus, *J.* and Hofer, F. (2011) Silicon: the key element in early stages of biocalcification, *J Str Biol*, **174**(1), 180–186.

Meyers, M.A., Chen, P-Y., Yu-Min Lin, A. and Seki, Y. (2008) Biological materials: structure and mechanical properties, *Progr Mater Sci*, **53**, 1–206.

Müller, F.A., Müller, L., Hofmann, I., Greil, P., Wenzel, M.M. and Staudenmaier, R. (2006) Cellulose-based scaffold materials for cartilage tissue engineering, *Biomaterials*, **27**, 3955–3963.

Nimni, M.E. (1988), *Collagen*, CRC Press, Boca Raton, FL.

O'Driscoll, S.W., Fitzsimmons, J.S. and Commisso, C.N. (1997) Role of oxygen tension during cartilage formation by periosteum, *J Orth Res*, **15**, 682–687.

Olszta, M.J., Cheng, X., Jee, S.S., Kumar, *R.*, Kim, Y-Y., Kaufman, M.J., Douglas, E.P. and Gower, L.B. (2007) Bone structure and formation: a new perspective, *Mater Sci Eng R*, **58**(3–5), 77–116.

Ozawa, C.R., Banfi, A., Glazer, N., Thurston, G., Springer, M.L., Kraft, P.E., McDonald, D.M. and Blau, H.M. (2004) Microenvironmental VEGF concentration, not total dose, determines a threshold between normal and aberrant angiogenesis, *J Clin Invest*, **113**, 516–527.

Patel, Z.S., Young, S., Tabata, Y., Jansen, J.A., Wong, M.E.K. and Mikos, A.G. (2008) Dual delivery of an angiogenic an osteogenic growth factor for bone regeneration enhances in a critical size defect model, *Bone*, **43**, 931–940.

Pietak, A.M., Reid, J.W., Stott, M.J. and Sayer, M. (2007) Silicon substitution in the calcium phosphate bioceramics, *Biomaterials*, **28**, 4023–4032.

Pool, A.R., Rizkalla, G., Ionescu, M., Reiner, A., Brooks, E., Rorabeck, C., Bourne, R. and Bogoch, E. (1993) Osteoarthritis in the human knee: a dynamic process of cartilage matrix degradation, synthesis and reorganisation, *Agents Actions Suppl*, **39**, 3–13.

Rey, C. (1998), 'Calcium phosphates for medical applications', in Amjad, Z., *Calcium Phosphates in Biological and Industrial Systems*, Kluwer Academic Publishers, Dordrecht, The Netherlands.

Robinson, R.A. and Watson, M.L. (1952) Collagen–crystal relationships in bone as seen in the electron microscope, *Anat Rec*, **114**, 383–409.

Schieker, M., Seitz, H., Drosse, I., Seitz, S. and Muetshler, W. (2006) Biomaterials as scaffold for bone tissue engineering, *Eur J Trauma*, **32**, 114–124.

Schmaljohann, D. (2006) Thermo- and pH-responsive polymers in drug delivery. *Adv Drug Deliv Rev*, **58**(15), 1655–1670.

Schoof, H., Apel, *J.*, Heschel, I. and Rau, G. (2001) Control of pore structure and size in freeze-dried collagen sponges, *J Biomed Mater Res (Appl Biomater)*, **58**, 352–357.

Schwarz, K. (1973) A bound form of Si in glycosaminoglycans and polyuronides, *Proc Natl Acad Sci USA*, **70**, 1608–1612.

Singh, N., Jenking, G.J.S., Asadi, R. and Doak, S.H. (2010) Potential toxicity of superparamagnetic iron oxide nanoparticles (SPION), *Nano Reviews*, **1**, 53–58.

Sionkowska, A., Wisniewski, M., Skopinska, J., Kennedy, C.J. and Wess, T.J. (2004) Molecular interactions in collagen and chitosan blends, *Biomaterials*, **25**, 795–801.

Sprio, S., Tampieri, A., Landi, E., Sandri, M., Martorana, S., Celotti, G. and Logroscino, G. (2008) Physico-chemical properties and solubility behaviour of multi-substituted hydroxyapatite powders containing silicon, *Mat Sci Eng C*, **28**, 179–187.

Sprio, S., Sandri, M., Panseri, S., Cunha, C. and Tampieri, A. (2012) Hybrid scaffolds for tissue regeneration: chemotaxis and physical confinement as sources of biomimesis. *J Nanomater*, **2012**(2012), Article ID 418281, 10 pages.

Sung, H-W., Lin, Y-H., Chen, M-C. and Tu, H. (2007) Nanoparticles for drug delivery. US 20070237827.

Svensson, A., Nicklasson, E., Harrah, T., Panilaitis, B., Kaplan, D.L., Brittberg, M. and Gatenholm, P. (2005) Bacterial cellulose as a potential scaffold for tissue engineering of cartilage, *Biomaterials*, **26**, 419–431.

Tampieri, A., Celotti, G., Landi, E., Sandri, M., Roveri, N. and Falini, G. (2003) Biologically inspired synthesis of bone-like composite: self-assembled collagen fibers/hydroxyapatite nanocrystals, *J Biomed Mater Res A*, **67**(2), 618–625.

Tampieri, A., Celotti, G. and Landi, E. (2005) From biomimetic apatites to biologically inspired composites, *Anal Bioanal Chem*, **381**(3), 568–576.

Tampieri, A., Sandri, M., Landi, E., Pressato, D., Francioli, S., Quarto, R. and Martin, I. (2008) Design of graded biomimetic osteochondral composite scaffolds, *Biomaterials*, **29**, 3539–3546.

Tampieri, A., Sprio, S., Sandri, M. and Valentini, F. (2011). Mimicking natural biomineralization processes: a new tool for osteochondral scaffold development, *Trends Biotech*, **29**(10), 526–535.

Tampieri, A., D'Alessandro, T., Sandri, M., Sprio, S., Landi, E., Bertinetti, L., Panseri, S., Pepponi, G., Goettlicher, J., Bañobre-López, M. and Rivas, J. (2012) Intrinsic magnetism and hyperthermia in bioactive Fe-doped hydroxyapatite, *Acta Biomater*, **8**, 843–851.

Tang, M.X., Redemann, C.T. and Szoka, F.C., Jr. (1996) *In vitro* gene delivery by degraded polyamidoamine dendrimers, *Bioconj Chem*, **7**, 703–714.

Tang, M.X. and Szoka, F.C. (1997) The influence of polymer structure on the interactions of cationic polymers with DNA and morphology of the resulting complexes, *Gene Therapy*, **4**, 823–832.

Tangsadthakun, C., Kanokpanont, S., Sanchavanakit, N., Banaprasert, T. and Damrongsakkul, S. (2006) Properties of collagen/chitosan scaffolds for skin tissue engineering, *J Met Mat Min*, **16**, 37–44.

Tomalia, D.A., Reyna, L.A. and Svenson, S. (2007) Dendrimers as multi-purpose nanodevices for oncology drug delivery, *Biochem Soc Trans*, **35**, 61–67.

Traub, W., Arad, T. and Weiner, S. (1989) Three-dimensional ordered distribution of crystals in turkey tendon collagen fibers, *Proc Natl Acad Sci*, **86**, 9822–9826.

Venkatesan, J. and Kim, S-K. (2010) Chitosan composites for bone tissue engineering – an overview, *Mar Drugs*, **8**, 2252–2266.

Wang, X., Tan, Y., Zhang, B., Gu, Z. and Li, X. (2009) Synthesis and evaluation of collagen–chitosan– hydroxyapatite nanocomposites for bone grafting, *J Biomed Mater Res*, **89A**, 1079–1087.

Ward, M.A. and Georgiou, T.K. (2011) Thermoresponsive polymers for biomedical applications, *Polymers*, **3**, 1215–1242.

Weiner, S. and Price, P.A. (1986) Disaggregation of bone into crystals, *Calcif Tissue Int*, **39**, 365–375.

Wernike, E., Li, Z., Alini, M. and Grad, S. (2008) Effect of reduced oxygen tension and long-term mechanical stimulation on chondrocyte-polymer constructs, *Cell Tis Res*, **331**(2), 473–483.

Whitaker, M.J., Quirk, R.A., Howdle, S.M. and Shakesheff, K.M. (2001) Growth factor release from tissue engineering scaffolds, *J Pharm Pharmacol*, **53**, 1427–1437.

Yao, J., Zhou, J.P., Ping, Q.N., Lu, Y. and Chen, L. (2008) Distribution of nobiletin chitosan-based microemulsions in brain following i.v. injection in mice, *Intern J Pharmaceut*, **352**, 256–262.

Ylä-Herttuala, S., Rissanen, T.T., Vajanto, I. and Hartikainen, J. (2007) Vascular endothelial growth factors. Biology and current status of clinical applications in cardiovascular medicine, *J Am Coll Cardiol*, **49**(10), 1015–1026.

Zaborowska, M., Bodin, A., Bäckdahl, H., Popp, J., Goldstein, A. and Gatenholm, P. (2010) Microporous bacterial cellulose as a potential scaffold for bone regeneration, *Acta Biomater*, **6**(7), 2540–2547.

Zhou, S., Cui, Z. and Urban, J.P. (2004) Factors influencing the oxygen concentration gradient from the synovial surface of articular cartilage to the cartilage-bone interface: a modelling study, *Arthritis Rheum*, **50**, 3915–3924.

2
Experimental quantification of bone mechanics

P. BHATTACHARYA and G. H. VAN LENTHE,
KU Leuven – University of Leuven, Belgium

DOI: 10.1533/9780857099037.1.30

Abstract: Design of bone substitute biomaterials requires an understanding of how bone behaves mechanically at different scales. In this chapter, common experimental methods that are used to measure bone mechanical properties at the whole-bone scale, micrometre scale and sub-micrometre scale are detailed. Special attention is given to the governing principles behind the methods, the properties that are measured by these methods, the measurement protocols and the associated limitations. Current scientific knowledge of bone mechanical characteristics at each scale is reviewed. Results are presented for the wrist, spine and hip, which are sites of high fracture risk and relevant for bone substitute biomaterials design.

Key words: bone, mechanical properties, architecture, multi-scale, experimental methods, wrist, spine, hip.

2.1 Introduction

This chapter presents a review of the mechanical characterization of bone. Following the multi-scale organization of bone, mechanical characterization of bone is also resolved at different scales (Augat and Schorlemmer, 2006). In particular, mechanical properties at three length scales are provided: nano-scale, micro- (or tissue) scale and macro- (or whole-bone) scale. The organization of bone at these scales is depicted in Fig. 2.1. The term 'structure' refers to an identifiable physical unit that performs a mechanical function. Depending on the scale of investigation, structure can therefore be identified as the entire bone, or a single osteon, or a lamella. The mechanical properties of the structure at a given length scale depend on the constitutive properties of the 'material' at the lower scale, as well as the 'architectural' properties that determine how the same material is organized (Fig. 2.2a). For example, the mechanical properties of micro-scale structures (osteons, trabeculae, etc.) will depend on the constitutive properties of the mineralized collagen fibres (the material) and their layered arrangement

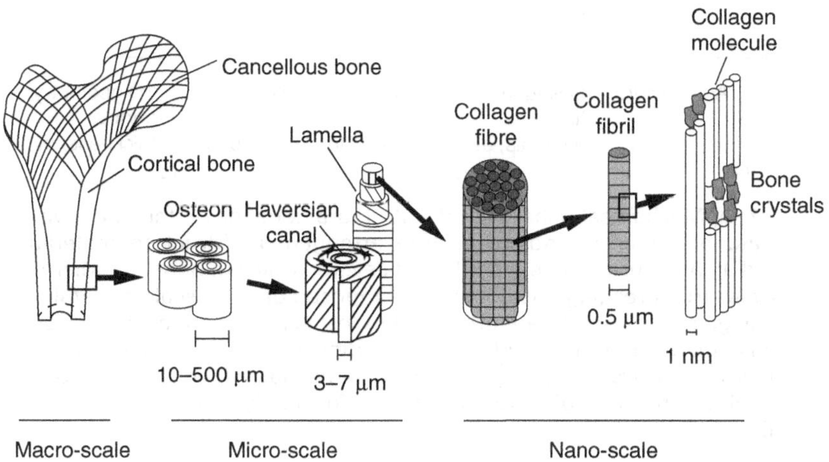

2.1 Multi-scale and multi-material characteristics of bone. Structural units of bone material are shown for several scales. At each scale the architectural and mechanical properties of the structural units can influence the stiffness and strength of the whole bone. (*Source*: Reprinted from *Medical Engineering and Physics*, **20** (2), Rho *et al.*, Mechanical properties and the hierarchical structure of bone, 92–102, Copyright (1998), with permission from Elsevier.)

(the architecture) at the sub-micron scale (Fig. 2.2b). In the same way as mineralized collagen fibres are the 'material' with respect to the micro-scale, osteons, trabeculae, interstitial lamellae, etc., constitute the 'material' with respect to the macro-scale. This material is identified as bone tissue in the literature. The architectural arrangement of bone tissue, together with the extracellular matrix (ECM) and network of vascular canals, determines the mechanical characteristics at the whole-bone level (Fig. 2.2c).

The chapter is organized as follows. In Section 2.2 some basic knowledge of bone biology and composition is reviewed. The understanding of how bone is created and lost is used to motivate the design of bone substitute biomaterials. Epidemiology and the general loss in mechanical competence of bone are further reviewed with particular focus at sites of clinical interest, for example, the wrist, the spine and the hip.

In Sections 2.3–2.5, the mechanical response and architectural properties are considered at the macro-, micro- and nano-scale respectively. In each section, the motivation to assess these properties is briefly highlighted, experimental methods of assessment are detailed, with a focus on principles, protocols, and drawbacks associated with them. In particular, Section 2.3 considers the whole bone as a structure. Stiffness, strength and toughness are highlighted as relevant mechanical characteristics. For the wrist, the spine and the hip, results from experiments conducted on human bones

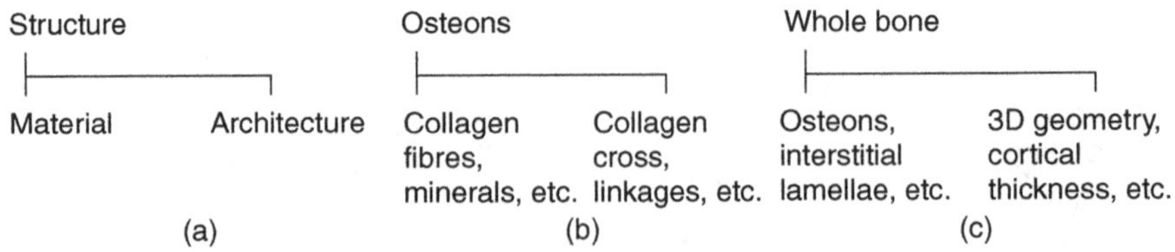

2.2 (a) Mechanical behaviour of a structure depends on the underlying material properties and architectural arrangement. Structure, material and architecture can be identified at several scales in bone. (b) Micro-scale structures (e.g., osteons) are composed of nano-scale mineralized collagen fibres, which are organized to form specific nano-scale cross-collagen linkages. (c) Whole bone is composed of osteons and other micro-scale structures (e.g., interstitial lamellae), which are organized to allow a network of vascular canals to form only along specific directions.

are summarized. We review widely used bone densitometric techniques which assess bone composition as a substitute for bone mechanical properties. Results obtained from these methods, and important limitations can be found in this section. Separately, methods of assessing bone architecture at the macro-scale are reviewed.

In Section 2.4 bone tissue is considered as the structure of interest. A review is conducted for the mechanical characterization of bone tissue as obtained from indentation techniques. The underlying principles of bone indentation and assumptions and limitations of current analyses are highlighted. The architectural properties at the micro-scale as obtained by high-resolution computed tomography are summarized. The reader is provided with an overview of possible future directions in micro-scale analysis of bone.

In Section 2.5 methods of assessment of mechanical and architectural properties of nano-scale structures in bone are reviewed. At this scale, the structure under investigation is typically an individual lamella. Isolation of such small structures requires careful preparation of specimens. Details of preparation procedures are presented in this section. Results from analyses are summarized for nano-scale bone structures originating from several anatomical sites.

In Section 2.6 two frameworks of assessment that combine methods from multiple length scales (or hierarchical approaches) are introduced. Recent studies employing these cutting-edge techniques are revisited. In a hierarchical approach, the success depends on the effectiveness of the assessment techniques at all the different length scales considered. Future possibilities are discussed, in view of the state of the art achieved independently in the assessment techniques at different scales. Section 2.7 concludes with a summary of current understanding of mechanical characterization of bone at different length scales.

The target audience for this chapter is expected to possess a basic understanding of mechanical engineering and material science concepts (elastic modulus, anisotropy, etc.). Some knowledge of general physics will help to better appreciate the experimental methods, but is not otherwise necessary. This chapter should be useful for scientists and engineers, who are interested in bone mechanical properties at different scales, and need an overview of associated measurement techniques. Even this requirement is quite broad; hence this chapter provides information typically useful for designing bone substitute biomaterials.

2.2 Bone biology and mechanical function

Bone tissue is a multiphase material comprising about 70% mineral (mainly hydroxyapatite), 22% proteins (mainly type I collagen) and 8% water by weight (Rho *et al.*, 1998; Augat and Schorlemmer, 2006). To form bone as a composite material, these constituent phases combine following a dominant underlying arrangement, even at the molecular level. As a living tissue, bone is actively remodelled: collagen and mineralized matrix are produced by particular bone cells (osteoblasts) while bone is lost through resorption by other cells (osteoclasts). A particular concern of advanced age is the prevalence of osteoporosis and other bone-degenerative diseases. As it is more difficult to restore bone strength at a later stage of disease, early detection and quantification of loss of bone competence is the aim of most diagnostic techniques (Lochmuller *et al.*, 2002b). Because of the complexity in determining what contributes to the strength of bone, as well as its failure mechanisms, it remains a challenge to establish parameters that govern the etiology of age-related fractures (see Fig. 2.3).

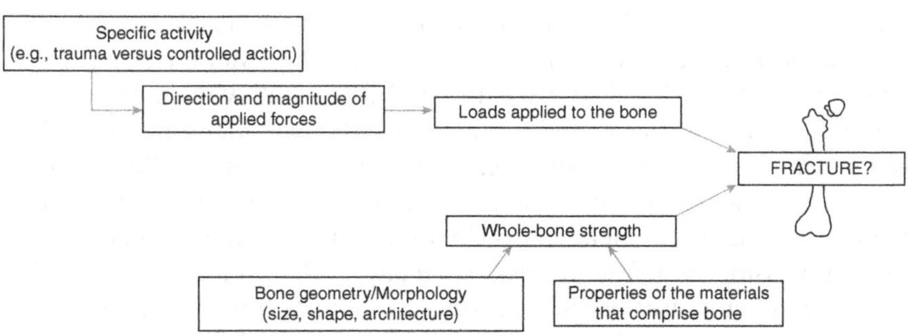

2.3 Etiology of age-related fractures. (*Source*: Reprinted from *Best Practice & Research in Clinical Rheumatology*, **19**, Bouxsein, M. L., Determinants of skeletal fragility, 897–911, Copyright (2005), with permission from Elsevier.)

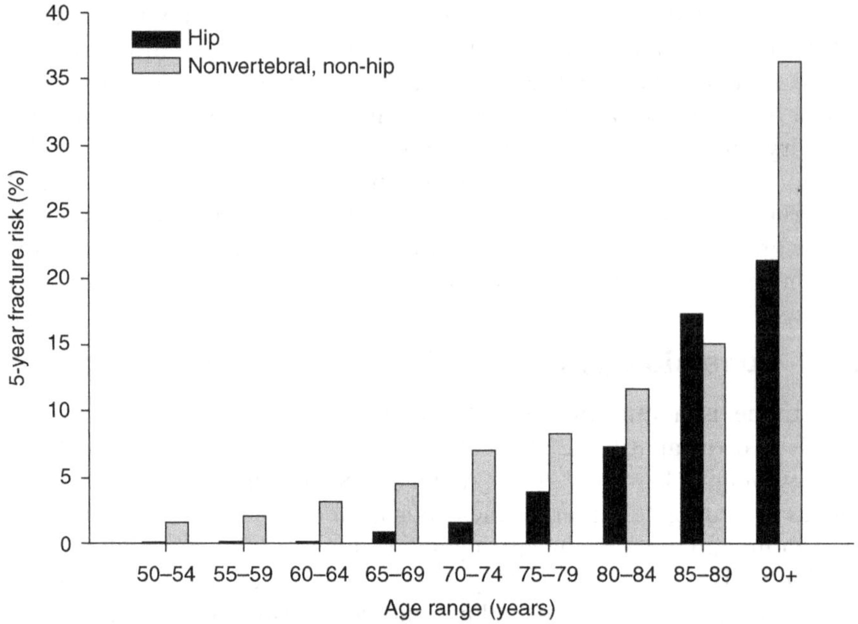

2.4 Age-stratified, model-based estimates of the 5-year fracture risk for Australian women above 50 years of age (*Source*: data from Doherty *et al.*, 2001). (*Source*: Reprinted from *American Journal of Medicine,* **120**, Delmas *et al.* Beyond hip: Importance of other nonspinal fractures, 381–387, Copyright (2007), with permission from Elsevier.)

This chapter focuses on the specific anatomical sites that seem to attract a high incidence of bone fracture in humans, in particular in the elderly. Development of bone substitute materials is vital so as to target such sites for a satisfactory clinical outcome. These sites are the distal radius (forearm), thoracolumbar vertebrae (spine) and the proximal femur (hip) (Eckstein *et al.*, 2004). The incidence of fracture rises sharply after menopause (Jensen *et al.*, 1982; Riggs and Melton, 1986; Melton *et al.*, 2005; Delmas *et al.*, 2007) (Fig. 2.4). An estimated 20% of all women will sustain a clinical fracture of the wrist by the age of 70 years (Jensen *et al.*, 1982; Owen *et al.*, 1982), the most common type of fracture being Colles' fracture (7.3 per 100 person years in women; 1.7 per 100 person years in men). The dominant loading modes leading to Colles' fracture are dorsal bending and axial compression, with fracture risk being largely determined by trabecular bone strength. Fractures of the forearm may indicate a high relative risk of subsequent fractures at the spine, hip and other sites (Klotzbuecher *et al.*, 2000). The spine is required to sustain compressive and shear forces during daily activities. It is known that the margin of safety between compressive strength of vertebrae and physiological loads is alarmingly small (Hutton *et al.*, 1979). Further the loading at this site is usually a complex combination of shear and compression (Brinckmann *et al.*, 1989a; van Dieën *et al.*, 2006). The particular loading scenario depends not only on posture but also on the relative health of the load-bearing components such as the neural arches, intervertebral discs, end-plates and trabecular core of the vertebral body. Hip fractures remain the most deleterious of all cases (Cumming *et al.*, 1997). About a fifth of the individuals who suffer a hip fracture die within the subsequent year. Rehabilitation to pre-fracture mobility levels is not common. Incidence of hip fractures is higher in women than in men, and the lifetime risk for women is three times that for men. Fall-related trauma combined with general loss in bone competence are the main causes of hip fractures in the elderly.

2.3 Whole-bone mechanical properties

At the whole-bone level, the operative interpretation of the term 'bone' is as a structure. External loads, as a result of physical activity (running, getting up from a chair) or traumatic events (falls), are applied on this structure by muscles and tendons at respective points of attachment and at the joints through the articular cartilage. Assessment of bone at this level is made either by direct investigation (usually invasive) of its mechanical behaviour or indirectly (usually non-invasive) through factors that are supposed to determine such behaviour.

2.3.1 Quantitative measures of mechanical and architectural properties

Stiffness, strength and toughness are the most common properties that are used to directly characterize the behaviour of bone to physiological loads and conditions. These properties help to quantify fracture risk and bone quality. For instance, for human vertebrae, applied load per unit ultimate compressive strength is a relevant measure of fatigue fracture risk (Brinckmann *et al.*, 1987). Another use of these properties relates to the design of bone-substitute biomaterials. Design of bone substitutes must be based on several criteria at once: biocompatibility, reduction of stress-shielding effects, performance under load-bearing situations and favourability to bone growth (Brydone *et al.*, 2010). Of these, reduction of stress-shielding effects and performance under load-bearing situations typically demand that the fabricated biomaterials have mechanical properties very similar to the bone being substituted. The determination of mechanical properties mentioned above also helps to quantify the effects of drugs, diet and micro-scale changes due to age or disease. However the definition of these properties is not obvious for a complex tissue such as bone. Conventional interpretations of terms such as 'stiffness', 'strength' and 'toughness' will be discussed in the following sections, and will be dependent on the scale of investigation (macro, micro or nano).

Bone morphology is known to affect bone competence and can serve as a predictor of direct mechanical properties such as stiffness, strength and toughness. Geometrical measures such as bone size, cross-sectional area or area moment of inertia can predict up to 70–80% of whole-bone strength (Augat and Schorlemmer, 2006). In this way, bone geometry aids in the identification of disease, quantification of disease progression and effect of drugs. Bone morphological assessment is the most appealing aspect which is frequently non-destructive and easier to use compared to mechanical testing. Hence, it is attractive for *in vivo* diagnosis at the whole-bone level. In the design of bone substitute materials, geometric details of bone must be accounted for to ensure compatibility within the body. Therefore, the development of bone morphological assessment techniques is directly relevant to the main topic of this book.

2.3.2 Destructive testing

For compression and bending tests, the behaviour of the whole-bone structure under loading can be characterized by a load-displacement diagram; a schematic is shown in Fig. 2.5 (Aspden, 2003; Cole and van der Meulen, 2011). Stiffness is normally equated with the slope of the initial linear regime (k). Yield is defined as the level of load at which the mechanical response deviates from the initial linear response (point A), and is generally determined following the 0.2% offset rule (Mosekilde and Danielsen, 1987; Bayraktar *et al.*, 2004). Bone strength is defined as the maximum load that can be sustained (point B). Toughness is defined as area under the load-deflection curve until failure (shaded region). The force-displacement curve is applicable to compression and tension, as well as bending tests. However, the strength and toughness values will depend on the particular mode of loading. For example, in the literature, bone strength obtained in tension is distinguished from bone strength obtained in compression.

Compression tests

At the whole-bone level, compression tests are used to determine bone mechanical properties at sites where compressive loads play an important role. Owing to their physiological role, compression tests at the vertebral body and the femoral neck are of immediate relevance. Though physiological activities might not impose predominantly compressive loads at the distal radius, if one accounts for the fact that most fractures at the wrist are

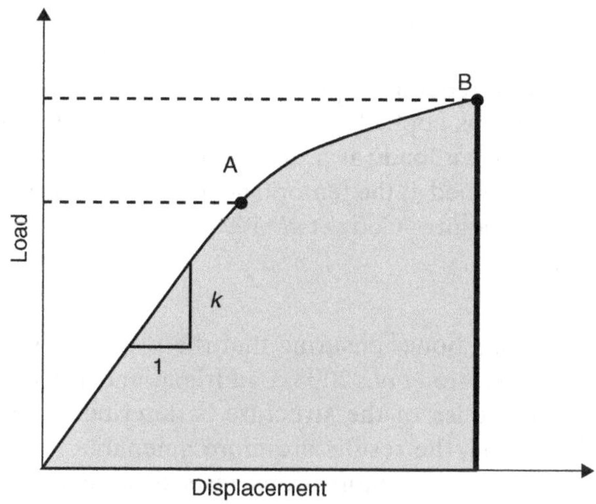

2.5 Typical load–displacement curve of a structure. Here the initial slope, *k*, is identified with the stiffness of the structure, points A and B locate yield and failure events, and the (shaded) area under the curve is identified with work to failure (or fracture toughness) of the structure.

due to falls sustained by the outstretched hand (Njeh *et al.*, 2000; Goldfarb *et al.*, 2001), it becomes clear why compression tests too are useful at this site.

Typically the whole anatomical region is tested at specific locations: spine, an entire vertebral motion segment (Brinckmann *et al.*, 1989b; Zhao *et al.*, 2009); wrist, the intact radius, ulna and carpal bones together (Njeh *et al.*, 2000; Pistoia *et al.*, 2002; Eckstein *et al.*, 2004); and hip, the proximal region of the femur (Cody *et al.*, 1999). Since the micro-scale architecture of bone (osteons and trabeculae) exhibits a preferred orientation, mechanical properties of the whole bone are also dependent on the loading direction. Thus, during mechanical testing the anatomical section (comprising a single or multiple bones) is oriented so as to simulate the typical scenario in which fracture occurs. For example, at the wrist, a 10° inclination of the radial shaft with respect to the vertical (loading direction) is used to simulate the outstretched hand in a fall scenario (Njeh *et al.*, 2000). At the vertebra, compression test fixtures are designed such that flexion, extension and side-bending modes of deformation cannot occur (Brinckmann *et al.*, 1989b). At the femoral neck, a 25° inclination of the femoral shaft with respect to the vertical (loading direction) simulates the one-legged stance phase of normal gait (Cody *et al.*, 1999), whereas a 60° inclination simulates a sideways fall (Keyak *et al.*, 1998; Lenaerts and van Lenthe, 2009). The combined loading gives rise to a complex state of stress, and often a wide variety of failure types are recorded (Brinckmann *et al.*, 1989b; Cody *et al.*, 1999; Lochmuller *et al.*, 2002b). Yet, such combined loading resembles the relevant real-life scenarios. In order to stabilize the structure in the desired orientation various degrees of preload are frequently applied. For instance, at the vertebral body, a preload of 300 N was applied for 15 min (Adams *et al.*, 2006), followed by 1500 N compressive load (at a displacement rate of 0.21 mm/s) and a 220 N preload was applied at the femoral head prior to a displacement control at 0.21 mm/s until failure (Cody *et al.*, 1999).

Three-point bending test

This test is suitable for long bones ensuring that the response is predominantly a flexural one (Torcasio *et al.*, 2008; Landrigan and Roeder, 2009). Since the mechanical response of the structure is described largely by a single mode of deformation, the results are more amenable to modelling studies, unlike those from the compressions tests described previously. Therefore three-point bending tests results have been widely used in fundamental bone mechanics research with small animals (Jamsa *et al.*, 1998; Silva *et al.*, 2005; Leppanen *et al.*, 2006; Margolis *et al.*, 2006). A testing protocol as applied to the human distal radius is provided by Lochmuller *et al.* (2002b). Briefly, the bone is supported at two locations equidistant from the site where the load is to be applied. The distance between the supports is usually chosen to be the largest possible span of bone for which the cross-section is more or less homogeneous. A displacement rate of 6.5 mm/s, applied at the 33% distal location is typical of most protocols (Lochmuller *et al.*, 2002b).

Static histomorphometry

Histomorphometry is the examination of a bone specimen to determine its geometric features. Histomorphometry provides direct information that can be used to quantify bone turnover, remodelling and architecture, which are otherwise only indirectly accessible by procedures such as bone densitometry. The bone specimen is obtained from the desired site (using a trephine for a biopsy) followed by fixation and processing in ethanol (Vedi and Compston, 2003) and then stained using specialized dyes to distinguish between the different bone architectural features.

A typical as-prepared two-dimensional (2D) section is investigated for resorption cavities, osteoids, bone remodelling units and mineralized bone. Following this identification it is possible to define measurements (Parfitt *et al.*, 1987) that quantify the whole-bone geometry and its microstructure with the aid of computer-assisted technologies. Some examples of commonly used primary measurements are bone area (B.Ar), bone volume (BV), wall thickness (W.Th) and trabecular thickness (Tb.Th). Derived indices in common usage are trabecular number (Tb.N), trabecular separation

(Tb.Sp) and others. However, it must be noted that histomorphometric indices are obtained from 2D sections and hence both primary and secondary measurements (e.g., Tb.Th, Tb.Sp) include assumptions regarding the three-dimensional (3D) architecture (Müller and van Lenthe, 2004). True 3D measurements can only be obtained using advanced imaging techniques such as micro-computed tomography (micro-CT) (see Section 2.3.1). Though histomorphometry can be applied to bone biopsies taken from live patients, it is also performed on specimens excised from cadavers. Histomorphometry serves as a benchmark for less invasive methods targeted towards quantifying bone architecture, for example, CT.

2.3.3 Non-destructive testing

Non-destructive testing procedures are non-invasive and are attractive to clinical diagnosis and prediction of bone disease. The loss of bone mass is the most popularly held cause of osteoporosis (McCreadie and Goldstein, 2000; Melton *et al.*, 2005). Indeed, reduction in bone mass at the distal radius has a strong correlation with fracture risk (Kelsey *et al.*, 1992; Cummings and Black, 1995; Nguyen *et al.*, 2001; Cummings *et al.*, 2002). The importance of assessing bone mass in determining bone fragility becomes clear in the light of significant challenges that exist in quantifying the effects of other factors (McCreadie and Goldstein, 2000; Bouxsein, 2005). This is why most non-destructive testing procedures are targeted at quantitative bone mass measurement.

Dual-energy X-ray absorptiometry (DXA)

DXA (or sometimes DEXA) is the primary quantitative bone mass measurement tool and, similar to other techniques in this category, is based on the principle of linear attenuation of radiation in tissue. Many reviews of quantitative bone mass measurement methods (DXA and others preceding it) exist in the literature (e.g., Augat *et al.*, 1998). DXA is based on the following principle: to distinguish between absorption in soft tissue and in bone, and thus eliminate the need for a constant path length radiation source where two different energy peaks are being used (Greenfield, 1992). A soft tissue equivalent – indispensable for previously used single-energy source methods and characteristically cumbersome to establish – is not necessary. This is due to the fact that two different energy peaks are used to calculate the contribution of soft tissue mass and bone mass. Regions of interest occupying bone and soft tissue are easily identifiable in DXA-generated images. The images are processed to compute the bone mineral content per projected area (in g/cm^2) which is commonly interpreted as the projected bone mineral density (BMD). Bone mineral content (BMC) is calculated by summing the BMD values over the projected area. Even though it measures only projected bone mass and not a mechanical property *per se*, bone quality is still largely defined by DXA in the clinical setting. This can be attributed to the ease of use of DXA and the strong correlation between bone mass and fracture risk. Another particular advantage is that DXA can be applied at any site (including radius, femur and spine).

Nevertheless, care needs to be taken to interpret measurements obtained from DXA (Jonson, 1993). Some researchers quote the BMC normalized to the width of the scanned region, resulting in bone mineral content per unit length (in g/cm). Typical DXA instruments are DPX-L scanner (GE Lunar, Madison, WI, USA) (Lochmuller *et al.*, 2002b) and QDR 4500 (Hologic, Waltham, MA) (Njeh *et al.*, 2000), among others. Care needs to be taken to ensure that the surrounding soft tissue does not introduce relevant measurement errors (Svendsen *et al.*, 1995; Lochmuller *et al.*, 2001, 2002b). A particular disadvantage of DXA is in its inability to discriminate between cortical and trabecular bone (Augat *et al.*, 1998). The most important limitation on the applicability of DXA remains the fact that bone mass is a very rudimentary measure of bone competence.

Computed tomography (CT)

Quantitative computed tomography (QCT) is also an imaging technique based on radiation absorption. However, compared to conventional X-ray imaging, CT scans have about 100× higher resolution (Hounsfield, 1973). Furthermore, QCT measures a volumetric density and the results are given in g/cm^3. The attenuation coefficients (or CT numbers) at each point of the cross-sectional image are reconstructed from the projected data (Herman, 1980). CT numbers can be converted to a measure of bone mineral quantity, by calibrating scans against phantoms, typically containing dipotassium hydrogen phosphate or hydroxyapatite (Genant and Boyd, 1977; Cann, 1987). Sectional slices varying in thickness between 1 and 9 mm are imaged, with a precision of 1–5% within a control population (Cody *et al.*, 1989; Sabatier and Guaydier-Souquieres, 1989). Three-dimensional reconstructions can be created by a computer program by overlaying projections of multiple slices (Hounsfield, 1973; Ruegsegger *et al.*, 1976). The multiple image slices may be obtained for example by moving a projection system relative to the subject (see Fig. 2.6). One main advantage of CT lies in its ability to provide separate measures for quantity of cancellous, cortical and integral bone (Genant and Boyd, 1977). Similar to other densitometry procedures, CT has the potential to be used *in vivo* to predict bone competence (Brinckmann *et al.*, 1989b). For excised human femurs, bone strength is better predicted by CT-based finite element (FE) models compared to DXA and QCT (Cody *et al.*, 1999).

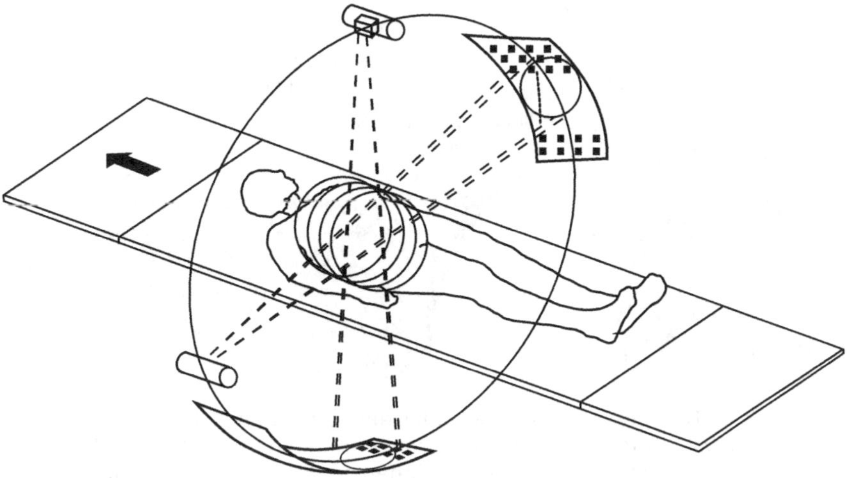

2.6 The movement of the projection system relative to the patient in a spiral scanner. (*Source*: Reproduced from *X-ray computed tomography in biomedical engineering*, Cierniak, R., Technical concepts of X-ray computed tomography scanners, 21–62, Copyright (2011), with kind permission from Springer Science and Business Media.)

Peripheral quantitative computed tomography (pQCT)

Peripheral QCT scanners are targeted towards assessing bone quality at peripheral skeletal sites (see Fig. 2.7). Slice thicknesses are in the 165 μm range and thus an order of magnitude smaller than general-purpose CT scans (Pistoia *et al.*, 2002). This reduction in imaging requirement is reflected in the reduction in size, mobility and cost of pQCT instruments (compared to conventional CT scanners). Additionally, radiation shielding, unlike in conventional CT imaging, is also unnecessary thus substantially reducing complexity (Pistoia *et al.*, 2002). Energy sources for pQCT systems have over the years moved from ^{125}I photon sources to new generation X-ray sources (Hangartner *et al.*, 1985; Schneider and Borner, 1991; Lehmann *et al.*, 1992). Measurement precision of pQCT can be enhanced by employing computer-assisted location-matching procedures (Hangartner, 1993; Muller *et al.*, 1996b). In fact, the resolution capabilities of pQCT methods have increased to a great extent (Gordon *et al.*, 1996; Muller *et al.*, 1996a); these techniques, high-resolution pQCT (HR-pQCT) and nano-CT are discussed separately in later sections (Sections 2.3 and 2.4). Like CT, pQCT can distinguish between cortical and trabecular compartments (Eckstein *et al.*, 2004). A particular disadvantage related to the increase in resolution in pQCT compared to DXA is that radiation dose and scan times increase multifold (Augat *et al.*, 1998; Cummings *et al.*, 2002). A typical scanning instrument is the Norland/Stratec XCT-3000A. Details of protocols for testing human (and murine)

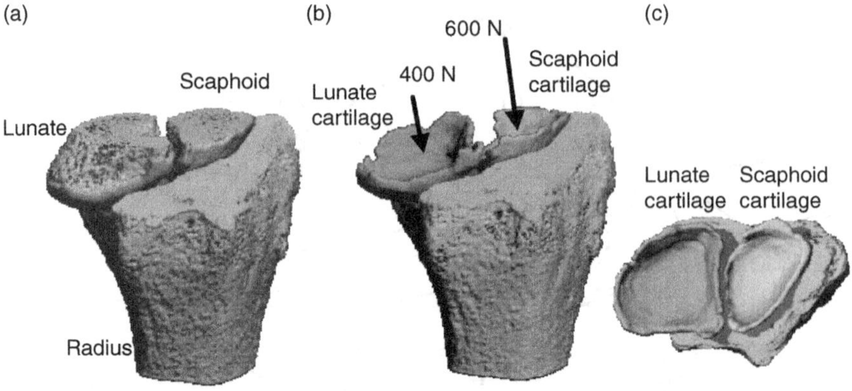

2.7 (a) The radius and the two adjacent carpal bones, together with artificially modelled cartilage (surface-rendered image from a 3D-pQCT measurement). (b) Typical three-dimensional representation of the radius and the two adjacent cartilage layers of the carpal bones as input for the FE analysis. The arrows symbolize the applied load on the radius representing a near-neutral hand position: 400 N force applied on the lunate side and 600 N force on the scaphoid side. (c) Top view of the same radius. (*Source*: Reprinted from *Bone*, **30**, Pistoia *et al.*, Estimation of distal radius failure load with micro-finite element analysis models based on three-dimensional peripheral quantitative computed tomography images, 842–848, Copyright (2002), with permission from Elsevier.)

bones are available in the literature (Groll *et al.*, 1999; Lochmuller *et al.*, 2002a; Gasser, 2003).

Quantitative ultrasonography (QUS)

The principle of QUS assessment of bone is characterized by transmission of sound through a bone specimen, essentially an elastic wave propagation phenomenon, related to architectural and constitutive properties of bone. Speed of sound (SOS) and broadband ultrasonic absorption (BUA) are sensitive to sound transmission characteristics of the architecture and composition of bone being typically measured. Thus, unlike the radiation absorption based assessment methods (McCreadie and Goldstein, 2000; Ammann and Rizzoli, 2003), the measured parameters in QUS contain information on the true mechanical properties of bone. These mechanical properties are inferred using inverse solution methods. A strong motivation for the use of QUS is its ability to sense the underlying trabecular details of cancellous bone, for example, orientation and angle-dependent tortuosity (Gluer *et al.*, 1993; Aygun *et al.*, 2009). The technique therefore has the potential to shed more light on the variation in cancellous bone compressive strength than an apparent density-based only assessment (about 36%). It is important to note that bone density varies linearly with strength only for cortical bone, while the dependence is non-linear for cancellous bone (typically power-law, e.g., Carter and Hayes, 1977; Gibson, 2005). QUS is often applied to the calcaneus or heel bone (Bouxsein and Radloff, 1997) but increasingly at other sites as well, for example, phalanges, femur, etc. (Ashman *et al.*, 1987; Gluer, 1997; Njeh *et al.*, 2000). Details of the underlying physics as well as the protocol as applied to human bone can be found in Njeh *et al.* (1999). Two important advantages of QUS are the complete lack of ionizing radiation and that it is an inexpensive tool (Hans *et al.*, 1997; Njeh *et al.*, 1997). This has led to the rapid development of numerous commercially available instruments to assess bone at peripheral sites (Gluer, 1997), for example, Achilles+ scanner (GE Lunar, Madison, WI, USA) (Lochmuller *et al.*, 2002b) and Omnisense (Sunlight/BeamMed) (Njeh *et al.*, 2000). However, the International Society for Clinical Densitometry (ISCD), as of 2007, states that for QUS bone density measurements from different devices are not directly comparable (Krieg *et al.*, 2008). The physics of how cortical/trabecular bone properties influence ultrasound measured parameters is still not clearly understood (Eckstein *et al.*, 2004).

2.3.4 Experimental findings

The complex nature of bone as a structure poses multiple challenges in interpreting results from the way testing procedures are conducted. Care needs to be taken when comparing results from two different mechanical tests (compression, bending or tension), even if an identical protocol is followed. A particularly complex concept is quantifying the risk of fracture (Bouxsein, 2005). Fracture occurs when the structure fails under a given load. A bone of low strength is not necessarily more susceptible to fracture, if the body it supports (physiological load) is sufficiently small. Considering measurement from non-destructive tests, several studies have noted that, when correlating with bone strength, it is incorrect to directly compare two parameters (bone mass and architecture) of which one is size-dependent (e.g., BMC) while the other is not (e.g., Tb.Th) (Seeman, 1999; Eckstein *et al.*, 2004).

Another common pitfall is the extraction of bone constitutive properties from structural behaviour. For instance, the elastic modulus of bone, assuming a perfect Euler–Bernoulli beam theory, is usually derived from the strength of bone during three-point bending. The case for such theories, and therefore the quantities so derived, is weak because in practice there are often significant deviations from the assumptions underlying the applicability of these theories.

Site-specificity is another confounding aspect in determining bone fragility/competence. Measurements (strength, BMC) in the spine may not

Table 2.1 Correlation coefficients (*r*) between failure loads at different sites and for various loading configurations

		Femur		Vertebra (axial compression only)		
		Vertical loading	Side-impact loading	L4	T10	L3
Radius	Fall simulation	0.81[a], 0.46[b]	0.39[b]	0.79[a]	0.44–0.47[b]	
	Axial compression	0.57[b]	0.45[b]	–	0.44[b]	0.50[b]
	3-point bending*	0.62[b]	0.61[b]	–	0.54[b]	0.56[b]

Notes: Tests on the radius were conducted at the distal end, except for (*) which was conducted at the diaphysis.
Sources: [a] Augat *et al.*, 1996; [b] Eckstein *et al.*, 2002.

correlate highly with those at the distal radius. Likewise, it is difficult to predict geometric parameters of long bones from those of vertebral segments. Nevertheless, investigation of correlations across sites remains active (Augat *et al.*, 1996) owing to significant disparities in operational costs between testing procedures at different sites. Indeed, the search for cross-site correlations can frequently be found in previous and current research. Table 2.1 cites some correlation coefficients for failure loads across different sites for various loading configurations.

Results summarized in Table 2.1 suggest that mechanical competence at different skeletal regions is highly heterogeneous. In particular, correlation coefficients between failure loads at different sites were observed to be significantly smaller than those between failure loads and bone mass measured at the same site (Augat *et al.*, 1998; Lochmuller *et al.*, 2002b). A strong biomechanical basis for such heterogeneity is attributed to the difference in load-bearing functions of bones at different sites, a known influence on bone architecture. An unusual case of histomorphometric biopsies is typically based upon specimens extracted from the iliac crest, a non-load bearing site. The implications on load-bearing long bones from histomorphometry alone are of questionable use (Augat and Schorlemmer, 2006). Future effort therefore needs to be directed at the site of interest, with focus on improving accuracy and removing technical challenges in measurement.

With respect to the above-mentioned complexities in bone mechanical quantification, it is expected that the review as presented here will not fulfil the standards of a meta-analysis. Consequently, throughout this chapter, numerical values reported for mechanical properties, whether constitutive (e.g. tissue modulus), architectural (e.g. Tb.Th) or due to both (e.g. bone failure strength) should only be taken as indicative, and must not in any case be used for comparative studies.

Wrist

Typically, fracture load at the distal radius was found to be about 2.6 kN (Myers *et al.*, 1993; Njeh *et al.*, 2000). However, within the elderly population (mean age 74 years) the failure load displays significant sex-dependence, with larger values reported for men than for women (Lochmuller *et al.*, 2002b). Among bone mass and histomorphometric parameters, variation in fracture load is best correlated with variation in BMC, cortical bone mass, cortical area and cortical width (Myers *et al.*, 1993; Augat *et al.*, 1996). Radius bone stiffness is found to agree satisfactorily with radius cortical geometric measures obtained from DXA. Compiling DXA measurements from previous studies, projected BMD (per unit area) at the ultradistal location for pre-menopausal women was found to be around 0.43 g/cm^2, whereas post-menopause this decreases to 0.35 g/cm^2 (Augat *et al.*, 1998). For osteoporotic women, the projected BMD was found to be even lower, at 0.28 g/cm^2. Employing pQCT, BMD values (measured per unit volume) for the above three groups (pre-menopause, post-menopause and osteoporotic) were found to be 0.36, 0.32 and 0.29 g/cm^3, respectively. Differentiating between cortical and trabecular bone using pQCT, the loss of bone in women with age (in particular beyond about 52 years) has been attributed most strongly to the loss in trabecular bone (Ferretti *et al.*, 1995).

Spine

Indicative results are as follows: BMC (normalized to width of scanned region) for L1–L4 ranged from 1.4 to 6.2 g/cm (Hansson *et al.*, 1980), compressive strength ranged from 2 to 12 kN (Hansson *et al.*, 1980; Brinckmann *et al.*, 1989b), but with a large variation among individuals across ages and sexes. There appears to be a very small margin of safety compared to physiological loading, which is well within the 2–12 kN range of compressive strength. Whether the small margin of safety is due to errors in the measurement of vertebral strength and why the vertebral body does not fracture more often are currently unresolved questions (Hutton *et al.*, 1979). Other researchers report that strength increases slightly when loading rate is increased, hinting at a viscoelastic response in the vertebral body (Hutton *et al.*, 1979). Within a vertebral body, bone density varies (Fig. 2.8) and this variation is thought to be the reason behind certain fracture types predominating over others. Across different vertebral sites, BMC and compressive strength increase inferiorly once again underscoring the heterogeneity in bone quality (Hansson *et al.*, 1980; Brinckmann *et al.*, 1989b). A statistically significant correlation between BMC and fracture load has been reported (Hansson *et al.*, 1980). The coefficient of correlation does not change with anatomical level but shows gender-dependence: the increase in fracture load

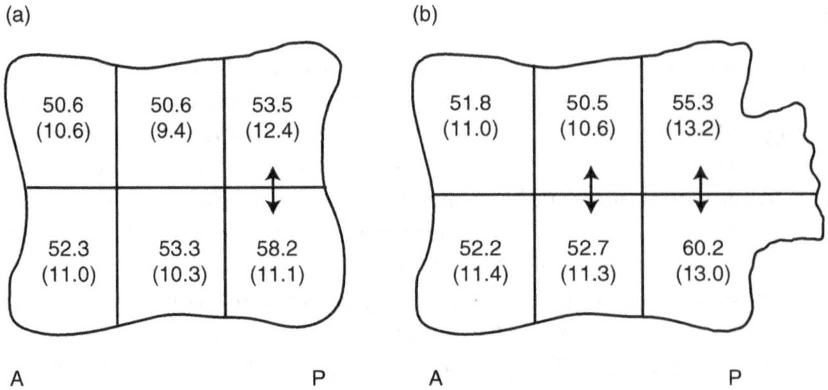

2.8 Diagrammatic representation of the optical densities of regions of trabecular bone in the vertebral body. Values represent the mean (SD) for (a) mid-sagittal and (b) pedicle slices. Numbers indicate optical density. Arrows indicate significant differences between cranial and caudal regions. (A: anterior, P: posterior). (*Source*: Reprinted from *Bone*, **44**, Zhao *et al.*, Vertebral fractures usually affect the cranial endplate because it is thinner and supported by less-dense trabecular bone, 372–379, Copyright (2009), with permission from Elsevier.)

for incremental increase in BMC is higher in men than in women. When vertebrae were imaged using CT, end-plate area was found to increase from T10 to L5 in a linear fashion, by 0.5 ± 0.19 cm^2 per end-plate (Brinckmann *et al.*, 1989b). BMD however decreased only marginally from T10 to L5. The linear correlation between the product, end-plate area × bone density, and compressive strength is excellent. Coefficients of correlation were not significantly different across age, sex or anatomical level (T10 to L5 were measured). Across sites, BMD values from spine and distal radius were not found to be strongly correlated (r = 0.4–0.7) even though trabecular bone predominates at both sites (Augat *et al.*, 1998). Histomorphometric results (Thomsen *et al.*, 2000) on the trabecular core of vertebral (L2) bodies indicate that trabecular bone volume ratio (BV/TV) drops from 15% to 7% over the ages 20–90 years. Over this same age span, trabecular pattern factor (TBPf) increases from 2.2 to 3.8. Low (even negative) TBPf values indicate well-connected 2D networks whereas high values characterize poorly connected networks. On the other hand, Tb.Th did not show a clear trend over age and any suggestion of sex-dependence was inconclusive.

Hip

For the femur, typical BMD measurements at different regions of interest (including the proximal femur) vary between 0.44 and 0.96 g/cm^2 (Cody *et al.*, 1999). Over the age range 21–93 years, the fracture load can vary from

3 to 17 kN, with values significantly larger for men than for women. For the femoral mid-shaft, the Young's modulus of cortical sections was found to be approximately 18.6 GPa (Cuppone *et al.*, 2004). At this location, the moment of inertia has been found to increase for men, but decrease for women with age (Martin and Atkinson, 1977). This is brought about by a thinning of the area of the cortical region in women whereas in men it tends to stay relatively constant. The increase in moment of inertia in men is explained by the fact that bone expands in the radial direction (simultaneous endosteal resorption and subperiosteal apposition). With age, the femoral shaft cross-section tends to become less circular due to remodelling. This causes the bone stiffness to vary with bending axis with up to a 50% difference seen between the minor and major axes. Among architectural parameters, cortical area, size of the femoral neck and area moment of inertia have been shown to be the strongest predictors of fracture load (Lochmuller *et al.*, 2002a). Coupling BMD measurements at the site of fracture with femoral neck geometry parameters (e.g., width and area moment of inertia), a correlation as high as $r = 0.93$ with fracture load has been reported (Augat *et al.*, 1996).

2.4 Micro-scale mechanical properties

Results presented in the previous section emphasize the need to use better quantification techniques than BMD and other bone mass related measures (McCreadie and Goldstein, 2000). Advanced imaging techniques (CT/pQCT) that provide architectural information improve bone competence prediction when coupled with BMD, but taken alone do not fare significantly better than BMD (Eckstein *et al.*, 2004). Although QUS combines architectural and constitutive effects within SOS and BUA measures, it is challenging to isolate these effects (Chao *et al.*, 2004) and this effectively limits a deeper understanding. There are indications that better insights can perhaps be gained at a microscopic scale rather than at the whole-bone level. For instance, considering the energy absorption, fracture toughness and ultimate tensile strain vary within 5–10% per decade of life whereas elastic moduli in tension or compression degrade by only about 2% per decade (Burstein *et al.*, 1976; Augat and Schorlemmer, 2006). Changes in relationships between macroscopic quantities with age, point towards modifications to bone architecture and constitution at the micro-scale. This notion is reinforced by the fact that non-trivial differences at the microstructural level have been shown to exist with age, sex, race and anatomic site (Yeni *et al.*, 2011) for trabecular regions of the human vertebral bone. Since bone is biologically an active hard tissue and remodels to account for loading patterns, diet and other external factors it is not straightforward to justify how a small number of whole-bone level parameters can capture all the complexity. It is well known that the bone remodelling process, due to injury or in response to the cyclic loading of daily activities, is characterized by the phenomenon of micro-cracking (Burr, 2002; Augat and Schorlemmer, 2006; Taylor, 2007). However, micro-cracks are almost impossible to capture separately by any of the whole-bone parameters. Equally challenging to capture using the previous methods is pore distribution, yet intracortical porosity accounts for about 70% of elastic modulus and 55% of yield stress. Another micro-scale feature, the collagen network, undergoes up to 50% loss in its capability to absorb energy during ageing and is a strong factor of bone competence, but is also not distinguished by macro-scale parameters (Wang *et al.*, 2001). Closely related to the collagen network, the mechanical characterization of bone tissue (or building blocks viz. lamellae, osteons, etc.) is not possible at the macro-scale. All these questions have motivated researchers to study the architecture of bone at smaller scales (including at the sub-micron scale).

2.4.1 Testing procedures

At the micro-scale mechanical testing is not destructive *per se* since the whole bone does not fail during the normal testing process. Thus the distinction between destructive and non-destructive tests is not made. The assessment methods at the micro-scale are instead focused on characterizing either the constitutive properties of bone tissue or its architecture.

Microhardness testing

Hardness of a substance is defined as its resistance to permanent deformation. In the context of bone, researchers study the deformation of the bone surface. For example, when pressed by an object that is itself much more resistant to deformation (much harder) the accompanying deformation is called indentation. The technique is performed such that inelastic (i.e., permanent) deformation occurs. In microhardness testing, the indented zone is sufficiently small (about 10–30 mm across and 0.2–12 mm deep (Zysset, 2009; Zwierzak *et al.*, 2009)) to ensure that the deformation is local. Preparation of bone surfaces prior to indentation (Yoon and Katz, 1976) and geometry of the indenter are tightly controlled factors. This ensures that the indented surface is visibly different from the unindented. The known indenter geometry makes it possible to compare with the indented surface thereby optically inferring the characteristics of indentation (depth, extent of plastic zone). The most common hardness measuring indenters are sphere (Brinell, Rockwell), four-sided pyramid (Vickers), three-sided pyramid (Berkovich) and asymmetric pyramid (Knoop). Berkovich tips are used mostly for nano-indentation (discussed in Section 2.5) and the others are widely used in microhardness testing. A typical protocol uses a Vickers diamond indenter (Leica VMHT), with a maximum load up to 1 N, applied over a span of about 15 s. Beyond maximum load, the indenter dwells for some time (5–60 s) to remove viscoelastic creep effects and is then retracted to the original position. Details on widely used measurement protocols are available in the literature (Ziv *et al.*, 1996; Huja *et al.*, 1998; Zioupos *et al.*, 2000; Johnson and Rapoff, 2007; Zwierzak *et al.*, 2009). To prepare the sample surface for indentation, fixation (usually in formalin) is necessary followed by dehydration. This procedure can be used only for larger samples. For indentation of individual trabeculae, embedding is therefore necessary for support. Hardness, H, defined as applied force, P, over indentation area, A, (Zysset, 2009) is the most commonly reported parameter for microhardness testing. The other test parameter is the relaxation modulus, E_r, from which the indentation modulus, E^{ind}, of the indented substance can be determined since the modulus of the indenter is known. Assuming that the substance is isotropic the elastic modulus, E, can be derived from the indentation modulus. If the substance is anisotropic, the indentation modulus would depend on the direction of indentation, and therefore on the elasticity tensor. The tensor can be derived by performing indentations in multiple directions (Zysset, 2009).

Reference point indentation (RPI)

RPI is essentially depth-sensing indentation. Therefore, from a mechanics perspective, the principle is similar to microhardness testing. Instead of a single indenter two coaxial probes are used. The inner probe indents the surface whereas the outer probe (which rests on the adjacent and unindented surface) serves as a reference for displacement of the inner probe (see Fig. 2.9). This depth-sensing technology is relatively new (see Randall *et al.*, 2009 for details), and in the context of bone testing, can simplify the analysis as well as the applicability of the procedure significantly. Since the instrument senses depth without the need of an optical measurement of the indented surface, associated systematic errors are avoided. Surface preparation is thereby less demanding, as a result of which RPI can be used *in vivo*. In microhardness measurements the surface should not recoil once the force is released, otherwise the optically measured indentation geometry would not correspond to the maximum force applied. Thus large loads that can induce plastic deformation of the surface are required. In RPI, since depth measurement is not dependent upon *a posteriori* optical measurement, large loads are not required, and damage to bone at organ level is not encountered. The test protocol comprises multiple indentation cycles, each of which includes a load, a dwell and an unloading phase. Typically, each phase spans one-third of the cycle. Randall *et al.* (2009), performed 10 cycles, each with a period of 0.5 s and the maximum applied load in each cycle varied from 2 to 10 N. The indentation distance increase (IDI) is the most common parameter in an RPI procedure. IDI is defined as the increase in

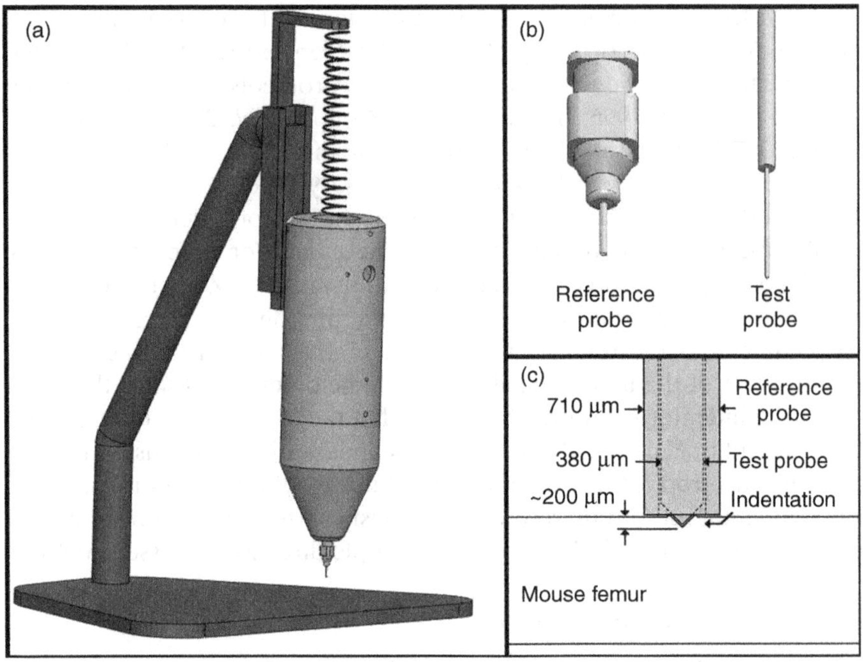

2.9 Schematic of the bone diagnostic instrument (BDI). (a) The BDI is supported by a spring to reduce the weight from 1.58 to 0.39 kg and held in place by a track, allowing only vertical movement. (b) The reference probe can be seen with a blunt end (left) and the test probe (right) has a 25 μm tip radius with a 90° conical end. (c) The blunt reference probe rests on the femur surface while the test probe indents into the bone to a total distance of approximately 200 μm. (*Source*: Reprinted with permission from *Review of Scientific Instruments*, **80**, Randall *et al.*, The bone diagnostic instrument III: Testing mouse femora, Copyright (2009), American Institute of Physics.)

the indentation distance in the last cycle relative to the indentation distance in the first cycle. It is a measure of the ability of the bone to resist additional deformation with repetitive loading (i.e., a local post-yield measure).

High-resolution pQCT (HR-pQCT) and micro-CT

HR-pQCT and micro-CT are based on the same principle as CT and pQCT (discussed earlier), but have substantially higher resolution. In HR-pQCT, an isotropic spatial resolution in the order of 82 μm is possible with the latest scanners, and is applied typically to peripheral sites. Micro-CT is considered the gold standard for microstructural imaging and can achieve resolutions in the order of 19 μm. Thus HR-pQCT has somewhat lower resolution than micro-CT, however the main advantage of HR-pQCT is that it can be applied *in vivo*. A good comparison between HR-pQCT and micro-CT is given by MacNeil and Boyd (2007). An excellent review of methods, capabilities, protocols and common instruments used in HR-pQCT is given by Patsch *et al.* (2011). Measures of 3D bone geometry, overall and compartment-specific bone density and bone microarchitecture can be acquired within a scan time of approximately 3 min. Radiation dose for a typical HR-pQCT application is of the order of 4 μSv, which is several orders of magnitude smaller than a typical 2 mSv dose during a clinical abdominal CT scan (Eckstein *et al.*, 2004; Patsch *et al.*, 2011). The HR-pQCT method has high accuracy compared to histomorphometry, high *in vivo* reproducibility and is relevant for micro-finite element (μFE) modelling (discussed in Section 2.6.2).

2.4.2 Experimental findings

The indentation moduli of bone tissue (Table 2.2) are found to vary from 7.5 to 25 GPa across anatomical sites. It is important to note that a micro-scale constitutive property at a particular location of the bone cannot be representative of the whole bone. Only data averaged over multiple locations within each bone can be meaningfully compared, instead of comparing data at individual locations (even if identical). Fixation of samples can increase hardness by up to 20%, whereas drying the sample also increases hardness by up to 54% (depending on how long the drying process is continued). Upon drying, a 12% increase in the hardness for cortical bone (Zwierzak *et al.*, 2009) and an increase up to 20% for cancellous bone has been noted (Amprino, 1958; Weaver, 1966). The effect of drying on the indentation modulus appears to be weaker and the increase can be as much as 23% (Hoffler *et al.*, 2005). Boivin *et al.* (2008) found that the organic collagen matrix had one-third the hardness of calcified tissue. Thereby it potentially accounts for as much as one-third of the variance in the stiffness of bone tissue. Embedding has been found to have a minor effect on hardness. However, the effect of indentation modulus can be substantial if the resin used infiltrates the pores in the trabeculae. The hardness value was also found to change significantly if the location of indentation was too close to a Haversian canal (less than 60 mm for a 1 N load) (Zwierzak *et al.*, 2009). Because of the additional challenges of using hardness values to determine mechanical properties of bone and its limited application *in vivo*, the relevance of classical microhardness testing of bone is being increasingly questioned.

In contrast, RPI is under active development and is currently restricted largely to animal studies (Proctor, 2010; Jast, 2011; Kaye *et al.*, 2012). Studies on tibial diaphysis (Diez-Perez *et al.*, 2010) and on human tibial plateau cartilage (Tang *et al.*, 2011) are some of the few measurements carried out on human tissue. IDI was found to increase significantly in osteoporotic women

Table 2.2 Indentation measurements of human bone at different anatomical sites, measured in the longitudinal direction

Site	BSU	E^{ind} (GPa)	Reference
Distal radius	Osteonal	17.8 ± 2.2	(Hoffler *et al.*, 2000)
Distal radius	Interstitial	18.3 ± 2.0	(Hoffler *et al.*, 2000)
Distal radius	Trabecular	15.0 ± 1.8	(Hoffler *et al.*, 2000)
Vertebra*	Trabecular	21.3 ± 2.5	(Rho *et al.*, 1999)
Vertebra	Enthesophytic	12.0 ± 1.4	(Hoffler *et al.*, 2000)
Vertebra	Circumferential lamellae	9.7 ± 1.1	(Hoffler *et al.*, 2000)
Vertebra	Trabecular	8.9 ± 1.4	(Hoffler *et al.*, 2000)
Femur neck	Osteonal	17.4 ± 5.8	(Zysset *et al.*, 1999)
Femur neck	Interstitial	19.2 ± 5.8	(Zysset *et al.*, 1999)
Femur neck	Trabecular	12.5 ± 6.2	(Zysset *et al.*, 1999)

Notes: For each anatomical site, indentation measurements are further grouped according to the bone structural unit (BSU) that was indented. Bone samples were analyzed under wet condition, and an indentation depth of 500 nm was used, except in the experiment marked (*), where the bone sample was fixed, and the indentation depth was 1000 nm.
Source: Adapted from *Osteoporosis International*, **20**, Zysset, Indentation of bone tissue: A short review, 1049–1055, Copyright (2009), with kind permission from Springer Science and Business Media.

(25 hip fractures and 2 multiple vertebral fractures) compared to controls (from 12 to 18 mm) (Diez-Perez *et al.*, 2010).

Using HR-pQCT, BV/TV and Tb.Th. at the ultradistal radius were reported to be significantly higher in young men than in young women (Khosla *et al.*, 2006). Furthermore, the rate of decline in trabecular bone volume with age is more or less sex-independent. In ageing but otherwise healthy men, trabecular thinning (measured by Tb.Th) appeared to predominate over actual loss of trabeculae (measured by Tb.N), the latter being characteristic of women (Lochmueller *et al.*, 2008). HR-pQCT measurements at the femoral mid-shaft showed the correlation between CT number and cortical bone modulus is not significant (Cuppone *et al.*, 2004). Microstructural changes with location within the vertebral body have also been investigated in detail (Adams and Dolan, 2011). There is a general agreement that these differences critically affect predisposition to certain fractures over others (Banse *et al.*, 2001; Adams *et al.*, 2006; Hulme *et al.*, 2007; Zhao *et al.*, 2009).

2.5 Nano-scale mechanical properties

The variation in the hardness response of BSUs at different sites (see Table 2.2) and due to disease (Zysset, 2009) implies that differences must exist in the underlying sub-micron scale architecture. Previous research has indicated that osteon microhardness is related to its mineral content (Carlstrom, 1954; Amprino, 1958; Weaver, 1966). Indeed, toughness and other post-yield deformation of the tissue are affected by collagen characteristics in the ECM such as collagen content and maturity or cross-linking (Burstein *et al.*, 1975; Burr, 2002). Bone tissue toughness and strength are also affected by both bound and unbound water present within the bone tissue. The distribution of water molecules (a sub-micron scale feature) is altered during the bone remodelling process (Timlin *et al.*, 2000; Nyman *et al.*, 2006; Ruppel *et al.*, 2006). These aspects are the motivation for current research on nanoscale properties of bone so as to understand their effect on whole-bone mechanical behaviour.

2.5.1 Imaging: synchrotron-radiation nano-CT

Nano-CT is the reference method for assessing trabecular microarchitecture parameters and can produce resolutions of around 1 μm and even below (Stampanoni *et al.*, 2002, 2003; Thurner *et al.*, 2006). It is based on the acceleration of electrons that generate a high amount of secondary radiation, resulting in parallel monochromatic radiation (photons with the same energy) and no beam hardening. Highly coherent beams are produced, compared to conventional X-ray sources leading to improvement in image resolution and contrast, and reduction in geometric distortion (Fig. 2.10). Some major drawbacks are that the system requires a high-cost infrastructure and the equipment can only be used *in vitro*.

2.5.2 Mechanical testing: nano-indentation

Nano-indentation is similar in concept to micro-indentation; however, the length scale is an order of magnitude smaller (Zysset, 2009). Due to the reduction in scale, sample preparation needs to be even more tightly controlled. Sections to be analysed have a thickness in the order of 100 μm. Sections are progressively ground and finally polished with a diamond suspension to achieve surface roughness below 1 μm (Boivin and Meunier, 2002). Typically, biopsy specimens are fixed in 70% alcohol, dehydrated in absolute alcohol and then embedded in methyl methacrylate (MMA). Similar to micro-indentation, fixing in resins remains a concern (Amprino, 1958), since the process can alter mechanical properties. The indenter is usually a Berkovich diamond probe (three-side pyramidal tip) with a total included angle of 142.3° and a nominal tip radius of 100 nm. Successive indentations on the same specimen must be separated by about 20 indentation diameters to prevent interference (Norman *et al.*, 2008). Some common commercially available instruments for this test are Nano Indenter II (Nano Instruments Inc., USA), Hysitron TriboIndenter (Hysitron Inc., Minneapolis MN) and Nanohardness tester (CSM Instruments, Switzerland). Testing

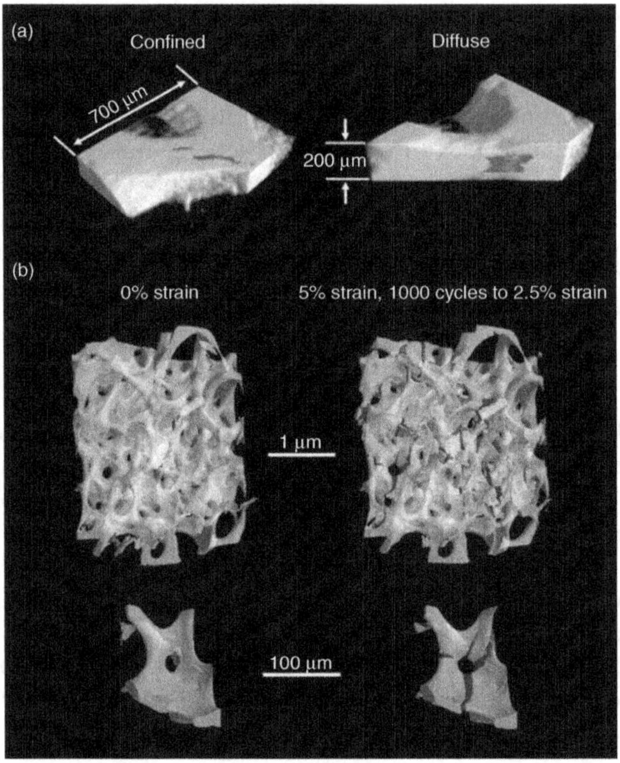

2.10 Illustration of a typical result of the crack detection algorithm using data from a cylindrical extract of bovine trabecular bone. The cracks appear in darker shades, whereas the bone appears transparent white. (*Source*: Reprinted from *Bone*, **39**, Thurner *et al.*, Time-lapsed investigation of three-dimensional failure and damage accumulation in trabecular bone using synchrotron light, 289–299, Copyright (2006), with permission from Elsevier.)

protocols typically comprise multiple cycles with loading and unloading phases. Each cycle is described by five successive steps (Fig. 2.11): loading to a maximum force which can vary from 2 to 500 mN (higher values are used to overcome the effects related to the heterogeneity of bone tissue at lamellar level) at a rate of 0.2–4 mN/s; holding for a period of 5–40 s (to limit the viscous behaviour of bone tissue); releasing the load to a fraction (around 10%) of maximum displacement; a hold period (20–50 s) for thermal drift calculation; and finally withdrawal to zero displacement (Chih Ling *et al.*, 2010; Bala *et al.*, 2011; Jirousek *et al.*, 2011).

Elastic modulus of the substance is obtained from the indentation response using the Oliver–Pharr method, and comparing it with a known substance (typically fused silica) (Oliver and Pharr, 1992). Analysis of nano-indentation

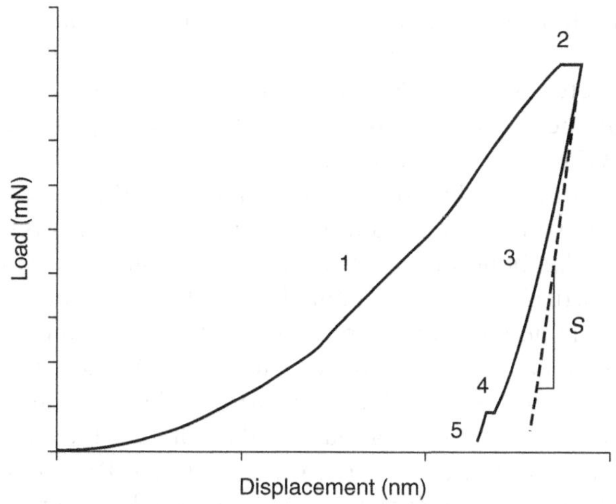

2.11 Typical load-depth curve obtained with indentation testing. The curve shows regions related to loading (1), holding segment (2), partial unloading (3), holding segment allowing the thermal drift correction (4) and complete unloading (5). Here *S* is the contact stiffness corresponding to the slope of the unloading curve tangent. (*Source*: Adapted from the *Journal of the Mechanical Behavior of Biomedical Materials*, **4**, Bala *et al*., Respective roles of organic and mineral components of human cortical bone matrix in micromechanical behavior: An instrumented indentation study, 1473–1482, Copyright (2011), with permission from Elsevier.)

results and troubleshooting during testing has been discussed in the literature (Oyen and Cook, 2009). Recognizing the underlying anisotropy of the bone tissue, a theory to quantify the indentation modulus from the stiffness tensor of a given substance has been formulated (Swadener and Pharr, 2001). An inverse solution can be used to determine the microstructural fabric-based stiffness tensor from the indentation moduli in different indentation orientations (Zysset, 2003). Viscoelastic response of bone during nano-indentation requires specialized analysis techniques (Oyen, 2008). Although it is known that tissue hydration can significantly modify elastic properties, technical challenges exist in conducting nano-indentation on wet bones (Hengsberger *et al.*, 2002).

2.5.3 Experimental findings

From nano-indentation measurements on the vertebral trabeculae in the transverse direction it was found that the average elastic modulus was around 14 GPa, and average hardness was approximately 470 MPa (Rho *et al.*, 1997). For trabecular bone segments sourced from the proximal femur, the indentation modulus was found to be 14–18 GPa (Jirousek *et al.*, 2011). In the trabecular bone of the proximal femur micromechanical properties were found to be heterogeneous (Norman *et al.*, 2008). The elastic modulus dropped by 17% in the transverse direction of a trabecular spicule, from the inner medial (14 GPa) to the outer margin. Along the same axis, hardness dropped by 32% (inner medial, 0.90 GPa). Significant differences across age (37–93 years) have not been found for micromechanical properties of cortical and trabecular bone obtained from the femoral head (Chih Ling *et al.*, 2010). For cortical segments obtained from the distal femoral shaft, secondary osteons (lamellar bone) were found to be anisotropic as a result of orientation of lamellae. The anisotropy factor (E_L/E_T) was found to vary within the same bone and reach values close to 1.5. Elastic modulus in the axial direction (E_L) was found to be about 25 GPa, whereas the elastic modulus in the circumferential direction (E_T) was about 17 GPa (Hengsberger *et al.*, 2001, 2002, Franzoso and Zysset, 2009). Similar values were obtained for the tibia (Rho *et al.*, 1997; Fan *et al.*, 2002).

Nano-indentation tests have been conducted on iliac biopsies obtained from pre- and post-menopausal women to determine frequency-dependent (dynamic) response of the specimens (Polly *et al.*, 2012). No significant difference in indentation modulus and hardness values between pre- and post-menopausal samples have been found. The dynamic storage modulus, E', was larger for the pre-menopausal group but the loss modulus, E'', and loss tangent, tan δ, showed the opposite trend at all frequencies. However, these differences were not significant, at any given frequency. Reduced modulus and hardness of the tibial interstitial lamellae were significantly higher than those of tibial osteons and vertebral trabeculae (Rho *et al.*, 1997).

2.6 Hierarchical or multi-scale methods of bone quality assessment

Methods considered in this section combine the assessment of bone at multiple scales, unlike those discussed in the foregoing sections. The advantage in these novel techniques is that they build upon more mature individual methods described above, and can be considered the cutting-edge of bone research. Presently used mostly for research purposes, it is expected that these methods will be employed clinically in the future.

2.6.1 Mechanical testing combined with functional micro-imaging

The basic principles of whole-bone mechanical testing have been discussed in Section 1.3.2. Functional micro-imaging is the use of high-resolution imaging (see Section 1.5.1) to guide the assessment of micro-scale mechanical behaviour (described in Section 1.4) during whole-bone testing. In a typical set-up, mouse femurs are mechanically tested and simultaneously imaged at three different length scales (Voide *et al.*, 2006). At the macro-level, three-point bending and axial compression of the femoral neck were conducted until failure. Images (at 10 mm resolution) were captured by a high-speed camera. This provided insight into how cracks initiate locally and propagate further, leading to failure of the entire structure. The understanding gained was used to determine a smaller field of view relevant to the crack propagation process. The mechanical tests were repeated but with micro-CT (3.5 mm resolution) focused at the above-mentioned region. At this micrometre scale, micro-scale structures (e.g., for cortical void spaces) were tracked throughout the process and statistically analysed. The insights gained were used to design mechanical tests conducted on a smaller section of the mouse femur, with a notch created to localize crack initiation. Nano-CT (700 nm resolution) was used to assess the response of sub-micron structures (e.g., for canaliculi) during cracking and eventual failure. This and similar time-lapsed functional-imaging research have been employed to examine differences in the crack propagation process in in-bred murine strains that have known differences in bone phenotype (Nazarian *et al.*, 2005; Voide *et al.*, 2009). Phenotype differences can be induced into in-bred mouse models by controlling factors like disease, diet, etc. Therefore, the above method can be used to reveal how and why external factors cause bones to be predisposed to failure under a given type of loading (see Fig. 2.12).

2.6.2 Image-based based μFE analysis

In earlier sections we have discussed how advances in the field of imaging have made quantifying bone microstructural parameters possible. Recent advances in computational modelling permit such parameters to serve as input into continuum mechanics-based μFE models of whole bone (Pistoia *et al.*, 2003; Homminga *et al.*, 2004; Eswaran *et al.*, 2007; van Lenthe *et al.*, 2008; Fields *et al.*, 2009; Lenaerts and van Lenthe, 2009; Burghardt *et al.*, 2011; Donnelly, 2011). This framework can be exercised in different investigative modes. In one approach, the objective is to understand the effect of microstructure on mechanical response (e.g., stiffness, toughness, post-yield behaviour, etc.) of whole bones, especially at clinically important sites such as the distal radius, vertebral body, proximal femur and femoral shaft. High-resolution images of the intact bone are used to define the macro- and micro-scale architectural detail of the computational model. Absolute values of constitutive properties (Young's modulus of bone tissue, Poisson's ratio, etc.) are either assumed, or scaled by the external load. Computer

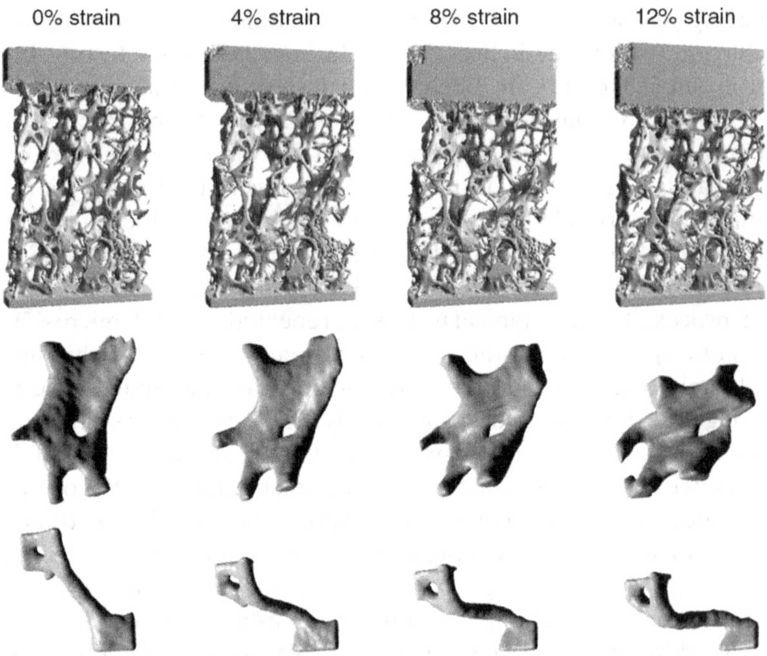

2.12 Failure assessment in human spine sample using time-lapsed tomographic imaging. Upper row shows a compressed specimen, imaged in steps of 4% strain. Middle and lower rows show how micro-compression can be used to non-invasively monitor the deformation of individual plates and rods respectively. (*Source*: Reproduced from *Mechanics of Biological Tissue* (Holzapfeland Ogden, eds.), Müller *et al.*, Functional microimaging at the interface of bone mechanics and biology, 473–487, Copyright (2006), with kind permission from Springer Science and Business Media.)

simulations reveal micro-scale details of the structural response to an applied load.

In another approach, the combined imaging and computational modelling framework is used to determine mechanical properties of bone tissue. Mechanical tests are conducted on the physical specimen to determine whole-bone level properties. Computer simulations are performed with micromechanical property parameter(s) (e.g., Young's modulus or elasticity tensor) being varied parametrically. The choice of parameter for which the macro-level response (e.g., stiffness) matches between the FE model and physical specimen is interpreted to characterize the tissue (van Rietbergen *et al.*, 1995). The freedom to analyse the same model repeatedly but under different types of loading is a distinct and major advantage of μFE methods. In using FE models that incorporate only whole-bone level geometry (e.g., beam models for bending), approximations on imposed boundary conditions frequently contribute to errors (Keaveny *et al.*, 1999). In contrast, the ability to easily apply natural boundary conditions in μFE modelling of whole bones is a significant advantage.

2.6.3 Experimental findings

In Section 2.3.4, correlations between bone microstructure and bone competence at the macro-scale were reviewed. The distinguishing advantage of hierarchical models, in particular image-based μFE presented above, is that these allow for physics-based explanations (instead of statistical indices) to relate microarchitecture and whole-bone properties. For example, it has been determined that load-distribution from articular surfaces of long bones to the cortical shell, causes individual trabeculae to predominantly deform in bending mode (van Rietbergen *et al.*, 1995). Other workers have found that modulus and strength of bone tissue found in trabecular bone are only slightly lower than in cortical bone, and that microstructure plays a more important role in distinguishing macro-scale properties in these bones (Bayraktar *et al.*, 2004). It has been reported that trabecular bone tissue modulus, extracted from the proximal tibia, lies within the range 2.23–10.1 GPa (van Rietbergen *et al.*, 1995), whereas for the vertebra, the mean modulus is 6.6 GPa (Ladd *et al.*, 1998). Research using image-based μFE models has demonstrated that to predict trabecular bone failure properties at the macro-scale, strength asymmetry in tension and compression must be incorporated (Niebur *et al.*, 2000).

Image-guided failure assessment has shown macro-scale failure of trabecular bone can be caused by failure of an individual trabecula within the structure (Muller *et al.*, 1998). This has led to using a 'weakest link of the chain' approach to predict strength and failure at the whole-bone scale. The identification of the 'weakest link' is based on investigation of local bone trabecular microarchitecture. It has been reported that, prior to catastrophic collapse at the macro-scale, a band-like local failure behaviour of trabecular bone occurred. Within this band, maximum local strain values were 5–8 times greater than the externally applied strain. It is important to note that the above insights are obtainable exclusively by employing a multi-scale approach to modelling.

2.7 Conclusions

At the whole-bone level mechanical testing remains the gold standard in determining bone competence. At the macro-scale, bone mechanical properties vary significantly dependent upon anatomical site. Furthermore, the correlations between mechanical properties at different sites are usually only moderate, and do not have high predictive capability. At a given site, mechanical properties show strong sex-dependence, especially in the elderly. In adult women, age (specifically, years after menopause) also plays an important role in determining bone quality. These results strongly suggest that specimen-specific approach design of bone substitute biomaterials can lead to much improved performance compared to a generic approach to design. However, designing specimen-specific bone substitute biomaterials to match macro-scale bone mechanical properties is non-trivial. This is because mechanical testing of whole bone is destructive, and cannot be included within the design process. On the other hand, non-destructive measures at the macro-scale (e.g., bone densitometric and macro-scale geometric measures) have failed to show clear trends with macro-scale bone competence, making design based solely on these non-destructive measures unreliable. The absence of such clear trends can be attributed to differences in constitutive and architectural properties of bone at the micro-scale which densitometry and histomorphometry are unable to capture in sufficient detail. Therefore direct assessment of the micro-scale properties of bone has the potential to guide specimen-specific design of bone substitute biomaterials.

Research and development of technology over the past few decades has made the determination of bone microstructure and tissue properties possible. Methods of high-resolution imaging have matured over the years and can deliver microstructural information with quality comparable to histomorphometry. HR-pQCT, which has a low radiation dose, can be used *in vivo* in man; its usage is expected to become widespread once its additive benefit in fracture risk prediction has been proven. Recent developments in depth-sensing micro-indentation testing have opened up significant opportunities to assess bone tissue mechanical properties. Though restricted to research use at present, these methods are expected to come into *in vivo* use in the recent future. QUS has the potential to distinguish differences in bone microstructure or bone tissue mechanical properties simultaneously. However, further development is needed in order for QUS results to be interpreted with clarity and thus become useful in diagnosis. Presently, extracting the separate effects of microstructure and bone tissue properties from QUS output remains a challenge.

Computational models of whole-bone behaviour can be developed by combining bone microarchitecture imaging with assessment of bone tissue constitutive properties. Advances in computational methods, and capabilities of modern computing hardware, point towards a promising future for such a hierarchical approach. Nano-scale assessment of bone provides interesting insights into bone mechanical characteristics. The major challenge in this approach is the integration (computationally) of nano-scale constitutive and architectural detail that is required to deliver a quantitative prediction of bone at the macro-scale. Integrative approaches like functional micro-imaging have demonstrated that computational models supplied with architectural and constitutive information correspond well with scale-matched experimental models.

2.8 References

Adams, M. A. and Dolan, P. (2011). Biomechanics of vertebral compression fractures and clinical application. *Archives of Orthopaedic and Trauma Surgery,* **131**, 1703–1710.

Adams, M. A., Pollintine, P., Tobias, J. H., Wakley, G. K. and Dolan, P. (2006). Intervertebral disc degeneration can predispose to anterior vertebral fractures in the thoracolumbar spine. *Journal of Bone and Mineral Research,* **21**, 1409–1416.

Ammann, P. and Rizzoli, R. (2003). Bone strength and its determinants. *Osteoporosis International,* **14**, S13–S18.

Amprino, R. (1958). Investigations on some physical properties of bone tissue. *Acta Anatomica,* **34**, 161–186.

Ashman, R. B., Corin, J. D. and Turner, C. H. (1987). Elastic properties of cancellous bone: Measurement by an ultrasonic technique. *Journal of Biomechanics,* **20**, 979–986.

Aspden, R. M. (2003). Mechanical testing of bone ex-vivo. In: Helfrich, M. H. and Ralston, S. H. (eds.) *Bone Research Protocols.* Totowa, NJ, USA, Humana Press Inc.

Augat, P., Fuerst, T. and Genant, H. K. (1998). Quantitative bone mineral assessment at the forearm: A review. *Osteoporosis International,* **8**, 299–310.

Augat, P., Reeb, H. and Claes, L. E. (1996). Prediction of fracture load at different skeletal sites by geometric properties of the cortical shell. *Journal of Bone and Mineral Research,* **11**, 1356–1363.

Augat, P. and Schorlemmer, S. (2006). The role of cortical bone and its microstructure in bone strength. *Age and Ageing,* **35**, 27–31.

Aygun, H., Attenborough, K., Postema, M., Lauriks, W. and Langton, C. M. (2009). Predictions of angle dependent tortuosity and elasticity effects on sound propagation in cancellous bone. *Journal of the Acoustical Society of America,* **126**, 3286–3290.

Bala, Y., Depalle, B., Douillard, T., Meille, S., Clement, P., Follet, H., Chevalier, J. and Boivin, G. (2011). Respective roles of organic and mineral components of human cortical bone matrix in micromechanical behavior: An instrumented indentation study. *Journal of the Mechanical Behavior of Biomedical Materials,* **4**, 1473–1482.

Banse, X., Devogelaer, J. P., Munting, E., Delloye, C., Cornu, O. and Grynpas, M. (2001). Inhomogeneity of human vertebral cancellous bone: Systematic density and structure patterns inside the vertebral body. *Bone,* **28**, 563–571.

Bayraktar, H. H., Morgan, E. F., Niebur, G. L., Morris, G. E., Wong, E. K. and Keaveny, T. M. (2004). Comparison of the elastic and yield properties of human femoral trabecular and cortical bone tissue. *Journal of Biomechanics,* **37**, 27–35.

Boivin, G., Bala, Y., Doublier, A., Farlay, D., Ste-Marie, L. G., Meunier, P. J. and Delmas, P. D. (2008). The role of mineralization and organic matrix in the microhardness of bone tissue from controls and osteoporotic patients. *Bone,* **43**, 532–538.

Boivin, G. and Meunier, P. J. (2002). The degree of mineralization of bone tissue measured by computerized quantitative contact microradiography. *Calcified Tissue International,* **70**, 503–511.

Bouxsein, M. L. (2005). Determinants of skeletal fragility. *Best Practice and Research in Clinical Rheumatology,* **19**, 897–911.

Bouxsein, M. L. and Radloff, S. R. (1997). Quantitative ultrasound of the calcaneus reflects the mechanical properties of calcaneal trabecular bone. *Journal of Bone and Mineral Research,* **12**, 839–846.

Brinckmann, P., Biggemann, M. and Hilweg, D. (1989a). Prediction of the compressive strength of human lumbar vertebrae. *Spine,* **14**, 606–610.

Brinckmann, P., Biggemann, M. and Hilweg, D. (1989b). Prediction of the compressive strength of human lumbar vertebrae. *Clinical Biomechanics,* **4**, iii–iv,1–27.

Brinckmann, P., Johannleweling, N., Hilweg, D. and Biggeman, M. (1987). Fatigue fracture of human lumbar vertebrae. *Clinical Biomechanics,* **2**, 94–97.

Brydone, A. S., Meek, D. and Maclaine, S. (2010). Bone grafting, orthopaedic biomaterials, and the clinical need for bone engineering. *Proceedings of the Institution of Mechanical Engineers, Part H: Journal of Engineering in Medicine,* **224**, 1329–1343.

Burghardt, A. J., Link, T. M. and Majumdar, S. (2011). High-resolution computed tomography for clinical imaging of bone microarchitecture. *Clinical Orthopaedics and Related Research,* **469**, 2179–2193.

Burr, D. B. (2002). The contribution of the organic matrix to bone's material properties. *Bone,* **31**, 8–11.

Burstein, A. H., Reilly, D. T. and Martens, M. (1976). Aging of bone tissue: Mechanical properties. *Journal of Bone and Joint Surgery-American Volume,* **58**, 82–86.

Burstein, A. H., Zika, J. M., Heiple, K. G. and Klein, L. (1975). Contribution of collagen and mineral to elastic-plastic properties of bone. *Journal of Bone and Joint Surgery-American Volume,* **57**, 956–961.

Cann, C. E. (1987). Quantitative CT applications: Comparison of current scanners. *Radiology,* **162**, 257–261.

Carlstrom, D. (1954). Micro-hardness measurements on single Haversian systems in bone. *Experientia,* **10**, 171–172.

Carter, D. R. and Hayes, W. C. (1977). Compressive behavior of bone as a 2-phase porous structure. *Journal of Bone and Joint Surgery-American Volume,* **59**, 954–962.

Chao, E. Y. S., Inoue, N., Koo, T. K. K. and Kim, Y. H. (2004). Biomechanical considerations of fracture treatment and bone quality maintenance in elderly patients and patients with osteoporosis. *Clinical Orthopaedics and Related Research,* **425**, 12–25.

Chih Ling, L., Han, H., Cribb, B. W. and Russell, A. (2010). A study of mechanical properties of human femoral heads using nanoindentation. *Advanced Materials Research,* **126–128**, 957–962.

Cierniak, R. (2011). Technical concepts of X-ray computed tomography scanners. In: *X-ray Computed Tomography in Biomedical Engineering.* Berlin, Springer-Verlag, 21–62.

Cody, D. D., Flynn, M. J. and Vickers, D. S. (1989). A technique for measuring regional bone-mineral density in human lumbar vertebral bodies. *Medical Physics,* **16**, 766–772.

Cody, D. D., Gross, G. J., Hou, F. J., Spencer, H. J., Goldstein, S. A. and Fyhrie, D. P. (1999). Femoral strength is better predicted by finite element models than QCT and DXA. *Journal of Biomechanics,* **32**, 1013–1020.

Cole, J. H. and van der Meulen, M. C. H. (2011). Whole bone mechanics and bone quality. *Clinical Orthopaedics and Related Research,* **469**, 2139–2149.

Cumming, R. G., Nevitt, M. C. and Cummings, S. R. (1997). Epidemiology of hip fractures. *Epidemiologic Reviews,* **19**, 244–257.

Cummings, S. R., Bates, D. and Black, D. M. (2002). Clinical use of bone densitometry: Scientific review. *Journal of the American Medical Association,* **288**, 1889–1897.

Cummings, S. R. and Black, D. (1995). Bone mass measurements and risk of fracture in Caucasian women: A review of findings from prospective studies. *American Journal of Medicine,* **98**, S24–S28.

Cuppone, M., Seedhom, B. B., Berry, E. and Ostell, A. E. (2004). The longitudinal Young's modulus of cortical bone in the midshaft of human femur and its correlation with CT scanning data. *Calcified Tissue International,* **74**, 302–309.

Delmas, P. D., Marin, F., Marcus, R., Misurski, D. A. and Mitlak, B. H. (2007). Beyond hip: Importance of other nonspinal fractures. *American Journal of Medicine,* **120**, 381–387.

Diez-Perez, A., Guerri, R., Nogues, X., Caceres, E., Jesus Pena, M., Mellibovsky, L., Randall, C., Bridges, D., Weaver, J. C., Proctor, A., Brimer, D., Koester, K. J., Ritchie, R. O. and Hansma, P. K. (2010). Microindentation for in vivo measurement of bone tissue mechanical properties in humans. *Journal of Bone and Mineral Research,* **25**, 1877–1885.

Doherty, D. A., Sanders, K. M., Kotowicz, M. A. and Prince, R. L. (2001). Lifetime and five-year age-specific risks of first and subsequent osteoporotic fractures in postmenopausal women. *Osteoporosis International,* **12**, 16–23.

Donnelly, E. (2011). Methods for assessing bone quality: A review. *Clinical Orthopaedics and Related Research,* **469**, 2128–2138.

Eckstein, F., Kuhn, V. and Lochmuller, E. M. (2004). Strength prediction of the distal radius by bone densitometry: Evaluation using biomechanical tests. *Annals of Biomedical Engineering,* **32**, 487–503.

Eckstein, F., Lochmuller, E. M., Lill, C. A., Kuhn, V., Schneider, E., Delling, G. and Muller, R. (2002). Bone strength at clinically relevant sites displays substantial heterogeneity and is best predicted from site-specific bone densitometry. *Journal of Bone and Mineral Research,* **17**, 162–171.

Eswaran, S. K., Gupta, A. and Keaveny, T. M. (2007). Locations of bone tissue at high risk of initial failure during compressive loading of the human vertebral body. *Bone,* **41**, 733–739.

Fan, Z., Swadener, J. G., Rho, J. Y., Roy, M. E. and Pharr, G. M. (2002). Anisotropic properties of human tibial cortical bone as measured by nanoindentation. *Journal of Orthopaedic Research,* **20**, 806–810.

Ferretti, J. L., Capozza, R. F., Salica, D., Tysarczyk-Niemeyer, G., Schiessl, H. and Steffens, M. (1995). Noninvasive estimation of bending or torsion strength of the human radius by peripheral quantitative computed-tomography (pQCT). *Journal of Bone and Mineral Research,* **10**, S372–S372.

Fields, A. J., Eswaran, S. K., Jekir, M. G. and Keaveny, T. M. (2009). Role of trabecular microarchitecture in whole-vertebral body biomechanical behavior. *Journal of Bone and Mineral Research,* **24**, 1523–1530.

Franzoso, G. and Zysset, P. K. (2009). Elastic anisotropy of human cortical bone secondary osteons measured by nanoindentation (vol. 131, art. no. 021001, 2009). *Journal of Biomechanical Engineering-Transactions of the ASME,* **131**, 021001-1-11.

Gasser, J. A. (2003). Bone measurements by peripheral quantitative computed tomography in rodents. In: Helfrich, M. H. and Ralston, S. H. (eds.) *Bone Research Protocols.* Totowa, NJ, USA, Humana Press Inc.

Genant, H. K. and Boyd, D. (1977). Quantitative bone-mineral analysis using dual energy computed tomography. *Investigative Radiology,* **12**, 545–551.

Gibson, L. J. (2005). Biomechanics of cellular solids. *Journal of Biomechanics,* **38**, 377–399.

Gluer, C. C. (1997). Quantitative ultrasound techniques for the assessment of osteoporosis: Expert agreement on current status. *Journal of Bone and Mineral Research,* **12**, 1280–1288.

Gluer, C. C., Wu, C. Y. and Genant, H. K. (1993). Broad-band ultrasound attenuation signals depend on trabecular orientation: An in-vitro study. *Osteoporosis International,* **3**, 185–191.

Goldfarb, C. A., Yin, Y., Gilula, L. A., Fisher, A. J. and Boyer, M. I. (2001). Wrist fractures: What the clinician wants to know. *Radiology,* **219**, 11–28.

Gordon, C. L., Webber, C. E., Adachi, J. D. and Christoforou, N. (1996). In vivo assessment of trabecular bone structure at the distal radius from high-resolution computed tomography images. *Physics in Medicine and Biology,* **41**, 495–508.

Greenfield, M. A. (1992). Current status of physical measurements of the skeleton. *Medical Physics,* **19**, 1349–1357.

Groll, O., Lochmuller, E. M., Bachmeier, M., Willnecker, J. and Eckstein, F. (1999). Precision and intersite correlation of bone densitometry at the radius, tibia and femur with peripheral quantitative CT. *Skeletal Radiology,* **28**, 696–702.

Hangartner, T. N. (1993). The osteoquant: An isotope-based CT scanner for precise measurement of bone-density. *Journal of Computer Assisted Tomography,* **17**, 798–805.

Hangartner, T. N., Overton, T. R., Harley, C. H., Vandenberg, L. and Crockford, P. M. 1985. Skeletal challenge: An experimental-study of pharmacologically induced changes in bone-density in the distal radius, using gamma-ray computed-tomography. *Calcified Tissue International,* 37, 19–24.

Hans, D., Fuerst, T. and Duboeuf, F. (1997). Quantitative ultrasound bone measurement. *European Radiology,* **7**, S43–S50.

Hansson, T., Roos, B. and Nachemson, A. (1980). The bone mineral content and ultimate compressive strength of lumbar vertebrae. *Spine,* **5**, 46–57.

Hengsberger, S., Kulik, A. and Zysset, P. (2001). A combined atomic force microscopy and nanoindentation technique to investigate the elastic properties of bone structural units. *European Cells and Materials,* **1**, 12–17.

Hengsberger, S., Kulik, A. and Zysset, P. (2002). Nanoindentation discriminates the elastic properties of individual human bone lamellae under dry and physiological conditions. *Bone,* **30**, 178–184.

Herman, G. T. (1980). *Image Reconstruction from Projections: The Fundamentals of Computerized Tomography,* New York, Academic press.

Hoffler, C. E., Guo, X. E., Zysset, P. K. and Goldstein, S. A. (2005). An application of nanoindentation technique to measure bone tissue lamellae properties. *Journal of Biomechanical Engineering-Transactions of the Asme,* **127**, 1046–1053.

Hoffler, C. E., Moore, K. E., Kozloff, K., Zysset, P. K., Brown, M. B. and Goldstein, S. A. (2000). Heterogeneity of bone lamellar-level elastic moduli. *Bone,* **26**, 603–609.

Homminga, J., Van-Rietbergen, B., Lochmuller, E. M., Weinans, H., Eckstein, F. and Huiskes, R. (2004). The osteoporotic vertebral structure is well adapted to the loads of daily life, but not to infrequent 'error' loads. *Bone,* **34**, 510–516.

Hounsfield, G. N. (1973). Computerized transverse axial scanning (tomography): Part I. Description of system. *British Journal of Radiology,* **46**, 1016–1022.

Huja, S. S., Katona, T. R., Moore, B. K. and Roberts, W. E. (1998). Microhardness and anisotropy of the vital osseous interface and endosseous implant supporting bone. *Journal of Orthopaedic Research,* **16**, 54–60.

Hulme, P. A., Boyd, S. K. and Ferguson, S. J. (2007). Regional variation in vertebral bone morphology and its contribution to vertebral fracture strength. *Bone,* **41**, 946–957.

Hutton, W. C., Cyron, B. M. and Stott, J. R. R. (1979). The compressive strength of lumbar vertebrae. *Journal of Anatomy,* **129**, 753–758.

Jamsa, T., Jalovaara, P., Peng, Z., Vaananen, H. K. and Tuukkanen, J. (1998). Comparison of three-point bending test and peripheral quantitative computed tomography analysis in the evaluation of the strength of mouse femur and tibia. *Bone,* **23**, 155–161.

Jast, J. A. (2011). *Investigation of Ultrastructural and Mechanical Properties of Rat Cortical Bone Using Micro-CT, Three-Point Bending Testing, and the Reference Point Indentation Technique.* M.S., University of Illinois at Urbana-Champaign.

Jensen, G. F., Christiansen, C., Boesen, J., Hegedus, V. and Transbol, I. (1982). Epidemiology of post-menopausal spinal and long-bone fractures: A unifying approach to post-menopausal osteoporosis. *Clinical Orthopaedics and Related Research,* **166**, 75–81.

Jirousek, O., Nemecek, J., Kytyr, D., Kunecky, J., Zlamal, P. and Doktor, T. (2011). Nanoindentation of trabecular bone: Comparison with uniaxial testing of single trabecula. *Chemicke Listy,* **105**, S668–S671.

Johnson, W. M. and Rapoff, A. J. (2007). Microindentation in bone: Hardness variation with five independent variables. *Journal of Materials Science-Materials in Medicine,* **18**, 591–597.

Jonson, R. (1993). Mass attenuation coefficients, quantities and units for use in bone-mineral determinations. *Osteoporosis International,* **3**, 103–106.

Kaye, B., Randall, C., Walsh, D. and Hansma, P. (2012). The effects of freezing on the mechanical properties of bone. *The Open Bone Journal,* **4**, 14–19.

Keaveny, T. M., Pinilla, T. P., Crawford, R. P., Kopperdahl, D. L. and Lou, A. (1999). Systematic and random errors in compression testing of trabecular bone (vol 15, pg 101, 1997). *Journal of Orthopaedic Research,* **17**, 151.

Kelsey, J. L., Browner, W. S., Seeley, D. G., Nevitt, M. C. and Cummings, S. R. (1992). Risk-factors for fractures of the distal forearm and proximal humerus. *American Journal of Epidemiology,* **135**, 477–489.

Keyak, J. H., Rossi, S. A., Jones, K. A. and Skinner, H. B. (1998). Prediction of femoral fracture load using automated finite element modeling. *Journal of Biomechanics,* **31**, 125–133.

Khosla, S., Riggs, B. L., Atkinson, E. J., Oberg, A. L., Mcdaniel, L. J., Holets, M., Peterson, J. M. and Melton, L. J. (2006). Effects of sex and age on bone microstructure at the ultradistal radius: A population-based noninvasive in vivo assessment. *Journal of Bone and Mineral Research,* **21**, 124–131.

Klotzbuecher, C. M., Ross, P. D., Landsman, P. B., Abbott, T. A. and Berger, M. (2000). Patients with prior fractures have an increased risk of future fractures: A summary of the literature and statistical synthesis. *Journal of Bone and Mineral Research,* **15**, 721–739.

Krieg, M.-A., Barkmann, R., Gonnelli, S., Stewart, A., Bauer, D. C., Barquero, L. D. R., Kaufman, J. J., Lorenc, R., Miller, P. D., Olszynski, W. P., Poiana, C., Schott, A.-M., Lewiecki, E. M. and Hans, D. (2008). Quantitative ultrasound in the management of osteoporosis: The 2007 ISCD Official Positions. *Journal of Clinical Densitometry,* **11**, 163–187.

Ladd, A. J. C., Kinney, J. H., Haupt, D. L. and Goldstein, S. A. (1998). Finite-element modeling of trabecular bone: Comparison with mechanical testing and determination of tissue modulus. *Journal of Orthopaedic Research,* **16**, 622–628.

Landrigan, M. D. and Roeder, R. K. (2009). Systematic error in mechanical measures of damage during four-point bending fatigue of cortical bone. *Journal of Biomechanics,* **42**, 1212–1217.

Langton, C. M. and Njeh, C. F. (eds.) (2004). *The Physical Measurement of Bone,* Bristol: Institute of Physics Pub.

Lehmann, R., Wapniarz, M., Kvasnicka, H. M., Baedeker, S., Klein, K. and Allolio, B. (1992). Reproducibility of measurements of bone-mineral density of the distal radius with a special-purpose computed-tomography scanner. *Radiologe,* **32**, 177–181.

Lenaerts, L. and Van Lenthe, G. H. (2009). Multi-level patient-specific modelling of the proximal femur. A promising tool to quantify the effect of osteoporosis treatment. *Philosophical Transactions of the Royal Society a-Mathematical Physical and Engineering Sciences,* **367**, 2079–2093.

Leppanen, O., Sievanen, H., Jokihaara, J., Pajamaki, I. and Jarvinen, T. L. N. (2006). Three-point bending of rat femur in the mediolateral direction: Introduction and validation of a novel biomechanical testing protocol. *Journal of Bone and Mineral Research,* **21**, 1231–1237.

Lochmueller, E.-M., Matsuura, M., Bauer, J., Hitzl, W., Link, T. M., Mueller, R. and Eckstein, F. (2008). Site-specific deterioration of trabecular bone architecture in men and women with advancing age. *Journal of Bone and Mineral Research,* **23**, 1964–1973.

Lochmuller, E. M., Groll, O., Kuhn, V. and Eckstein, F. (2002a). Mechanical strength of the proximal femur as predicted from geometric and densitometric bone properties at the lower limb versus the distal radius. *Bone,* **30**, 207–216.

Lochmuller, E. M., Krefting, N., Burklein, D. and Eckstein, F. (2001). Effect of fixation, soft-tissues, and scan projection on bone mineral measurements with dual energy X-ray absorptiometry (DXA). *Calcified Tissue International,* **68**, 140–145.

Lochmuller, E. M., Lill, C. A., Kuhn, V., Schneider, E. and Eckstein, F. (2002b). Radius bone strength in bending, compression, and falling and its correlation with clinical densitometry at multiple sites. *Journal of Bone and Mineral Research,* **17**, 1629–1638.

Macneil, J. A. and Boyd, S. K. (2007). Accuracy of high-resolution peripheral quantitative computed tomography for measurement of bone quality. *Medical Engineering and Physics,* **29**, 1096–1105.

Margolis, D. S., Kim, D., Szivek, J. A., Lai, L.-W. and Lien, Y.-H. H. (2006). Functionally improved bone in Calbindin-D28k knockout mice. *Bone,* **39**, 477–484.

Martin, R. B. and Atkinson, P. J. (1977). Age and sex-related changes in structure and strength of human femoral-shaft. *Journal of Biomechanics,* **10**, 223–231.

Mccreadie, B. and Goldstein, S. A. (2000). Biomechanics of fracture: Is bone mineral density sufficient to assess risk? *Journal of Bone and Mineral Research,* **15**, 2305–2308.

Melton, L. J., Chrischilles, E. A., Cooper, C., Lane, A. W. and Riggs, B. L. (2005). How many women have osteoporosis? (Reprinted from *Journal Bone and Mineral Research*, vol **7**, pg 1005, 1992). *Journal of Bone and Mineral Research,* **20**, 886–892.

Mosekilde, L. and Danielsen, C. C. (1987). Biomechanical competence of vertebral trabecular bone in relation to ash density and age in normal individuals. *Bone,* **8**, 79–85.

Muller, R., Gerber, S. C. and Hayes, W. C. (1998). Micro-compression: A novel technique for the nondestructive assessment of local bone failure. *Technology and Health Care: Official Journal of the European Society for Engineering and Medicine,* **6**, 433–444.

Muller, R., Hahn, M., Vogel, M., Delling, G. and Ruegsegger, P. (1996a). Morphometric analysis of noninvasively assessed bone biopsies: Comparison of high-resolution computed tomography and histologic sections. *Bone,* **18**, 215–220.

Muller, R., Hildebrand, T., Hauselmann, H. J. and Ruegsegger, P. (1996b). In vivo reproducibility of three-dimensional structural properties of noninvasive bone biopsies using 3D-pQCT. *Journal of Bone and Mineral Research,* **11**, 1745–1750.

Myers, E. R., Hecker, A. T., Rooks, D. S., Hipp, J. A. and Hayes, W. C. (1993). Geometric variables from DXA of the radius predict forearm fracture load in-vitro. *Calcified Tissue International,* **52**, 199–204.

Müller, R. and Van Lenthe, G. H. (2004). Microarchitectural aspects of quality and competence of bone. *Advances in Osteoporotic Fracture Management,* **3**, 2–12.

Müller, R., Nazarian, A., Schneider, P., Stauber, M., Thurner, P., Van Lenthe, G. H. and Voide, R. (2006). Functional microimaging at the interface of bone mechanics and biology. *In:* Holzapfel, G. A. and Ogden, R. W. (eds.) *Mechanics of Biological Tissue.* Berlin, Springer-Verlag, 473–487.

Nazarian, A., Stauber, M. and Muller, R. (2005). Design and implementation of a novel mechanical testing system for cellular solids. *Journal of Biomedical Materials Research Part B-Applied Biomaterials,* **73B**, 400–411.

Nguyen, T. V., Center, J. R., Sambrook, P. N. and Eisman, J. A. (2001). Risk factors for proximal humerus, forearm, and wrist fractures in elderly men and women: The Dubbo osteoporosis epidemiology study. *American Journal of Epidemiology,* **153**, 587–595.

Niebur, G. L., Feldstein, M. J., Yuen, J. C., Chen, T. J. and Keaveny, T. M. (2000). High-resolution finite element models with tissue strength asymmetry accurately predict failure of trabecular bone. *Journal of Biomechanics,* **33**, 1575–1583.

Njeh, C. F., Boivin, C. M. and Langton, C. M. (1997). The role of ultrasound in the assessment of osteoporosis: A review. *Osteoporosis International,* **7**, 7–22.

Njeh, C. F., Hans, D., Fuerst, T., Glueer, C.-C. and Genant, H. K. (1999). *Quantitative Ultrasound: Assessment of Osteoporosis and Bone Status.,* London, Martin Dunitz.

Njeh, C. F., Wu, C., Fan, B., Hans, D., Fuerst, T., He, Y. and Genant, H. K. (2000). Estimation of wrist fracture load using phalangeal speed of sound: An in vitro study. *Ultrasound in Medicine and Biology,* **26**, 1517–1523.

Norman, J., Shapter, J. G., Short, K., Smith, L. J. and Fazzalari, N. L. (2008). Micromechanical properties of human trabecular bone: A hierarchical investigation using nanoindentation. *Journal of Biomedical Materials Research Part A,* **87A**, 196–202.

Nyman, J. S., Roy, A., Shen, X. M., Acuna, R. L., Tyler, J. H. and Wang, X. D. (2006). The influence of water removal on the strength and toughness of cortical bone. *Journal of Biomechanics,* **39**, 931–938.

Oliver, W. C. and Pharr, G. M. (1992). An improved technique for determining hardness and elastic-modulus using load and displacement sensing indentation experiments. *Journal of Materials Research,* **7**, 1564–1583.

Owen, R. A., Melton, L. J., Johnson, K. A., Ilstrup, D. M. and Riggs, B. L. (1982). Incidence of Colles fracture in a North-American community. *American Journal of Public Health,* **72**, 605–607.

Oyen, M. L. (2008). Poroelastic nanoindentation responses of hydrated bone. *Journal of Materials Research,* **23**, 1307–1314.

Oyen, M. L. and Cook, R. R. (2009). A practical guide for analysis of nanoindentation data. *Journal of the Mechanical Behavior of Biomedical Materials,* **2**, 396–407.

Parfitt, A. M., Drezner, M. K., Glorieux, F. H., Kanis, J. A., Malluche, H., Meunier, P. J., Ott, S. M. and Recker, R. R. (1987). Bone histomorphometry: Standardization of nomenclature, symbols, and units. *Journal of Bone and Mineral Research,* **2**, 595–610.

Patsch, J. M., Burghardt, A. J., Kazakia, G. and Majumdar, S. (2011). Noninvasive imaging of bone microarchitecture. *Skeletal Biology and Medicine II: Bone and Cartilage Homeostasis and Bone Disease,* **1240**, 77–87.

Pistoia, W., Van Rietbergen, B., Lochmuller, E. M., Lill, C. A., Eckstein, F. and Ruegsegger, P. (2002). Estimation of distal radius failure load with micro-finite element analysis models based on three-dimensional peripheral quantitative computed tomography images. *Bone,* **30**, 842–848.

Pistoia, W., Van Rietbergen, B. and Ruegsegger, P. (2003). Mechanical consequences of different scenarios for simulated bone atrophy and recovery in the distal radius. *Bone,* **33**, 937–945.

Polly, B. J., Yuya, P. A., Akhter, M. P., Recker, R. R. and Turner, J. A. (2012). Intrinsic material properties of trabecular bone by nanoindentation testing of biopsies taken from healthy women before and after menopause. *Calcified Tissue International,* **90**, 286–293.

Proctor, A. (2010). In situ mechanical measurements of rat tibia. *Journal of Bone and Mineral Research,* **25**, S235.

Randall, C., Mathews, P., Yurtsev, E., Sahar, N., Kohn, D. and Hansma, P. (2009). The bone diagnostic instrument III: Testing mouse femora. *Review of Scientific Instruments,* **80**, 065108-1-3.

Rho, J. Y., Kuhn-Spearing, L. and Zioupos, P. (1998). Mechanical properties and the hierarchical structure of bone. *Medical Engineering and Physics,* **20**, 92–102.

Rho, J. Y., Tsui, T. Y. and Pharr, G. M. (1997). Elastic properties of human cortical and trabecular lamellar bone measured by nanoindentation. *Biomaterials,* **18**, 1325–1330.

Rho, J. Y., Zioupos, P., Currey, J. D. and Pharr, G. M. (1999). Variations in the individual thick lamellar properties within osteons by nanoindentation. *Bone,* **25**, 295–300.

Riggs, B. L. and Melton, L. J. (1986). Involutional osteoporosis. *New England Journal of Medicine,* **314**, 1676–1686.

Rogel, M. R., Qiu, H. and Ameer, G. A. (2008). The role of nanocomposites in bone regeneration. *Journal of Materials Chemistry,* **18**, 4233–4241.

Ruegsegger, P., Elsasser, U., Anliker, M., Gnehm, H., Kind, H. and Prader, A. (1976). Quantification of bone mineralization using computed tomography. *Radiology,* **121**, 93–97.

Ruppel, M. E., Burr, D. B. and Miller, L. M. (2006). Chemical makeup of microdamaged bone differs from undamaged bone. *Bone,* **39**, 318–324.

Sabatier, J. P. and Guaydier-Souquieres, G. (1989). Noninvasive methods of bone-mass measurement. *Clinical Rheumatology,* **8**, 41–46.

Schneider, P. and Borner, W. (1991). Peripheral quantitative computed-tomography for bone-mineral measurement with a new special purpose QCT-scanner: Method, normal references, comparison to osteoporotic patients. *Fortschritte Auf Dem Gebiete Der Rontgenstrahlen Und Der Neuen Bildgebenden Verfahren,* **154**, 292–299.

Seeman, E. (1999). The structural basis of bone fragility in men. *Bone,* **25**, 143–147.

Silva, M. J., Brodt, M. D. and Hucker, W. J. (2005). Finite element analysis of the mouse tibia: Estimating endocortical strain during three-point bending in SAMP6 osteoporotic mice. *Anatomical Record Part a-Discoveries in Molecular Cellular and Evolutionary Biology,* **283A**, 380–390.

Stampanoni, M., Borchert, G., Wyss, P., Abela, R., Patterson, B., Hunt, S., Vermeulen, D. and Ruegsegger, P. (2002). High resolution X-ray detector for synchrotron-based microtomography. *Nuclear Instruments and Methods in Physics Research, Section A: Accelerators, Spectrometers, Detectors and Associated Equipment,* **491**, 291–301.

Stampanoni, M., Borchert, G., Abela, R. and Ruegsegger, P. (2003). Nanotomography based on double asymmetrical Bragg diffraction. *Applied Physics Letters,* **82**, 2922–2924.

Svendsen, O. L., Hassager, C., Skodt, V. and Christiansen, C. (1995). Impact of soft-tissue on in-vivo accuracy of bone-mineral measurements in the spine, hip, and forearm: A human cadaver study. *Journal of Bone and Mineral Research,* **10**, 868–873.

Swadener, J. G. and Pharr, G. M. (2001). Indentation of elastically anisotropic half-spaces by cones and parabolae of revolution. *Philosophical Magazine a-Physics of Condensed Matter Structure Defects and Mechanical Properties,* **81**, 447–466.

Tang, S. Y., Souza, R. B., Ries, M., Hansma, P. K., Alliston, T. and Li, X. (2011). Local tissue properties of human osteoarthritic cartilage correlate with magnetic resonance T(1)rho relaxation times. *Journal of Orthopaedic Research,* **29**, 1312–1319.

Taylor, D. (2007). Fracture and repair of bone: A multiscale problem. *Journal of Materials Science,* **42**, 8911–8918.

Thomsen, J. S., Ebbesen, E. N. and Mosekilde, L. (2000). A new method of comprehensive static histomorphometry applied on human lumbar vertebral cancellous bone. *Bone,* **27**, 129–138.

Thurner, P. J., Wyss, P., Voide, R., Stauber, M., Stampanoni, M., Sennhauser, U. and Müller, R. (2006). Time-lapsed investigation of three-dimensional failure and damage accumulation in trabecular bone using synchrotron light. *Bone*, **39**, 289–299.

Timlin, J. A., Garden, A., Morris, M. D., Rajachar, R. M. and Kohn, D. H. (2000). Raman spectroscopic imaging markers for fatigue-related microdamage in bovine bone. *Analytical Chemistry*, **72**, 2229–2236.

Torcasio, A., Van Oosterwyck, H. and Van Lenthe, G. H. (2008). The systematic errors in tissue modulus of murine bones when estimated from three-point bending. *Journal of Biomechanics*, **41**, S14.

Van Dieën, J. H., Van Der Veen, A., Van Royen, B. J. and Kingma, I. (2006). Fatigue failure in shear loading of porcine lumbar spine segments. *Spine*, **31**, E494–E498.

Van Lenthe, G. H., Voide, R., Boyd, S. K. and Muller, R. (2008). Tissue modulus calculated from beam theory is biased by bone size and geometry: Implications for the use of three-point bending tests to determine bone tissue modulus. *Bone*, **43**, 717–723.

Van Rietbergen, B., Weinans, H., Huiskes, R. and Odgaard, A. (1995). A new method to determine trabecular bone elastic properties and loading using micromechanical finite-element models. *Journal of Biomechanics*, **28**, 69–81.

Vedi, S. and Compston, J. (2003). Bone histomorphometry. *In:* Helfrich Mh and Sh, R. (eds.) *Bone Research Protocols*. Totowa, NJ, USA, Humana Press Inc.

Voide, R., Schneider, P., Stauber, M., Wyss, R., Stampanoni, M., Sennhauser, U., Van Lenthe, G. H. and Muller, R. (2009). Time-lapsed assessment of microcrack initiation and propagation in murine cortical bone at submicrometer resolution. *Bone*, **45**, 164–173.

Voide, R., Van Lenthe, G. H., Schneider, P., Thurner, P. J., Wyss, P., Sennhauser, U., Stampanoni, M., Stauber, M., Snedeker, J. G. and Muller, R. Functional micro-imaging: An integrated approach for advanced bone biomechanics and failure analysis – art. no. 61430X. Medical Imaging 2006 Conference, 12–14 February 2006 San Diego, CA. X1430-X1430.

Wang, X. D., Bank, R. A., Tekoppele, J. M. and Agrawal, C. M. (2001). The role of collagen in determining bone mechanical properties. *Journal of Orthopaedic Research*, **19**, 1021–1026.

Weaver, J. K. (1966). Microscopic hardness of bone. *Journal of Bone and Joint Surgery-American Volume*, A **48**, 273–288.

Yeni, Y. N., Zinno, M. J., Yerramshetty, J. S., Zauel, R. and Fyhrie, D. P. (2011). Variability of trabecular microstructure is age-, gender-, race- and anatomic site-dependent and affects stiffness and stress distribution properties of human vertebral cancellous bone. *Bone*, **49**, 886–894.

Yoon, H. S. and Katz, J. L. (1976). Ultrasonic wave-propagation in human cortical bone – II. Measurements of elastic properties and microhardness. *Journal of Biomechanics*, **9**, 459–464.

Zhao, F. D., Pollintine, P., Hole, B. D., Adams, M. A. and Dolan, P. (2009). Vertebral fractures usually affect the cranial endplate because it is thinner and supported by less-dense trabecular bone. *Bone*, **44**, 372–379.

Zioupos, P., Currey, J. D. and Casinos, A. (2000). Exploring the effects of hypermineralisation in bone tissue by using an extreme biological example. *Connective Tissue Research*, **41**, 229–248.

Ziv, V., Wagner, H. D. and Weiner, S. (1996). Microstructure-microhardness relations in parallel-fibered and lamellar bone. *Bone,* **18**, 417–428.

Zwierzak, I., Baleani, M. and Viceconti, M. (2009). Microindentation on cortical human bone: Effects of tissue condition and indentation location on hardness values. *Proceedings of the Institution of Mechanical Engineers Part H-Journal of Engineering in Medicine,* **223**, 913–918.

Zysset, P. K. (2003). A review of morphology-elasticity relationships in human trabecular bone: Theories and experiments. *Journal of Biomechanics,* **36**, 1469–1485.

Zysset, P. K. (2009). Indentation of bone tissue: A short review. *Osteoporosis International,* **20**, 1049–1055.

Zysset, P. K., Guo, X. E., Hoffler, C. E., Moore, K. E. and Goldstein, S. A. (1999). Elastic modulus and hardness of cortical and trabecular bone lamellae measured by nanoindentation in the human femur. *Journal of Biomechanics,* **32**, 1005–1012.

3 Osteoinductivization of dental implants and bone-defect-filling materials

E. B. HUNZIKER, University of Bern, Switzerland

DOI: 10.1533/9780857099037.1.72

Abstract: Novel tools and techniques are continually being sought to improve the osseointegration of implants in clinical dentistry. This chapter addresses the functionalization of biomimetic calcium phosphate coatings.

Key words: biomimetic, calcium phosphate, osteogenic agent, degradation.

Note: This chapter is a reproduction of Chapter 17 'Osteoinductivization of dental implants and bone-defect-filling materials' by E. B. Hunziker, originally published in *Preprosthetic and maxillofacial surgery*, ed. Joël Ferri and Ernst B. Hunziker, Woodhead Publishing Limited, 2011, ISBN 978-1-84569-589-7.

3.1 Introduction

In the fields of clinical dentistry and maxillofacial surgery, novel tools and techniques are continually being sought to improve and expedite the osseointegration of dental implants and the bony remodelling of defect-filling materials. To achieve this goal, the implanted materials must be rendered not only osteoconductive but also osteoinductive. The osteoconductivity of a material can be enhanced by coating its surface with a bone-mineral-like layer of calcium phosphate, which can be rendered osteoinductive by functionalizing it with an osteogenic agent.[1,2]

3.2 Biomimetic coating technique

During the past 30 years, activities in clinical dentistry have focused mainly on the coating of metallic implants. About a decade ago, the technique whereby such coatings are produced underwent a revolutionary change,[3,4] which has had a profound impact on the potential of these layers to act as carriers for an osteogenic agent. Formerly, layers of calcium phosphate could be deposited upon the surfaces of metallic implants only at highly unphysiological temperatures (>10 000 °C), which precluded the co-precipitation of a biologically-active agent. Such agents could be only superficially adsorbed onto pre-formed layers of calcium phosphate. In this adsorbed state, the protein molecules are released too rapidly[5] to exert a sustained osteoinductive response at the site of implantation. But with the advent of the so-called biomimetic coating process, it became possible to deposit layers of calcium phosphate under physiological conditions of temperature. The coating process can be achieved in the presence of an osteogenic agent, which thereby becomes incorporated into the crystalline latticework that comprises the calcium phosphate layer.[6–8] By virtue of their integration into the inorganic latticework, the molecules of the osteogenic agent are released gradually at the site of implantation, as the coating undergoes osteoclast-mediated degradation.

The structure of the crystals that are formed under biomimetic conditions of coating is more akin to that of bone mineral than are the partially molten hydroxyapatite particles that are formed at high temperatures. Consequently, biomimetically-prepared coatings are more readily degraded within a biological milieu.

The biomimetic technique involves the nucleation and growth of calcium phosphate crystals upon an amorphous seeding layer under physiological conditions of temperature (37 °C) and pH (7.4). The method was developed by Kokubo *et al.* in 1990,[3] and has since been refined by several groups of investigators.[7,9–13]

The implant is first immersed for 24 hours (pH 7.4) within a solution that resembles bodily fluids in terms of its ionic composition, but within which the anions and cations are present at five-fold-higher concentrations. To inhibit nucleation and crystal growth, the concentration of Mg^{2+} is further raised. A thin, 1-to-3 μm thick, amorphous layer of calcium phosphate is thereby uniformly deposited upon the implant surface. This layer serves as a seeding substratum for the growth of a more substantial crystalline one, which is prepared by immersion for 4–8 hours (pH 7.4) within a supersaturated solution of calcium phosphate. Under these conditions, the crystalline latticework is composed of octacalcium phosphate (Fig. 3.1). It is during the deposition of this crystalline layer that an osteogenic agent can be co-precipitated. The agent is simply introduced, at the appropriate concentration, into the supersaturated solution of calcium phosphate.[14]

Apart from its amenability to a process of biological functionalization, another great asset of the biomimetic coating technique is that it can be applied not only to metallic implants but also to heat-sensitive, non-conductive, porous and bulky materials with complex surface geometries. Consequently, not only dental implants but also alloplastic bone-defect-filling materials can be coated. Thus far, this has been achieved using Helistat® (a sponge-like material that is manufactured from natural collagen), Polyactive® (a sponge-like synthetic co-polymer of ethyleneoxide

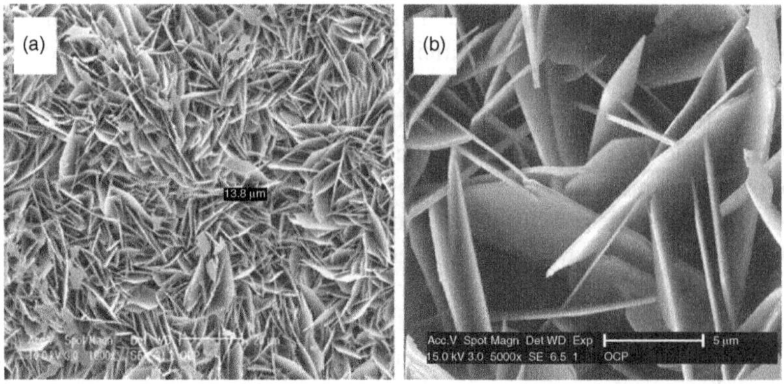

3.1 (a) Low- and (b) high-resolution scanning electron micrographs illustrating the latticework of octacalcium-phosphate crystals that constitute a biomimetically-prepared coating

terephthalate and butylene terephthalate), Ethisorb™ (a fibrous co-polymer of glactin and ρ-dioxanone), PLGA (a fibrous co-polymer of lactic and glycolic acids)[15] and Bio-Oss® (deproteinized bovine bone). In each case, the crystalline coating could be unproblematically co-precipitated with the model protein bomvine serum albumin.[8]

3.2.1 Bone morphogenetic proteins (BMPs)

Given that biomimetically-prepared calcium-phosphate coatings can serve as a vehicle for the slow and sustained delivery of a protein, which has been demonstrated *in vitro* for bovine serum albumin, the next issue to be addressed is the choice of a suitable osteogenic agent. Members of the transforming growth factor beta superfamily (TGF-βs), particularly the bone morphogenetic proteins (BMPs), probably hold the most promise for osteoinduction.[17]

BMP-2 has been successfully co-precipitated with calcium-phosphate coatings, and has been shown to be incorporated into the crystalline latticework.[18] Using an *in vitro* assay, which involved monitoring the alkaline-phosphatase activity of osteoprogenitor cells that had been cultured directly upon a co-precipitated layer of calcium phosphate and BMP-2, the process of incorporation, far from compromising the response, was shown to enhance it greatly.[18]

This BMP-2-carrier system has also been tested at an ectopic ossification site (the subcutaneous space) in rats.[19] In this study, the co-precipitated layers of calcium phosphate and BMP-2 were deposited upon discs of titanium. The histomorphometric findings revealed that a coating-incorporated depot of BMP-2 was capable not only of inducing ectopic bone formation at a very

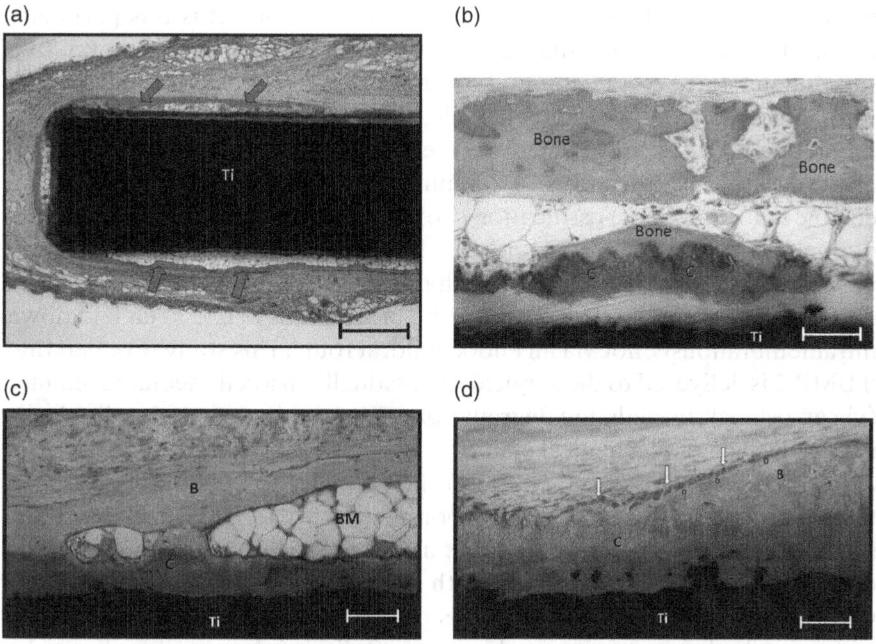

3.2 Light micrographs of discs in the incorporated-BMP-2 group, retrieved together with the surrounding tissue five weeks after implantation, and stained with McNeal's Tetrachrome, basic Fuchsine and Toluidine Blue O. (a) Low-magnification view of a disc (Ti) which is almost completely surrounded by bone tissue (arrows). This has been deposited both directly upon the coating and at some distance therefrom within the connective tissue capsule. Scale bar = 0.5 mm. (b) Higher-magnification view of the disc illustrated in (a) showing bone tissue in direct contact with the coating (C) and within the surrounding connective tissue. Scale bar = 20 μm. (c) Along this stretch of the disc surface, both bone tissue (B) and bone marrow (BM) contact the coating (C) directly. Ti = disc. Bar = 100 μm. (d) Osteoblasts (arrows) are here seen to be aligned in a characteristic epithelioid fashion upon the surface of osteoid material (o). Ti = disc; C = coating; B = bone tissue. Scale bar = 200 μm.

low concentration level, but of sustaining this process for a considerable period of time (minimally five weeks: the extent of the monitoring period) (Fig. 3.2). A depot of BMP-2 that had been superficially adsorbed onto a pre-formed calciumphosphate coating was unable to induce more than a very transient and sporadic osteogenic response.

Furthermore, bone tissue was laid down directly, namely, by intramembranous growth, rather than via an endochondral mechanism. Intramembranous ossification is generally believed to occur only within a mechanically stable field, in the absence of shear stress, which was obviously furnished by the rigid titanium discs. However, it now appears that it is this particular mode of BMP-2-carriage that favours the direct mode of ossification. In a subsequent study, co-precipitated layers of calcium phosphate and BMP-2 were deposited upon discs of a flexible, alloplastic bone-defect-filling material, namely, Ethisorb™, which were either left unsupported or mechanically stabilized by affixation to titanium plates prior to implantation at an ectopic (subcutaneous) ossification site in rats. Two weeks later, the volumes of bone that were associated with mechanically-unstabilized and titanium-plate-stabilized discs of Ethisorb™ that bore a coating-incorporated depot of BMP-2 were similar. In each case, the osseous tissue had been laid down intramembranously, not via an endochondral route. This study revealed that if BMP-2 is delivered to the surgical site gradually, in a cell-mediated manner (via an osteoclast-mediated degradation of the coating), then the alloplastic material's lack of mechanical stability becomes a matter of indifference.[20]

In addition to Ethisorb™, three other alloplastic materials have been tested as carriers for a coating-incorporated deposit of BMP-2 at an ectopic site in rats.[16] This study revealed that also these materials can be rendered osteoinductive by coating them with a co-precipitated layer of calcium phosphate and BMP-2. Moreover, as gauged by the local volume density of foreign-body giant cells, the biocompatibility of these materials can be improved by the coating process. Furthermore, since the coatings must be degraded before the underlying material can be degraded, the inorganic layers protect the host from the potentially deleterious effects of the organic polymers' degradation products at a crucial juncture, namely during the initial phase of osteoblast mobilization and bone formation.

3.2.2 Mechanical stability of functionalized coatings

One other aspect that must be considered in a clinical setting is the mechanical stability of the functionalized coatings themselves. The implantation of coated alloplastic materials within large bony defects is not likely to compromise the integrity of the functionalized calcium-phosphate layers. However, the shear forces that are generated during the insertion of screw-like dental implants could render the coatings susceptible to fragmentation. Coatings into which bovine serum albumin was incorporated have been shown to be characterized by higher torque values than non-protein-bearing ones, and to be less susceptible to fragmentation.[21] Hence, the incorporation of a protein increases the resistance of the coatings to the shear forces that are generated during implant insertion. This attribute reflects the greater compactness of the protein-bearing layers (Fig. 3.3).

However, even protein-bearing coatings are subject to fragmentation at vulnerable points, namely, at the tips of the implant threads. Notwithstanding

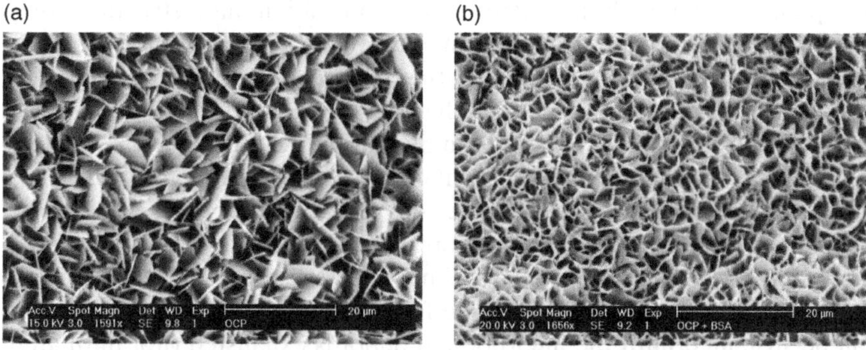

3.3 Scanning electron micrographs of implant coatings that either lacked (a) or bore (b) an incorporated depot of bovine serum albumin (BSA). In the absence of BSA, the coating is composed of distinct, squarish, plate-like crystals of calcium phosphate. In the presence of BSA, the crystals are smaller, have a more rounded profile, and are less readily distinguishable as distinct entities.

this circumstance, the loss of coating material may not be a disadvantage. The delaminated portions remain within the peri-implant space, and will retain their incorporated protein deposit. In a clinical setting, the incorporated protein would be an osteogenic agent, the osteoinductive potential of which would not be compromised by the fragmentation process. Indeed, it may even be enhanced: scattered fragments of coating not only expose a larger protein-bearing surface to the environment than does an intact layer, but they also decrease the osteogenic 'jumping distance' between the thread valleys and the native bone wall. The process of osseointegration is thus not likely to be compromised in any way by coating fragmentation. Ongoing studies in which coated, BMP-2-functionalized screw-like implants were inserted into the partially edentulous mandibles of adult miniature pigs bear out this expectation: no inflammatory reactivity was evoked, bone-formation activity was promoted, and osseointegration was expedited.[21]

3.3 Conclusions

In summary, biomimetically-prepared layers of calcium phosphate can improve the osteoconductivity of both dental implants and bone-defect-filling materials. These osteoconductive layers can be rendered osteoinductive by co-precipitating them with an osteogenic agent. The beauty of this co-precipitation process lies in the fact that the osteogenic agent is truly incorporated into the inorganic crystalline latticework during its growth, with the consequence that it can be liberated only as a result of coating degradation. Since these biomimetically-prepared coatings of octacalcium phosphate resemble bone mineral in composition and structure, they are degraded in a similar manner, namely, via the activity of osteoclasts. The similarity of this process of coating degradation to the physiological remodelling of bone is borne out by the total absence of foreign-body-giant-cell reactivity at orthotopic sites of implantation. As the coating undergoes degradation, the osteogenic agent is liberated from the inorganic matrix at a steady rate which keeps pace with the activity of the osteoclasts. Once freed, it stimulates the recruitment of osteoprogenitor cells and their differentiation into bone-forming osteoblasts. Since the osteogenic agent is liberated gradually from the coating, the bone-formation activity of the osteoblasts can be sustained until the incorporated depot is exhausted (not until after the elapse of several weeks). This sequence of events mimics the physiological process of bone formation in that the osteoclast-mediated degradation of bone mineral likewise liberates osteogenic growth factors. Furthermore, the coating-incorporated depot of the osteogenic agent is efficacious at physiological – not at pharmacological – levels.

Hence, the osteoinductivization of prosthetic and bone-defect-filling materials with biomimetically-prepared, functionalized coatings is likely to have far reaching consequences for the osseointegration of dental implants and for the repair of bony defects in the maxillofacial region, as well as at other bodily sites.

3.4 References

1. Liu Y, Enggist L, Kuffer AF, Buser D, Hunziker EB (2007) The influence of BMP-2 and its mode of delivery on the osteoconductivity of implant surfaces during the early phase of osseointegration. *Biomaterials* **28** (16): 2677–86.
2. Liu Y, Li JP, Hunziker EB, de Groot K (2006) Incorporation of growth factors into medical devices via biomimetic coatings. *Philos Transact A Math Phys Eng Sci* **364** (1838): 233–48.
3. Kokubo T, Kushitani H, Sakka S, Kitsugi T, Yamamuro T (1990) Solutions able to reproduce in vivo surface-structure changes in bioactive glass-ceramic A-W. *J Biomed Mater Res* **24** (6): 721–34.
4. Kokubo T (1991) Bioactive glass ceramics: properties and applications. *Biomaterials* **12** (2): 155–63.
5. Agrawal CM, Best J, Heckman JD, Boyan BD (1995) Protein release kinetics of a biodegradable implant for fracture non-unions. *Biomaterials* **16** (16): 1255–60.
6. Wen HB, de Wijn JR, van Blitterswijk CA, de Groot K (1999) Incorporation of bovine serum albumin in calcium phosphate coating on titanium. *J Biomed Mater Res* **46** (2): 245–52.
7. Liu Y, Layrolle P, de Bruijn J, van Blitterswijk C, de Groot K (2001) Biomimetic coprecipitation of calcium phosphate and bovine serum albumin on titanium alloy. *J Biomed Mater Res* **57** (3): 327–35.
8. Liu Y, Hunziker EB, Layrolle P, Van Blitterswijk C, Calvert PD, de Groot K (2003) Remineralization of demineralized albumin-calcium phosphate coatings. *J Biomed Mater Res A* **67** (4): 1155–62.

9. Barrere F, van Blitterswijk CA, de Groot K, Layrolle P (2002) Influence of ionic strength and carbonate on the Ca-P coating formation from SBFx5 solution. *Biomaterials* **23** (9): 1921–30.
10. Barrere F, van Blitterswijk CA, de Groot K, Layrolle P. Nucleation of biomimetic Ca-P coatings on Ti6A14V from a SBFx5 solution: influence of magnesium. *Biomaterials* **23** (10): 2211–20.
11. Barrere F, Layrolle P, van Blitterswijk CA, De Groot K (2001) Biomimetic coatings on titanium: a crystal growth study of octacalcium phosphate. *J Mater Sci Mater Med* **12** (6): 529–34.
12. Barrere F, Layrolle P, van Blitterswijk CA, de Groot K (1999) Biomimetic calcium phosphate coatings on Ti6AI4V: a crystal growth study of octacalcium phosphate and inhibition by Mg 2+ and HCO3. *Bone* **25** (2 Suppl): S107–11.
13. Gondolph-Zink B (1998) Effect of hydroxyapatite layering on the osteointegration of weightbearing and non-weightbearing implants. Comparison to other microporous surfaces in animal experiments. *Orthopade* **27** (2): 96–104.
14. Liu Y, Hunziker EB, Randall NX, de Groot K, Layrolle P (2003) Proteins incorporated into biomimetically prepared calcium phosphate coatings modulate their mechanical strength and dissolution rate. *Biomaterials* **24** (1): 65–70.
15. Wu G, Liu Y, Iizuka T, Hunziker EB (2010) Biomimetic coating of organic polymers with a protein-functionalized layer of calcium phosphate: The surface properties of the carrier influence neither the coating characteristics nor the incorporation mechanism or release kinetics of the protein. *Tissue Eng Part C Methods*, **16**(6): 1255–65.
16. Wu G, Liu Y, Iizuka T, Hunziker EB (2010) The effect of a slow mode of BMP-2 delivery on the inflammatory response provoked by bone-defect-filling polymeric scaffolds. *Biomaterials* **31**(29): 7485–93.
17. Liu Y, de Groot K, Hunziker EB (2004) Osteoinductive implants: the mise-en-scene for drug-bearing biomimetic coatings. *Ann Biomed Eng* **32** (3): 398–406.
18. Liu Y, Hunziker EB, Layrolle P, De Bruijn JD, De Groot K (2004) Bone morphogenetic protein 2 incorporated into biomimetic coatings retains its biological activity. *Tissue Eng* **10** (1–2): 101–8.
19. Liu Y, de Groot K, Hunziker EB (2005) BMP-2 liberated from biomimetic implant coatings induces and sustains direct ossification in an ectopic rat model. *Bone* **36** (5): 745–57.
20. Hägi TT, Wu G, Liu Y, Hunziker EB (2010) Cell-mediated BMP-2 liberation promotes bone formation in a mechanically unstable implant environment. *Bone* **46** (5): 1322–7.
21. Hägi TT, Enggist L, Michel D, Ferguson SJ, Liu Y, Hunziker EB (2010) Mechanical insertion properties of calcium-phosphate implant coatings. *Clin Oral Implants Res* **21**(11): 1214–22.

4 Bioresorbable bone graft substitutes

T. J. BLOKHUIS, University Medical Center
Utrecht, The Netherlands

DOI: 10.1533/9780857099037.1.80

Abstract: Bone graft substitutes are frequently used as void fillers. Ideally, the material will resorb and participate in bone remodelling, gradually replacing the bone graft substitute over time. However, all available materials either resorb faster or slower than predicted, and it is therefore important to understand the mechanisms of resorption and the properties that determine the resorption of a specific material. In this chapter these items are discussed. The mechanism of resorption, the properties of frequently used materials, and the use of resorbable materials as a source for other substances will be discussed, including future perspectives and challenges.

Key words: resorption, calcium phosphate ceramics, mechanical strength, bioglass, bone morphogenetic proteins, osteoconduction.

4.1 Introduction

Bioresorbable bone graft substitutes are materials that are used as bone void fillers and gradually resorb over time. These materials should be bioinert, at least osteoconductive, and have the capacity to disappear over time, either through active bioresorption or through passive dissolution. The advantages of resorbable materials are the disappearance of a local foreign body as well as the reconstitution of the normal bone properties. The local bone will not, after healing, be influenced in remodelling by the presence of artificial substances, resulting in a stronger and more biological outcome. The effects of these advantages are less risk of secondary infection and a stronger bone area. Ideally, the resorption process takes place at a similar rate as the physiological bone remodelling. In this way, the material is gradually replaced over time with normal bone, and bone restoration is not hampered by the presence of the material.

The resorption of biomaterials is mainly a cellular process. It involves foreign body giant cells, osteoclasts and possibly macrophages[1–4] that adhere to the material and phagocytose it over time. This makes the process of resorption fundamentally different from dissolution, where an unstable material dissolves over time, for example in an aqueous solution. Resorption is influenced by the composition of the material in two ways: its geometry and the chemical composition of the material. The geometrical properties that influence resorption are porosity and interconnectivity of the pores.[5] These properties can be altered by the manufacturing process, for example by changing the sintering temperature. The chemical composition depends on the manufacturer as well, and adding substances or other materials will have a direct influence on the resorption rate. In most materials, resorption takes place in a sealed extracellular environment created by the multinuclear cells. The cells excrete acid into this environment, and the acid dissolves the material, such as a calcium phosphate. The release of calcium into the extracellular microenvironment causes the osteoclast to stop the resorption and migrate. Through this mechanism the resorption rate can be influenced, as the amount of calcium released will cause the osteoclast to migrate. Changing a material, for example changing its stability or the relative amount of calcium, will thus change the rate of resorption. It is also important to realize in this perspective that the activity of osteoblasts and osteoclasts is interlinked.[6] This means that a change in activity of one will automatically influence the other, and thus the rate of bone remodelling.

Ideally, a resorbable bone graft substitute takes part in the bone remodelling process. However, the speed of bone remodelling is variable. The rate of bone turnover is higher in cancellous compared to cortical bone,[7] and bone turnover is furthermore influenced by other factors such as loading. As a consequence the already variable resorption rate of bone graft substitutes will become even more unpredictable. The concept of using different materials for different indications and areas is emerging in the clinical application of bone graft substitutes, but many clinicians tend to use one single material for each indication. Given the aforementioned variability this seems illogical, and a more differentiated approach should be taken.

4.2 Materials that allow resorption

4.2.1 Calcium phosphates

Calcium phosphates are the largest group of artificial bone graft substitutes. This is mainly due to their close resemblance to the mineral components of bone. Numerous forms, compositions, and application methods are available. The number of commercially available products is large and growing, leading to a difficult area to make an educated choice. Choosing specific calcium phosphate products should be based on their properties in combination with the desired indication. One important property is the resorption rate. The resorption rate of calcium phosphates can be influenced in two different ways. First, the geometrical properties can be changed, and second, the chemical composition influences the resorption rate.

The geometrical properties of calcium phosphate ceramics are mainly the porosity and the interconnectivity of the pores. In various studies, the optimal pore size for ingrowth of new tissue was determined to be between 150 and 500 μm.[8] This pore size facilitates the ingrowth of fibrous tissue and formation of vessels, and also the invasion of cells involved in the resorption of the material. Ingrowth of tissue is also determined by interconnectivity of the pores.

The chemical composition of the calcium phosphates has changed significantly over the past decades. Initially the calcium phosphates consisted of hydroxyapatite (HA) ($CA_{10}(PO_4)_6(OH)_2$), an inert material which is not resorbed over time. With the introduction of β-tricalcium phosphate (β-TCP) (β-$Ca_3(PO_4)_2$) as well as other compositions such as α-TCP,[9] octocalcium phosphate, dicalcium phosphate, and others, the field of calcium phosphates has changed significantly. Generally speaking HA is not resorbable, whereas the other compositions are resorbed quickly. By combining HA and TCP, resulting in biphasic calcium phosphates (BCP) the rate of resorption of a material can be influenced, and the other calcium phosphate compositions all aim to establish different resorption rates. In theory, adding more TCP will make resorption more quickly, adding HA more slowly. Recently a combination of HA and calcium sulphate has been introduced. This material, Cerament™, aims to combine the mechanical behaviour of HA and the fast resorption of calcium sulphate. This injectable paste combines easy handling properties with good biomechanical behaviour after implantation, as well as excellent osteointegration. It is therefore an example of the newest generation of calcium phosphates, and others will probably emerge.

Another form of calcium phosphate is the amorphous calcium phosphate (ACPs), which may have a role in the precipitation and initial assembly of more structured HA particles.[10] Therefore the ACPs are considered promising as a source of initial bone formation. However, due to the manufacturing process of most biomaterials, their application is limited,[11] and in fact the injectable Biobon cement (α-BSM) is the only commercially available calcium phosphate for orthopaedic application that contains a portion of ACP.

Various forms of resorbable calcium phosphates have been on the market for two decades, and their development is still ongoing. The difficulty in resorption of calcium phosphate ceramics is that the resorption rate is not only dependent on the properties of the bone substitute but, as stated before, also on the host characteristics. It is therefore the case that the behaviour of a given material in clinical conditions is, in terms of resorption, unpredictable.

Mechanical strength

The mechanical strength of calcium phosphates depends on their composition and their geometrical properties. The mechanical properties of bone are different from calcium phosphates, as bone displays some elasticity due to the presence of collagen in the bone. Also, different bones have different properties, and different bone tissues have different properties. The longitudinal compression strength of (femoral) cortical bone is 205 MPa, whereas the strength of human trabecular bone ranges from 0.1 to 30 MPa, depending on the site and the age of the patient.[12,13] It is, however, an aspect to be taken into account when choosing a material for a specific indication. For example, compression strength of a calcium phosphate should resemble that of cancellous bone when implanting it in a tibial plateau fracture, but implantation in impaction grafting in a total hip revision case poses specific demands on the applied material. Also, a material should not exceed the strength of the surrounding bone in order to avoid stress shielding. It is for these reasons that compression strength after implantation (and hardening) is important to consider. In general, HA has a high compression strength, and TCP a lower compression strength. Compact materials will in general yield a higher strength than porous materials. It is important to realize that bone graft substitutes will behave differently from bone. Compression strength is one aspect of the mechanical behaviour of a given material, but tensile strength is another. In contrast to bone tissue, which not only contains minerals but also collagen, calcium phosphates will behave poorly when exposed to shear forces. Therefore the combination of bone graft substitutes and an adequate fixation or mechanical environment is required, as the graft materials do not have the elasticity of bone. In a recent overview, van Lieshout *et al.* have summarized several important mechanical and geometrical properties of bone substitute materials.[14] This overview includes compression strength and Young's modulus of various materials and illustrates the wide variety in characteristics of materials. For example, the range in compression strength varies between 47.3 MPa for CortOss (Synthes) and 0.81 MPa for ChronOS (Orthovita), whereas the Young's modulus ranges between 790 MPa for Calcibon (Biomet) and 6 MPa for Ostim (Heraus). Caution must be taken in interpretation of mechanical data, however, as different testing conditions will significantly influence the outcome. For example, dry or wet testing conditions and size of the material tested will lead to variations in the outcome. It is therefore generally acknowledged that only data from tests directly comparing materials under similar circumstances are acceptable. The data provided in Table 4.1 are merely shown to illustrate the range of existing materials.

Table 4.1 Examples of mechanical properties of bone substitute materials

Material	Compression strength (MPa)	Young's modulus (MPa)	Manufacturer
Cortical bone*	205	18.6 GPa	–
Cancellous bone*	0.1–30	10.4 GPa	–
Cerament™**	30–50 (dry)	0.2–0.5 GPa	BoneSupport
CalciBon™***	34 ± 6.8	790 ± 132	Biomet
Norian™***	25.6 ± 7.4	674 ± 146	Synthes
BoneSource™***	6.3–34	475	Stryker
Hydroset™***	14–24	375	Stryker
ChronOS™***	0.81 ± 0.32	54 ± 20	Synthes

**Source*: Rho *et al., J Biomech* 1993.[42]
***Source*: Cerament™ brochure, BoneSupport.
****Source*: van Lieshout *et al., BMC Musculoskelet Disord* 2011.[14]

Clinical evidence

Clinical evidence regarding the use of calcium phosphate ceramics is available for different indications. The aim of most clinical or pre-clinical studies using calcium phosphates is evaluation of the material as void filler. For example in metaphyseal fractures calcium phosphates are used as support for the joint surface after an intraarticular fracture, and the primary outcome of these studies is often prevention of secondary subsidence.[15,16] Resorption of the calcium phosphate is usually not the scope of the study, or even evaluated. The available evidence in the literature is therefore not conclusive regarding the aforementioned advantages of the resorbability of the calcium phosphate ceramics, but the clinical outcome of patients treated with these materials is at least equal, and in most studies superior, to autograft. Given the favorable outcome of artificial bone void fillers compared to autograft, the indication for using these materials is clear for tibial plateau fractures. Although their use in distal radius fractures, proximal humerus fractures, and calcaneus fractures has shown good results in clinical studies so far,[17] application of calcium phosphate ceramics in these fractures is not as widespread as in tibial plateau fractures. An example in the current discussion regarding the use of calcium phosphates is the treatment of distal radius fractures, as recently reviewed by Suhm *et al.*[18] Initially, augmentation of fracture reduction with skeletal repair system (SRS) was shown to be better than K-wire fixation (operative stabilization of the fracture with metal wires), thereby avoiding the risks of an operation. However, later studies showed that a void filler alone led to inferior results compared to K-wire fixation. The used material was different from SRS. As an example, the discussion surrounding the use of a cement as augmentation in the distal radius fractures illustrates that the variation in initial treatment of a given fracture, with plates or K-wires, in combination with the use of different materials, obscures the final analysis of studies. Well-designed clinical trials with long follow-up will be necessary to provide the correct answers, but the current studies mostly show benefits in using resorbable calcium phosphate ceramics for augmentation of fracture stabilization.

4.2.2 Other materials

Bioglass

Bioactive glasses are inorganic materials with clear bone bonding and osteoconductive properties. The surface characteristics of these materials lead to rapid bonding and colonization of bone forming cells.[19] Because of the release of ions after contact with body fluids, especially Cu, Zn, and SR, specific activation and differentiation of bone forming cells occurs. Moreover, they are effective in treatment of osteomyelitis[20] and are associated with the release of angiogenic factors.[8,21,22] By varying their chemical composition and structure, both degradation and bioreactivity can be influenced.

Although the surface characteristics and the mechanical properties of bioactive glasses are favorable for application in bone replacement, clinical evidence is surprisingly limited. A recent clinical trial in tibia plateau fractures showed outcomes comparable with autologous bone,[23] but this study was rather small and the follow-up of one year does not describe the outcome in terms of arthrosis. Most of the publications regarding bioactive glass describe results of pre-clinical studies and periodontology. However, recent developments are promising, and the combination of bioactive glass and other materials appears to be a logical next step.

Synthetic polymers

Synthetic polymers such as polylactic acid (PLA) and polyglycolic acid (PGA) are suitable as a carrier material. Because of the manufacturing process, these polymers can be adjusted or designed according to the needs. By changing molecular weights, configuration, and conformation of polymer chains, degradation rate and application form can be determined, for example by combining PLA and PGA into poly(lactic-co-glycolic) acid (PLGA).[24] The initial application form for these polymers was their use as resorbable fixation materials (screws, plates), because a resorbable material would save the patient a second operative procedure for hardware removal.[25] When used as a block or as granules, the polymers can be used as a graft material. These materials degrade into lactic and glycolic acids, depending on their manufacturing process.[26] Also, degradation times *in vitro* depend on the degradation media used[27] and this makes translation from *in vitro* results to the clinical implementation of this wide range of materials difficult.

Degradation of synthetic polymers involves hydrolysis and enzymatic cleavage. The hydrolysis and enzymatic cleavage of these polymers leads to inflammatory responses, which may have a negative influence on the bone repair process.[28] The inflammatory response, or tissue reaction, is often mentioned as a drawback in the use of PLGA implants. In a large study on 2528 patients who received implants for bone fixation made of PGA or PLA, 108 (4.3%) were affected by a clinically significant tissue reaction.[29] It becomes clear from their analysis that the risk for tissue reaction is much smaller in PLA implants than in PGA implants, with an incidence of tissue reaction of 0.2% (one of 491) versus 5.3% (107 of 2037). Today implants made of polymers consist of a mixture of PLA, PGA, and other polymers. The purpose of this third generation of implants is to maintain stability over a longer period of time to achieve bone union, but also to diminish the tissue reaction to the released ingredients during degradation. However, recent studies indicate that the tissue reaction to these third generation implants is not less frequent, but only postponed.[30] Tissue reaction therefore remains an issue to take into account when considering polymers as bone graft substitutes.

4.3 Bioresorbable materials as a source of other substances

Bone graft substitutes are used to fill bone defects. Their function in treatment of bone defects is to act as void filler, provide initial mechanical support, and to enhance bone formation by osteoconduction and some stimulation of biological activity in specific materials. In larger bone defects, osteoconduction alone will not be sufficient and additional ingredients are necessary to achieve healing. As described in a phenomenon known as the Diamond Concept[31,32] an osteoconductive scaffold needs to be applied in combination with growth factors, osteogenic cells, adequate mechanical stability, and sufficient vascularization in order to achieve healing. These factors may be endogenous or exogenous, but are requirements in the treatment of these often challenging conditions. When the biological condition of the patient is considered insufficient, addition of growth factors or osteogenic cells to an osteoconductive scaffold should be considered.

Combinations of treatment concepts are currently finding their way into clinical application. Still, development of a clinically applicable combination is difficult, with many factors interacting and leading to surprising outcomes in translational research.[33] The successful application of combined implants, especially those with osteogenic cells, depends largely on the biology of the host. Vascularization of the implant, and consequently the survival of the cells, is a concern in larger implants. So far this has hampered the translation from pre-clinical concepts to the clinical situation, in particular for scaffolds seeded with osteogenic cells. Other factors that influence survival of cells in these implants are porosity and scaffold characteristics.

The combination of resorbable materials with drugs or other substances has theoretical advantages. First, the purpose of the material can be extended: for example, in a bone defect after an infected non-union osteoconduction of the implanted material can be combined with antibiotics to treat the infection simultaneously. Second, the release of a drug can be determined by the resorption rate of the material, leading to a slow release of the material and a longer mean residual time (MRT) of the drug at the site of interest. By adding growth factors to osteoconductive materials, osteoinduction is added to the treatment protocol. In this manner, tissue engineering is taken from bench to bedside, and the contributing factors and circumstances are difficult to understand for clinicians who are used to working with screws, plates and nails. The use of resorbable materials makes decisions for specific products even more difficult. Still, several aspects have to be taken into account for each group of materials.

The degradation rate determines the release of attached growth factors,[34,35] and this is therefore an important factor to consider in choosing a carrier material. Binding of growth factors depends on the materials used. For example, binding to calcium phosphates is excellent, but binding to polymers remains the subject of investigation. As bone morphogenetic proteins (BMPs) are water-soluble and polymers are hydrophobic, an even or stable attachment is difficult to achieve. Several methods have been investigated, such as solution dipping, but the ideal method has not yet been developed. Recently, a study by Shen *et al.* reported a good bonding of rhBMP-2 on PLGA after oxygen plasma treatment[36] *in vitro*, and these and other methods will eventually lead to a good binding affinity.

In order to combine positive effects of different materials, combinations of synthetic polymers and growth factors have been studied as well. For example, a recent study[37] described the combination of gelatin, PLGA microspheres, and poly(propylene fumarate) (PPF), a linear polyester which displays mechanical properties suitable for bone replacement applications and encapsulation of PLGA microspheres.[38] These materials were then combined with BMP-2. In this study, the *in vitro* release of BMP-2 was determined as 42 days, but the *in vivo* implantation displayed a shorter and more variable release pattern. The *in vivo* behaviour of combinations of materials, natural or synthetic, will be difficult to predict, as shown in this study as well as by others, because of the mechanisms by which animals, or humans, will physiologically respond to implantation of synthetic polymers. Another combination that has been investigated is the mixture of PLGA and bioactive glass particles.[39] This combination aims to increase local vascular supply, and the first results confirm this effect in experimental studies.

As bone graft substitutes are often applied in revision cases, infection is a commonly encountered problem. The infection can either be active or already treated and under control. Obviously, in case of an active infection debridement will be the first step, followed by bone reconstruction. In case of a history of infection things become more difficult, as many cases of (post-traumatic) osteomyelitis have a tendency for recurrence. In these cases, combination of a bone graft substitute and antibiotics should be considered. The combination of a variety of calcium phosphates and antibiotics is on the market, where the spectrum of the antibiotics will often determine the choice for a given product. However, new products combine the handling and biomechanical properties of a calcium phosphate with broad spectrum antibiotics. A new example in this field is Cerament-G^{TM}, a calcium phosphate consisting of HA and calcium sulphate, to which gentamycin is added. The combination of polymers with antibiotics is feasible as well. Generally, the antibiotics are added to a polymer during the preparation using a solvent/non-solvent separation process, resulting in particulate delivery systems. However, using other manufacturing techniques, solid implants containing antibiotics can be produced as shown recently.[40] This combination is promising for use in infected defects, but largely depends on the composition and properties of the polymer. Their effectiveness has yet to be proven in clinical trials.

Synthetic polymers alone or in combination with bioactive glass are suitable vehicles for drug or growth factor delivery, as described above. Also calcium phosphate ceramics are useful and widely investigated for this purpose.[41] Promising results have been described in numerous publications. The combination of gentamycin or tobramycin in polymethyl methacrylate (PMMA) cement is common in coated implants, but other combinations are used widely as well. Still, the combination of calcium phosphates and living cells remains one of the challenges today. Surface characteristics, degradation rate, and porosity are some of the factors that influence cell survival and thus the success of such an implant. The final solution is yet to emerge.

4.4 Challenges

The broad use of bioresorbable graft substitutes suggests that all is well in this field. The use of these materials is safe and many clinicians will find their use beneficiary. However, several challenges remain. These challenges concern the behaviour of the materials and the evidence for their clinical use.

The behaviour of these materials, as outlined in the above sections, is influenced by the composition of the material, both physical and chemical. Ideally, the material will behave similarly to the surrounding bone, which is either cancellous or cortical. In cancellous bone, the compression strength of the material should be limited to avoid stress shielding. Also, bone turnover and remodelling in cancellous bone is faster than in cortical bone, and this should be reflected in the resorption rate of the applied material. Changes in the composition of the material and the physical properties of the material can be used to mould the behaviour of the material in this manner. Also, surface characteristics are important to consider. Small changes in surface characteristics can influence, for example, the attachment of multinuclear giant cells or osteoclasts, and thereby inhibit or improve the resorption rate. Also, several proteins have a high binding affinity for calcium phosphates, such as the BMPs. By changing the surface, binding affinity for these proteins could be improved, and when the threshold for osteoinduction is reached, bone formation by these substitute materials would be achieved. At this moment, however, only a few studies report on osteoinduction by calcium phosphates to indicate this working mechanism exists in clinical application. The behaviour of a given material will also affect the effectiveness of a combination between a bone graft substitute and other substances, such as antibiotics or growth factors. Local concentration and release rates of these drugs are closely related to the resorption rate of the material, and as indicated in the above, small changes will affect these properties directly. Finding the ideal material is therefore one of the challenges for the near future.

The evidence for the use of bioresorbable bone graft substitutes is only substantial for specific indications, such as the tibial plateau fracture. One of the main reasons for the lack of evidence in many fields, as well as for many materials, is due to the legal status of these materials. Their registration is not as a drug, but as an implant. This implies that their registration is obtained by demonstration of their safety, and not by demonstration of their efficacy. This changes the incentive for many producers, in contrast to the pharmaceutical industry. Many studies on the effectiveness of the use of bioresorbable bone graft substitutes are therefore hospital driven, and not industry driven. Although this approach may limit bias in some perspective, it also makes the design and implementation of these studies more difficult, as the funding and resources are limited. Gathering proper evidence on all possible indications is a major challenge, which requires close cooperation between centres to guarantee a sufficient sample size.

4.5 Conclusions

Resorbable bone graft substitutes have taken their place in clinical applications. The indication for their use is large and growing. Still, the resorption or degradation is influenced by factors from both the material and the recipient. As a consequence, the rate of resorption remains variable. This means that the resorption rate of these materials will not always coincide with the remodelling of the newly formed bone. Also, in case of combinations of materials with medication or growth factors, release patterns will remain somewhat unpredictable. Regulating the resorption rate or degradation speed will remain one of the biggest challenges in this field.

4.6 References

1. Dennis JE, Haynesworth SE, Young RG and Caplan AI. (1992) Osteogenesis in marrow-derived mesenchymal cell porous ceramic composites transplanted subcutaneously: effect of fibronectin and laminin on cell retention and rate of osteogenic expression. *Cell Transplant* **1**(1):23–32.
2. Frankenburg EP, Goldstein SA, Bauer TW, Harris SA and Poser RD. (1998) Biomechanical and histological evaluation of a calcium phosphate cement. *J Bone Joint Surg Am* August; **80**(8):1112–1124.
3. Gross UM, Muller-Mai CM and Voigt C. (1990) The interface of calcium-phosphate and glass-ceramic in bone, a structural analysis. *Biomaterials* July; **11**:83–85.
4. van Blitterswijk CA, Grote JJ, Kuijpers W, Daems WT and de GK. (1986) Macropore tissue ingrowth: a quantitative and qualitative study on hydroxyapatite ceramic. *Biomaterials* March; **7**(2):137–143.
5. Bohner M and Baumgart F. (2004) Theoretical model to determine the effects of geometrical factors on the resorption of calcium phosphate bone substitutes. *Biomaterials* August; **25**(17):3569–3582.
6. Katagiri T and Takahashi N. (2002) Regulatory mechanisms of osteoblast and osteoclast differentiation. *Oral Dis* May; **8**(3):147–159.
7. Ritzel H, Amling M, Posl M, Hahn M and Delling G. (1997) The thickness of human vertebral cortical bone and its changes in aging and osteoporosis: a histomorphometric analysis of the complete spinal column from thirty-seven autopsy specimens. *J Bone Miner Res* January; **12**(1):89–95.
8. Daculsi G and Passuti N. (1990) Effect of the macroporosity for osseous substitution of calcium phosphate ceramics. *Biomaterials* July; **11**:86–87.
9. Merten HA, Wiltfang J, Grohmann U and Hoenig JF. (2001) Intraindividual comparative animal study of alpha- and beta-tricalcium phosphate degradation in conjunction with simultaneous insertion of dental implants. *J Craniofac Surg* January; **12**(1):59–68.
10. Tao J, Pan H, Zeng Y, Xu X and Tang R. (2007) Roles of amorphous calcium phosphate and biological additives in the assembly of hydroxyapatite nanoparticles. *J Phys Chem B* 29 November; **111**(47):13410–13418.
11. Dorozhkin SV. (2010) Amorphous calcium (ortho)phosphates. *Acta Biomater* December; **6**(12):4457–4475.
12. Mosekilde L and Mosekilde L. (1986) Normal vertebral body size and compressive strength: relations to age and to vertebral and iliac trabecular bone compressive strength. *Bone* **7**(3):207–212.
13. Reilly DT and Burstein AH. (1975) The elastic and ultimate properties of compact bone tissue. *J Biomech* **8**(6):393–405.
14. Van Lieshout EM, Van Kralingen GH, El-Massoudi Y, Weinans H and Patka P. (2011) Microstructure and biomechanical characteristics of bone substitutes for trauma and orthopaedic surgery. *BMC Musculoskelet Disord* **12**:34.
15. Welch RD, Zhang H and Bronson DG. (2003) Experimental tibial plateau fractures augmented with calcium phosphate cement or autologous bone graft. *J Bone Joint Surg Am* February; **85-A**(2):222–231.

16. Trenholm A, Landry S, McLaughlin K, Deluzio KJ, Leighton J, Trask K and Leighton RK (2005) Comparative fixation of tibial plateau fractures using alpha-BSM, a calcium phosphate cement, versus cancellous bone graft. *J Orthop Trauma* November; **19**(10):698–702.
17. Larsson S and Bauer TW. (2002) Use of injectable calcium phosphate cement for fracture fixation: a review. *Clin Orthop Relat Res* February; (395):23–32.
18. Suhm N and Gisep A. (2008) Injectable bone cement augmentation for the treatment of distal radius fractures: a review. *J Orthop Trauma* September; **22**(8 Suppl):S121–S125.
19. Xynos ID, Edgar AJ, Buttery LD, Hench LL and Polak JM. (2001) Gene-expression profiling of human osteoblasts following treatment with the ionic products of Bioglass 45S5 dissolution. *J Biomed Mater Res* May; **55**(2):151–157.
20. Lindfors NC, Hyvonen P, Nyyssonen M, Kirjavainen M, Kankare J, Gullichsen E and Salo J (2010) Bioactive glass S53P4 as bone graft substitute in treatment of osteomyelitis. *Bone* August; **47**(2):212–218.
21. Day RM. (2005) Bioactive glass stimulates the secretion of angiogenic growth factors and angiogenesis in vitro. *Tissue Eng* May; **11**(5–6):768–777.
22. Keshaw H, Forbes A and Day RM. (2005) Release of angiogenic growth factors from cells encapsulated in alginate beads with bioactive glass. *Biomaterials* July; **26**(19):4171–4179.
23. Heikkila JT, Kukkonen J, Aho AJ, Moisander S, Kyyronen T and Mattila K. (2011) Bioactive glass granules: a suitable bone substitute material in the operative treatment of depressed lateral tibial plateau fractures: a prospective, randomized 1 year follow-up study. *J Mater Sci Mater Med* April; **22**(4):1073–1080.
24. Kaushiva A, Turzhitsky VM, Darmoc M, Backman V and Ameer GA. (2007) A biodegradable vascularizing membrane: a feasibility study. *Acta Biomater* September; **3**(5):631–642.
25. Peltoniemi H, Ashammakhi N, Kontio R, Waris T, Salo A, Lindqvist C, Gratz K and Suuronen R (2002) The use of bioabsorbable osteofixation devices in craniomaxillofacial surgery. *Oral Surg Oral Med Oral Pathol Oral Radiol Endod* July; **94**(1):5–14.
26. Oh SH, Kang SG and Lee JH. (2006) Degradation behavior of hydrophilized PLGA scaffolds prepared by melt-molding particulate-leaching method: comparison with control hydrophobic one. *J Mater Sci Mater Med* February; **17**(2):131–137.
27. Perron JK, Naguib HE, Daka J, Chawla A and Wilkins R. (2009) A study on the effect of degradation media on the physical and mechanical properties of porous PLGA 85/15 scaffolds. *J Biomed Mater Res B Appl Biomater* November; **91**(2):876–886.
28. Miyamoto S, Takaoka K, Okada T, Yoshikawa H, Hashimoto J, Suzuki S and Ono K. (1992) Evaluation of polylactic acid homopolymers as carriers for bone morphogenetic protein. *Clin Orthop Relat Res* May; (278):274–285.
29. Bostman OM and Pihlajamaki HK. (2000) Adverse tissue reactions to bioabsorbable fixation devices. *Clin Orthop Relat Res* February; (371):216–227.
30. Givissis PK, Stavridis SI, Papagelopoulos PJ, Antonarakos PD and Christodoulou AG. (2010) Delayed foreign-body reaction to absorbable implants in metacarpal fracture treatment. *Clin Orthop Relat Res* December; **468**(12):3377–3383.
31. Giannoudis PV, Einhorn TA and Marsh D. (2007) Fracture healing: the diamond concept. *Injury* September; **38** Suppl 4:S3–S6.

32. Giannoudis PV, Einhorn TA, Schmidmaier G and Marsh D. (2008) The diamond concept--open questions. *Injury* September; **39** Suppl 2:S5–S8.
33. Meijer GJ, de Bruijn JD, Koole R and van Blitterswijk CA. (2008) Cell based bone tissue engineering in jaw defects. *Biomaterials* July; **29**(21):3053–3061.
34. Holland TA and Mikos AG. (2006) Biodegradable polymeric scaffolds. Improvements in bone tissue engineering through controlled drug delivery. *Adv Biochem Eng Biotechnol* **102**:161–185.
35. Lee SH and Shin H. (2007) Matrices and scaffolds for delivery of bioactive molecules in bone and cartilage tissue engineering. *Adv Drug Deliv Rev* 30 May; **59**(4–5):339–359.
36. Shen H, Hu X, Yang F, Bei J and Wang S. (2009) The bioactivity of rhBMP-2 immobilized poly(lactide-co-glycolide) scaffolds. *Biomaterials,* June **30**(18): 3150–3157.
37. Kempen DH, Lu L, Hefferan TE, Creemers LB, Maran A, Classic KL, Dhert WJ and Yaszemski MJ (2008) Retention of in vitro and in vivo BMP-2 bioactivities in sustained delivery vehicles for bone tissue engineering. *Biomaterials* August; **29**(22):3245–3252.
38. Kempen DH, Lu L, Kim C, Zhu X, Dhert WJ, Currier BL and Yaszemski MJ (2006) Controlled drug release from a novel injectable biodegradable microsphere/scaffold composite based on poly(propylene fumarate). *J Biomed Mater Res A* April; **77**(1):103–111.
39. Day RM, Maquet V, Boccaccini AR, Jerome R and Forbes A. (2005) In vitro and in vivo analysis of macroporous biodegradable poly(D,L-lactide-co-glycolide) scaffolds containing bioactive glass. *J Biomed Mater Res A* 15 December; **75**(4):778–787.
40. Park PI, Makoid M and Jonnalagadda S. (2011) The design of flexible ciprofloxacin-loaded PLGA implants using a reversed phase separation/coacervation method. *Eur J Pharm Biopharm* February; **77**(2):233–239.
41. Bose S and Tarafder S. (2012) Calcium phosphate ceramic systems in growth factor and drug delivery for bone tissue engineering: a review. *Acta Biomater* April; **8**(4):1401–1421.
42. Rho JY, Ashman RB and Turner CH. (1993) Young's modulus of trabecular and cortical bone material: ultrasonic and microtensile measurements. *J Biomech* February; **26**(2):111–119.

Part II

Biomaterial substitute scaffolds and implants for bone repair

5
Multifunctional scaffolds for bone regeneration

V. GUARINO, M. G. RAUCCI, A. RONCA, V. CIRILLO and L. AMBROSIO, Institute for Polymers, Composites and Biomaterials, National Research Council, Italy

DOI: 10.1533/9780857099037.2.95

Abstract: Biodegradable scaffolds are generally considered as indispensable elements for engineering living tissues as they are used as temporary templates with specific mechanical and biological properties similar to native extracellular matrix (ECM). They allow modulating cell adhesion, invasion, proliferation and differentiation, prior to the regeneration of biologically functional tissue or natural ECM.

In the case of bone regeneration, a current challenge is to conceive new process strategies to fabricate composite or hybrid scaffolds able to provide three-dimensional templates and synthetic ECM environments. Here, we will describe current processing technologies used to achieve structural features mimicking the ECM on various levels and successfully emulate cell–ECM interactions in order to promote the regeneration of mineralized tissues such as bone.

Key words: composite scaffolds, bioactivity, electrospinning, stereolithography (SLA), sol gel technique, bone regeneration.

5.1 Introduction

A typical scaffold-aided tissue engineering approach involves the monitoring of tissue formation in three-dimensional (3D) structures. In the case of bone, scaffold acts as the 3D support for tissue formation. It possesses peculiar morphological and functional properties which promote cell adhesion, differentiation and proliferation, and desirable mechanical integrity to maintain the predesigned tissue structure, non-cytotoxicity, and osteoconductivity.[1] The development of bio-inspired scaffolds can facilitate the natural repair of damaged or diseased tissues, aiding in the treatment of bone defects caused by loss of bone through disease or injury and where conventional treatments, for example, casting or immobilization, are inadequate.[2,3]

However, the design of highly porous scaffold is often critical. The scaffold should provide a microenvironment for regenerative cells, supporting cell attachment, proliferation, differentiation, and neo-tissue genesis – namely, acting as a synthetic, temporary ECM.[4,5] The human body is a complex and sensitive biological system, and the requirements of scaffold materials for tissue engineering and regenerative medicine are correspondingly numerous, diverse, and extremely challenging.[6] The scaffold design must combine several structural and functional properties through an appropriate selection of constituent materials in order to adapt the scaffold features to the requirements of the specific application. An example is the design of biomaterial scaffolds with adequate morphological and functional properties. Factors governing scaffold design are complex and involve matrix architecture, pore size, morphology, surface properties and degradation.[7] A precise control of scaffold architectural parameters (e.g., pore geometry, size, interconnectivity, orientation, and branching) is necessary to maximize nutrient diffusion,[8] control cell growth and function,[9] manipulate tissue differentiation,[10] and mimic the mechanical properties of regenerated tissue, as outlined below.[11]

- A suitable macrostructure allows the scaffold to promote cell proliferation and cell-specific matrix production.[12]
- An open-pore geometry with a highly porous surface and microstructure enables cell ingrowth.[13]
- A suitable surface morphology and physico-chemical properties may encourage intracellular signalling[14] and recruitment of cells.
- The scaffold should be made from a material with a predictable rate of degradation,[15] which is non-toxic once degraded.

In this context, it is crucial to choose an appropriate material which satisfies the criteria listed above. Several materials have been identified and tested as potential scaffold materials in recent years. However, as biodegradability is one of the most important requirements, the choice of materials is limited to a few ceramics and polymers.[16] Several natural and synthetic ceramics have been used in the field of bone tissue engineering and the results are explored in several works.[17–20] Ceramics are considered an ideal material because of their osteoinductive properties and have been shown to provide good support for bone regeneration. One of the major disadvantages of ceramic material is its brittleness, which results in poor mechanical stability. Also, the high degradation rate of ceramic materials can lead to increased concentrations of calcium and phosphorous, which are detrimental for bone growth.

Natural and synthetic biodegradable polymers have been used as alternatives to overcome the disadvantages of ceramics. Biodegradable polymers such as collagen, elastin or polysaccharides are those that break down due to macromolecular degradation by attack of biological elements[7,21] and show relevant benefits including good bioactivity and chemical versatility.[22] Synthetic polymers such as polyesters – for example, polyglycolide, polylactides,[23] polydioxanone, polycaprolactone (PCL) – and polyanhydrides – such as poly(methacrylated sebacic anhydride) and poly(methacrylated 1,6-bis(carboxyphenoxyhexane) – show considerable achievements in scaffold design for bone due to their specific degradation properties and high workability in several processes.[24] In particular, they have been shown to undergo surface degradation via acidic degradation and this mechanism may be well suited to bone tissue engineering applications. However, some limitations exist, due to pronounced inflammatory response and the inhibition of tissue formation. Therefore, organic polymers with inorganic phases are preferable, such as bioceramics, metal oxide, carbon nanotubes, and bioglasses. They may often be ascribable to unrecognized degradation products or residual components related to the process. Conventional techniques for preparing porous structures to use as tissue engineering scaffolds include porogen leaching,[25] phase separation/freeze drying and gas foaming[26] processes. However, in the last decade, other technologies based on solid freeform fabrication (SFF) and electric field assisted processing have increasingly been applied for this purpose.[27]

5.2 Bone structures and extracellular matrix (ECM) mimics

Learning from nature is an approach which may be successfully adopted in bone and mineralized tissue regeneration. Bone is a natural, anisotropic composite structure, with higher stiffness and tensile strength than soft tissues such as skin, cartilage or blood vessels. Human bone is essentially a three-component composite material, comprising cellular, organic, and inorganic phases and strictly assembled to form the natural tissue. The cellular phase – involving three different types of cells, namely osteoblasts, osteoclasts, and osteocytes – synthesizes and regulates ECM deposition and mineralization, modulating the complex mechanism of bone remodelling.

The extracellular matter embeds tissue-specific cells in a highly complex matrix consisting of the other two components – a non-mineralized phase and a mineralized (hydroxyapatite – HA) phase. The organic phase contains collagens, glycoproteins, proteoglycans, and sialoproteins which play an essential role both in the control of growth and differentiation of osteoblasts, osteocytes, and osteoclasts and in bone remodelling. Finally, the inorganic phase comprising HA (65–70% of the total matrix) provides adequate structural support.

For bone replacement, engineering solutions involve the use of smart structural components, which respond *in situ* to the pertinent stimuli triggering bone tissue formation mechanisms such as biomineralization. Ideally, mimesis of the living tissue, in its mechanical, biological and functional aspects, must be achieved. As part of this strategy, open-pore biocompatible and biodegradable porous scaffolds have been used to provide a temporary substitute for the ECM of natural bone. However, correct bone growth must be coupled with efficient angiogenesis, to ensure the neovascularization and to preserve an efficient nutrient supply and oxygen to cells, as well as removal of metabolic products. Therefore, cell biology, materials, and engineering aspects must be addressed together to create biohybrid scaffolds which combine porous matrices with harvested cells, recombinant signalling molecules and specific inductive factors (e.g., bone morphogenetic protein (BMP)-4, vascular endothelial growth factor (VEGF)).

Composite and hybrid materials may be successfully used to build multifunctional scaffolds if they satisfy the three key requirements as follows:

1. material components with time- and space-controlled biodegradability.
2. percolative architectures, with interconnecting pores of the right scale to promote tissue integration and vascularization.
3. bioactive soluble and insoluble signals able to promote the mineralization of ECM and osteogenesis.

Osteoconductive or osteoinductive stimuli may be induced by using insoluble signals, such as, calcium phosphates (CaPs) and HA[28] inside the implant – that is, within the bone scaffold – or by soluble signals represented by proteins such as growth factors[29] and peptide sequences.[30]

5.3 Micro/macroporous scaffolds with bioactive solid signals

A consolidated approach in bone engineering consists of the improvement of bioactivity of polymer scaffold through the employment of HA and CaPs. They resemble the mineral phase of bone, and are biocompatible, bioactive, and osteoconductive. Moreover, their inclusion in biodegradable polymers is an interesting strategy to modulate scaffold degradation kinetics. This allows the polymer to reach scaffold component resorption *in vivo* at the same rate as novel tissue formation, preserving structural function to withstand loading where necessary.[31] The driving force behind the development of polymer/bioactive glass composite scaffolds for bone tissue engineering is the need to confer bioactive behaviour upon the polymer matrix. The degree of bioactivity may be adjustable by the volume fraction, size, shape, and arrangement of inclusions.[6]

Kim *et al.* proposed the development of a gelatin–HA composite as a good alternative for hard tissue regeneration, in view of its combination of the bioactivity and osteoconductivity of HA with the flexibility and hydrogel characteristics of gelatin.[32] Unlike the hydrophobic synthetic polymers, gelatin, with its hydrophilic character, has a high affinity with HA and so mixes well with HA in an aqueous solution. The gelatin–HA composite is fabricated as a foam, through a novel mixing, freeze-drying, and cross-linking process, to furnish a system with a high specific surface area for vascularization and cell ingrowth. The progressive addition of bioactive HA gives rise to an increasing anisotropy of the structure with a reduction of pore size, and to an irregular pore shape. Mechanically, the gelatin composite responds efficiently to applied stress, absorbing energy, and dissipating stress without being fractured. This high flexibility of the foam is beneficial in a surgical operation, keeping the scaffold securely at the defect site, whilst the rigidity and stiffness of the HA component enable resistance to higher stress levels. Good growth and spreading of human osteoblast-like (HOS) cells, cultured on the foams by static seeding, suggest an enhanced functional activity of osteoblastic cells due to the presence of bioactive HA filler.[33,34]

Guarino *et al.* proposed alternative strategies to improve the mechanical performance of the scaffolds through the use of synthetic polymers with slow degradation, such as PCL.[15,35,36] PCL has been widely used for the effective entrapment of antibiotic drugs, and so a construct made with PCL can be considered as a drug-delivery system, used to enhance bone ingrowth and regeneration in the treatment of bone defects.[37] Its degradation proceeds via a remarkably slow mechanism, requiring up to three years for complete removal from the host body. The process entails random hydrolytic ester cleavage and weight loss by the diffusion of oligometric species from the bulk,[38] which is particularly useful in bone regeneration. However, poor mechanical properties of PCL, depending on preparation technique and molecular weight, impair its use as material scaffold for hard tissue replacement.[39]

Causa *et al.* proposed enhancing the mechanical properties of microporous PCL matrices with rigid HA particles, which would also improve osteoconductivity of the polymer. In particular, the addition of HA particles with an optimal polymer/ceramic volume ratio led to a significant improvement in mechanical performance in tension. However, the absence of large pores to host cells limits their use for bone regeneration. Indeed, the achievement of adequate porosity is crucial to the success of these materials as scaffolds for orthopaedic applications. Accordingly, Guarino *et al.* proposed the use of PCL–HA composite scaffolds in the form of macroporous networks, prepared by the phase inversion/salt leaching technique, which contain larger pores inside the matrix.[15, 35] In this case, mechanical properties are partially compromised from the higher porosity degree (of about 90%), reducing the reinforcing effect of HA particles. However, bimodal distribution of pore sizes – large pores ranging from 100 to 300 μm coexisting with smaller pores of few microns in size – is crucial to assure a correct sustainment of cells during the *in vitro* culture. Indeed, larger pores can efficiently assume an orientation favourable to cell adhesion and proliferation, whilst the smaller pores are able to promote the nutrient supply and waste removal by improving scaffold vascularization.

Although recent studies emphasized the crucial role of porosity features (i.e., pore size and shape) on cell–material interaction,[40] it is recognized that pore architectures with tailored porosity need to be complemented by bone-inducing agents to stimulate the native bone-formation activity.[41] Recently, it has been demonstrated that chemical modification of HA crystals can induce improvements in the biological response of bone-like cells, better triggering the osteogenic response under *in vitro* and *in vivo* conditions, due to Mg^{2+} and CO_3^{2-} ion substitution into porous scaffolds with optimized morphological features (i.e., pore size, surface curvature, interconnectivity).[36] Indeed, Mg^{2+}-doped HA is able to retain more water at its surface than stoichiometric HA, with water molecules coordinated to cations and absorbed as a multilayered molecule.[42] This favours cell anchorage and growth, due to the ability of CaP to adsorb several important ECM proteins.[43] More recently, Guarino *et al.* demonstrated that ionic substitution also stimulates cells to produce mineral extracellular matter through activation of early osteogenesis mechanisms.[36] Meanwhile, porous architecture with tailored porosity features (i.e., pore size, concavity) supports the *in vivo* growth of mature bone with hierarchical organization by imparting the natural osteon-like structure of bone (with lamellae centripetally assembled from the wall of the macropores toward the central bone marrow cavity) during the first six months of implantation.

5.4 Hybrid scaffolds by sol–gel technique

Bioactive glasses are able to bond to living bone, and this has been attributed to the formation of a hydroxyl carbonate apatite layer under physiological conditions. Sola *et al.* have confirmed the ability of the glasses, both in bulk and sintered form, to generate a HA surface layer when immersed in a simulated body fluid.[44] On the other hand, CaPs have been successfully used for bone repair and regeneration because of their biocompatibility and osteoconductivity or osteoinductivity, which promote the formation of chemical bonds with living tissue by mimicking the apatite phase of the natural bone tissue.[45] In particular, synthetic HA has been recognized as one of the most important bone substitute materials in orthopaedics and dentistry over the past few decades[46,47] because of its ability to form chemical bonds with surrounding hard tissues[48,49] through the formation of a HA interfacial layer.[50] Moreover, HA particles on a nanometric scale (nHA) have been proved to be an osteoconductive material that also chemically binds to enamel and dentine.[51]

Recently, hybrid materials, particularly nano-hydroxyapatite (nHA) reinforcement and polymer matrix,[52] have gained much recognition as bone scaffolds. This is due not only to their composition and structural similarity with natural bone but also to their unique functional properties, such as a larger surface area and superior mechanical strength than those of their single-phase constituents.[53]

Numerous HA synthesis techniques have been developed to suit the numerous applications of CaP compounds in biomedical fields. The most commonly used technique for the formation of HA is the precipitation technique, involving wet chemical reactions between the calcium and phosphate precursors under controlled temperature and pH conditions.[54,55] However, this method has some inherent disadvantages, primarily difficulty in keeping the pH value above 9. This is essential to avoid the formation of a Ca-deficient HA which, in sintering, undergoes easy decomposition, forming tricalcium phosphates. In this respect, the sol–gel technique is an elective method for the preparation of highly pure powder due to the possibility of strict control of the process parameters.[56]

This method involves a molecular mixing of calcium and phosphorus, which can improve the chemical homogeneity. Moreover, the low processing temperature of sol–gel technology makes it an ideal technology for the fabrication of bioactive materials, particularly when combined with its intrinsic biocompatibility (the ability of materials not to produce a significant rejection or immune response when they are inserted into the body) and environmental friendliness. A bioactive material is defined as a material that elicits a specific response at the interface of the material which results in the formation of a bond between the tissue and that material.[57] The ability of sol–gel technology to manipulate the structure of materials at the molecular level, as well as its ability to precisely control the nature of interfaces, makes it an interesting approach for a wide range of practical applications.[58] The sol–gel technology based on various alkoxides allows for the production of conventional silica glasses[59–62] as well as multicomponent materials, merging silicates with titanates, borates, and a variety of other oxides. The alkoxide gel method can also be used to produce certain non-silicate oxide glass-like materials (e.g., ZrO_2^-, etc.).[63]

Furthermore, the sol–gel product is characterized by nanosized dimensions of the primary particles, and this small domain is a very important parameter to improve the contact reaction and the stability at the artificial–natural bone interface. Moreover, the high reactivity of the sol–gel powder allows a reduction of the processing temperature to produce homogeneous hybrid materials at room temperature, with any degradation phenomena occurring during sintering.

Raucci *et al.* have demonstrated the possibility of finely dispersing CaP nanoparticles into a biodegradable polymer such as PCL matrix, expecting an improvement in the functional features (i.e., mechanical response, bioactivity) using the sol–gel method.[64] Moreover, presence of HA particles in the composite material compensates for the acidic release from the polymer because of the presence of alkaline CaP.[65] Acidosis caused by the (chemically unavoidable) release of acidic degradation products may be a problem with biodegradable polyesters, but careful *in vivo* and *in vitro* measurements of pH in bone chambers have shown that the pH drop is 0.2 units near the eroding polyesters.[66] Incorporating bioceramic particles into biodegradable polymers to form bioactive scaffolds overcomes some limits of physically embedded particles, through a more efficient physical distribution of the bioactive signal into the scaffold which improves the bone bonding. The scaffold interaction at the tissue interface may be further improved by the creation of tailored surface topographies with features in the same scale-range as that seen on the old bone surface at bony remodelling sites.[67] Several approaches also involve surface modification by surface adsorption and HA grafting onto the substrate.[68] It has been shown that bone-like apatite coating could form *in vitro* on the composite scaffolds,[69,70] indicating osteoconductivity of these materials.[66,71] This route of making bioactive bone tissue engineering scaffolds is currently being explored, with several manufacturing techniques being actively investigated.

5.5 3D printed scaffolds via laser sintering

All common SFF technologies are based on the creation of 3D scaffolds by a layering process of 2D repeating slices. A computer model is sliced into a number of layers by computer-aided design (CAD) software and each layer of the original model is produced in succession.[72,73] Currently, rapid prototyping (RP) using SFF techniques is common practice in the automotive industry, for jewellery making, and for designing end-user devices and appliances,[74] but these additive fabrication methods have also been used in designing surgical tools, implants, and other biomedical devices.[75,76] The desired implant area of the patient is scanned by X-ray or magnetic resonance and the data are imported into CAD software. This software enables the surgeon to design the implant according to the individual requirements. Information is transferred to a RP system, which produces the scaffold from a biocompatible and biodegradable material. Living cells are seeded onto the surface of the scaffold after production or integrated into the RP process. After a cell culture treatment and a cell number enlargement, the implant is placed into the human body and starts to fulfil its function. The material disappears after a time and is replaced by natural tissue.[72,73,77]

Advances in RP techniques have significantly improved control over the whole design of three-dimensional solid and porous structures.[78] Among these techniques are selective laser sintering,[79] fused deposition modelling,[80,81]

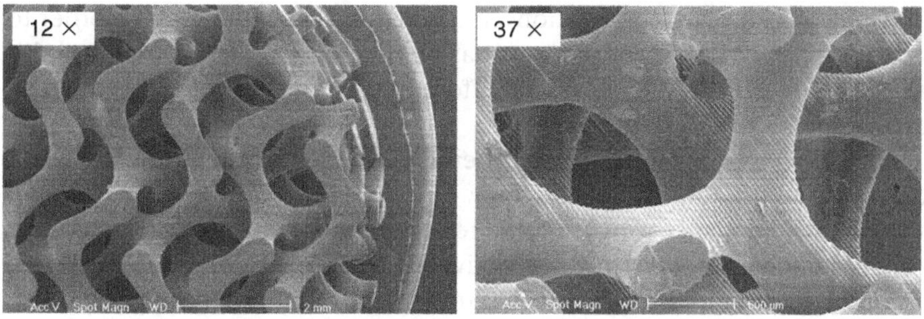

5.1 Scanning electron microscope (SEM) images at different magnification of porous structures with mathematically defined architecture prepared by stereolithography.

3D printing[82] and stereolithography (SLA).[83,77] SLA was developed by 3D systems in 1986 as the first commercially available SFF technique, and nowadays it is still superior to others in terms of accuracy and resolution.[73] SLA is a RP technique based on the polymerization of photopolymer resins that is initiated by energy from UV radiation to create 3D scaffolds from a CAD specification.[84–86] SLA is a useful strategy to fabricate precisely designed scaffolds (Fig. 5.1) with the external shape based on a 3D computed tomography (CT) scan of the patient, as well as parameters such as pore size, porosity, interconnectivity, and mechanical stiffness.[87] Like most SFF techniques, SLA allows fabrication of parts from CAD files. The CAD files describe the geometry and size of the parts to be built.[88] This designed structure is virtually sliced into layers and, using a computer-controlled laser beam or digital light projection and a computer-driven support platform, a 3D object can be constructed in a layer-by-layer fashion.[86,89] The sequential polymerization of layers with different illumination patterns leads to the creation of complex 3D objects, described by Griffith and Halloran.[90] Architectural cues for 3D porous scaffolds including porosity, pore size, interconnectivity, and channel orientation are also important design parameters that can affect osteogenic signal expression of a seeded cell population.[86] For example, pore size could be a factor controlling both the level of BMP2), mRNA expression and osteoblastic differentiation.[91]

Physical characteristics of the scaffold can be tailored using SLA systems to influence diffusion, fluid flow, and mechanical properties, as well as cell behaviour.[92] Ferry *et al.* showed that an open, accessible scaffold architecture, fabricated by SLA, improved cell seedability when compared with the tortuous pore architecture of scaffolds obtained by conventional methods.[83] Mathematically designed gyroid scaffolds can be more easily wetted and penetrated by a cell suspension, resulting in a more homogeneous cell distribution.[83,88] Besides the architectural features, scaffolds for load-bearing applications in bone regeneration should be mechanically stable in order to provide mechanical support, as well as being bioactive, facilitating or initiating proliferation and osteogenic differentiation of cells, ECM production, and eventually bone deposition.[93]

For this reason, great interest is being addressed to the optimization of the material used in SLA technology. Photopolymer resins are mixtures of simple low molecular weight monomers, capable of chain reaction to form solid, long-chain polymers when activated by radiant energy in a specific wavelength range.[94] However, the number of available biocompatible resins suitable for use in biomedical applications of SLA is limited,[95] because photocurable materials are predominantly acrylate-based resins which lack biodegradability and possess insufficient mechanical strength.[96] These materials are predominantly glassy, rigid, and brittle and very few resins have been described that allow the preparation of elastomeric objects by SLA.[73]

Ferry *et al.* developed a low molecular weight poly(D,L-lactide) (PDLLA) oligomer functionalized with methacrylate groups that enables the preparation of a mathematically defined porous structure for bone tissue engineering.[98] With values of the elasticity modulus of approximately 3 GPa, PDLLA polymers are among the few biodegradable polymers with mechanical properties that approach those of bone.[86] Cooke demonstrated the feasibility of using the SLA process to build and control 3D multilayer parts made from a biodegradable, biocompatible polypropylene fumarate (PPF) resin.[99] Lee *et al.* successfully investigated an optimal UV curable polymer resin, based on PPF, to fabricate 3D scaffolds with controlled microstructures for bone tissue engineering application.[96] The main disadvantage of methacrylates is their limited reactivity due to the sterical hindrance and inductive stabilization of the formed radical of the additional methyl group. Acrylates are considerably more reactive but also show a significant tendency toward Michael addition side reactions, with amino groups of proteins or DNA giving hydrolytically noncleavable aliphatic adducts.[100] This frequently results in some significant toxic effects, meaning that their use in biomedical applications can be controversial.

In order to create scaffolds that encompass the physical and chemical cues for bone tissue engineering, recent studies have demonstrated the feasibility of using a composite polymer/ceramic particles resin.[77,101] Ceramic particles (e.g., nHA) are homogeneously suspended in the photo-crosslinkable resin and used in SLA[102,103] to reproduce the natural composition of the bone. The presence of ceramic nanopowder improves the osteogenic potential, as well as the mechanical properties of the scaffold, and it also decreases the penetration depth of the incident light, improving the vertical resolution of the process. Ronca *et al.*[89] achieved composite structures using different amounts of nHA with a large freedom of design and complex internal architectures. Composite structures have demonstrated improved mechanical properties and cell compatibility compared with the conventional dimethacrylate structures, which showed greater amounts of alkaline phosphatase synthesis and deposition of calcium-containing mineral when human mesenchymal stem cells (hMSCs) were cultured.[89,104]

Bian *et al.* designed a bio-inspired biphasic osteochondral scaffold, composed of beta-tricalcium phosphate (β-TCP) and type-I collagen that included a bone phase with a 3D channel network, a cartilage phase, and a transitional structure, according to the biological transitional interface from cartilage to bone.[105] Arcaute *et al.* encapsulated cells in bioactive diacrylate polyethylene glycol (PEG) solution to fabricate complex 3D structures by SLA.[106] Human dermal fibroblasts were seeded within PEG hydrogel constructs by combining a cell suspension with the photopolymer solution, which was then used in the SLA fabrication process.

Looking towards the future, RP technologies hold great potential in the context of scaffold fabrication[107,72] and SLA, which has proved to be particularly versatile with respect to the freedom of design of the structures that are to be built, has strong potential for bone tissue engineering applications.

5.6 ECM-like scaffolds by electrospinning

ECM is composed of a network of nanometer-sized proteins and glycosaminoglycans.[109] The intricate complexities of this spatial and temporal environment dynamically influence phenotypic and other cellular behaviour by providing indirect and direct informational signalling cues.[109,110] Thus, the more closely the *in vivo* environment (i.e., chemical composition, morphology, surface functional groups) can be recreated, the more likely the success of the tissue engineering scaffold.[111,112]

Although the desired characteristics of a scaffold vary slightly with the tissue being recreated, there are general properties that are desirable. First and foremost, the scaffold should be biocompatible, meaning that it integrates with the host tissue without eliciting a major immune response.[113] The scaffold should also be porous, with a high surface/volume ratio to allow for cell attachment and ingrowth as well as exchange of nutrients during *in vitro* or *in vivo* culture.[114] Furthermore, the porous nature of the scaffold allows for angiogenesis upon implantation in a defect site (for vascularized tissues). Moreover, since the scaffold acts as a temporary support on which cells can adhere and proliferate, it should mimic native ECM both architecturally and functionally.[115] Finally, a tissue engineering scaffold should be biodegradable, so that further surgery is not required to remove the implant.[115] The rate of degradation should coincide with, or at least be controllable in order to mimic, the rate of neo-tissue formation.[109,113]

Electrospinning is a versatile technique that allows the production of continuous fibres, with diameters ranging from micro- to nanometres.

This process involves the use of a strong electric field applied to a droplet of fluid (polymer solution/melt) coming out from the tip of a needle, which acts as one of the electrodes. When the applied voltage overcomes the surface tension of the fluid, a charged jet is ejected from the cone-shaped tip of the droplet. The jet is accelerated towards the oppositely charged collector and the solvent evaporates, leaving behind a dry fibre on the collecting device.[116,117] The random collection of electrospun fibres on a plate generates connected porous mats with high porosity and high surface area which can mimic the ECM structure and, therefore, electrospinning is an excellent candidate for use in tissue engineering.

The diameter of electrospun fibres can be of similar magnitude to that of fibrils in ECM, mimicking the natural tissue environment, if the system and process parameters are tuned appropriately.[118] Electrospinning has now become the most extensively used fabrication method for preparation of these nanofibrous scaffolds. When compared to other fibre-forming processes such as self-assembly and phase separation techniques, electrospinning provides a simpler and more cost-effective means to produce fibrous scaffolds with an interconnected pore structure and fibre diameters in the sub-micron range.[119] A variety of polymeric fibres have been considered for use as scaffolds for engineering tissues such as cartilage,[120,121] dermal tissue,[122] bone,[123,124] arterial blood vessels,[125–127] heart,[128,129] and nerves.[130,131] Optimal choice of biomaterials is required, in terms of the necessary mechanical properties and degradation time, which depends upon the type of scaffold required, the type of tissues to be regenerated, and their regeneration time.

Several natural and synthetic polymers have been processed by electrospinning for bone tissue engineering applications. Synthetic biodegradable aliphatic polyesters, such as poly(lactic acid), poly(glycolic acid), and poly(ε-caprolactone) (PCL) are generally preferred due to their good processability, which allows for the development of micro- and/or nanostructured systems with good chemical stability and mechanical performance.[132–134]

Spasova *et al.*[135,136] have successfully electrospun nanofibrous polymer mats comprising poly(L-lactide) (PLLA)/PEG with fibre diameters ranging from 100 nm to 6 micron, and they observed in long-term cultures that osteoblast-like cells (MG-63) became organized in a tissue-like structure. Guarino *et al.* produced PCL fibrous membranes and tested them with human mesenchymal stem cells (hMSCs).[137] Results indicated that nanofibre scaffolds supported greater cell adhesion and viability in comparison with PCL casted film. Also, in osteogenic conditions, nanofibrous scaffolds cultured with hMSCs clearly displayed a positive alizarin red staining, indicating a more diffused mineral deposition on electrospun substrates. However, compared to natural polymers, synthetic materials show a relative lack of cell recognition sites on the scaffold surfaces, leading to poor cell affinity.[138]

Natural polymers are often used for preparing fibrous scaffolds because of their enhanced biocompatibility and bio-functional motifs such as collagen, alginate, silk protein, hyaluronic acid, fibrinogen, chitosan, starch, and others, and because their addition into synthetic polymers can improve the overall cytocompatibility of the scaffold.[139–142] It has been demonstrated that PCL nanofibres endowed with gelatin protein clearly encouraged cell recognition, directing the MSCs to adhere, proliferate, and form a mineralized matrix thanks to the synergistic effect of fibre architecture and biochemical signals that influence the basic processes of bone regeneration.[137] Moreover, Alvarez *et al.* showed that PCL/gelatin electrospun fibres provide an environment which markedly improves the osteogenic differentiation of hMSCs and mineralization, confirmed by the high tendency to CaP deposition and the overexpression of principal bone-related molecules and genes.[143]

Having the ability to process different kinds of materials by electrospinning (i.e., polymers, metals and ceramics), a novel approach has been proposed, involving the coupling of an inorganic component to improve bone-bioactivity with an organic phase that assures the fibrous architecture and easy processability. Biomimetic fibres containing HA crystallites have been produced, as shown in Fig. 5.2; these are similar to CaP, the inorganic component of bone ECM. Similarly, electrospun nanofibrous bioactive glass was thoroughly homogenized with collagen or a PLLA solution, with the aim of producing uniform and homogenous scaffolds for improving bone-bioactivity of the organic phase.[144–146]

An alternative method is the development of a biomimetic coating method, based on the immersion of polymeric electrospun membranes in

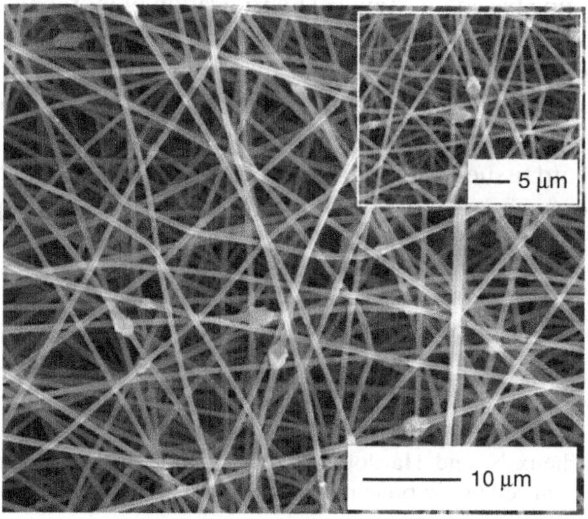

5.2 Electrospun gelatin fibres with HA inclusions.

solutions resembling body fluids, to functionalize polymeric scaffolds for bone tissue engineering by applying a thin CaP coating on their surface.

Recent advances focused on the development of novel 3D structures have shown that it is possible to produce tubular scaffolds with potential application as scaffolds for long bone treatment by collecting electrospun fibres onto a rotating mandrel instead of a metal conductive plate.[147] Nowadays, all the reported results on the evaluation of electrospun nano- or microfibrous matrices as scaffolds for bone tissue engineering clearly indicate that electrospun scaffolds represent suitable potential substrates for bone regeneration.

Along with the ability to produce a tissue-mimicking, nanofibrous structure, other factors such as the low cost of the process, the ease of set-up, and versatility in tuning fibre composition and morphology made the electrospun platforms attractive for broader purposes.

5.7 Conclusions and future trends

The ability to produce tissue-mimicking constructs depends on several factors, including versatility in tuning composition and morphology of scaffolds for target tissues. Firstly, the production of appropriate tissue-engineered platforms requires an appropriate resizing of pore structures to permit cellular migration and tissue perfusion. In this context, electrospinning allows the material to mimic the native microstructure of ECM, but significant intrinsic limitations relate to the creation of macropores – that is, pore sizes ranging from tens of nanometres or at most a few micrometres. Meanwhile, the bioactivation of fibres may be improved by finer control of bioactive phases in terms of particle sizes at the nanoscale, and spatial distribution. In this context, new strategies are emerging based on chemical hybridization of inorganic–organic compositions. These strategies will catalyse a significant interest due to their ability to synthesize nanoapatite at the molecular scale into organic–inorganic hybrids, with performance regarding their physicochemical and mechanical properties still to be explored. These approaches will show great potential for biomedical use in hard tissue regeneration, due to the opportunity to combine bioactive molecules such as native proteins and chemical drugs to produce multimaterial scaffolds with therapeutic efficacy and tissue mimesis.

5.8 References

1. Wang H., Li Y., Zuo Y., Li J., Ma S. and Cheng L. (2007) Biocompatibility and osteogenesis of biomimetic nano-hydroxyapatite/polyamide composite scaffolds for bone tissue engineering. *Biomaterials* **28**(22), 3338–3348.
2. Olivier V., Faucheux N. and Hardouin P. (2004) Biomaterial challenges and approaches to stem cell use in bone reconstructive surgery. *Drug Discov Today* **9**, 803–811.

3. Guarino V., Gloria A., Raucci M.G., De Santis R. and Ambrosio L. (2012) Bio-inspired composite and cell instructive platforms for bone regeneration. *Int Mater Rev* **57**(5), 256–275.
4. Langer R. and Vacanti J.P. (1993) Tissue engineering. *Science* **260**, 920–926. P.X. Ma (2004) Scaffolds for tissue fabrication, *Mater Today* **7** 30–40.
5. Liu X. and Ma P.X. (2004) Polymeric scaffolds for bone tissue engineering. *Ann Biomed Eng* **32,** 477–486.
6. Rezwan K., Chen Q.Z., Blaker J.J. and Boccaccini A.R. (2006) Biodegradable and bioactive porous polymer/inorganic composite scaffolds for bone tissue engineering. *Biomaterials* **27,** 3413–3431.
7. Hutmacher D.W., Sittinger M. and Risbud M.V. (2004) Scaffold-based tissue engineering: rationale for computer-aided design and solid free-form fabrication systems. *Trends Biotechnol* **22**(7), July (2004).
8. Kuboki Y., Takita H., Kobayashi D., Tsuruga E., Inoue M., Murata M., Nagai N., Dohi Y. and Ohgushi H. (1998) BMP-induced osteogenesis on the surface of hydroxyapatite with geometrically feasible and non-feasible structures: topology of osteogenesis. *J Biomed Mater Res* **39**(2), 190–199.
9. Blokhuis T.J., Termaat M.F., den Boer F.C., Patka P., Bakker F.C. and Haarman H.J. (2000) Properties of calcium phosphate ceramics in relation to their in vivo behavior. *J Traum Inj Inf Crit Care* **48**(1), 179.
10. Holmes R.E., Bucholz R.W. and Mooney V. (1986) Porous hydroxyapatite as a bone-graft substitute in metaphyseal defects – a histometric study. *J Bone Jt Surg Am* **68A**(6), 904–911.
11. Martin R.B., Chapman M.W., Sharkey N.A., Zissimos S.L., Bay B. and Shors E.C. (1993) Bone ingrowth and mechanical-properties of coralline hydroxyapatite 1 yr after implantation. *Biomaterials* **14**(5), 341–348.
12. Yeong W.-Y., Chua C.-K., Leong K.-F. and Chandrasekaran M. (2004) Rapid prototyping in tissue engineering: challenges and potential. *Trends Biotechnol* **22**(12), December.
13. Melchels F.P.W., Barradas A.M.C., van Blitterswijk C.A., de Boer J., Feijen J. and Grijpma D.W. (2010) Effects of the architecture of tissue engineering scaffolds on cell seeding and culturing. *Acta Biomater* **6,** 4208–4217.
14. Mangano C., De Rosa A., Desiderio V., d'Aquino R., Piattelli A., De Francesco F, Tirino V., Mangano F. and Papaccio G. (2010) The osteoblastic differentiation of dental pulp stem cells and bone formation on different titanium surface textures. *Biomaterials* **31,** 3543–3551.
15. Guarino V., Taddei P., Di Foggia M., Fagnano C., Ciapetti G. and Ambrosio L. (2009) The influence of hydroxyapatite particles on in vitro degradation behavior of poly ε-caprolactone-based composite scaffolds. *Tissue Eng Part A* November 2009, **15**(11), 3655–3668.
16. Meijer G.J., de Bruijn J.D., Koole R. and van Blitterswijk C. (2007) Cell-based bone tissue engineering. *PLoS Med* **4**(2): 260–264. Guarino V., Causa F., Ambrosio L. (2007) *Exp Rev Med Dev*, **4**(3), 405–418.
17. LeGeros R.Z. (2002) Properties of osteoconductive biomaterials: calcium phosphates. *Clin Orthop Relat R* **395**, 81–98.
18. Baksh D., Davies J.E. and Kim S. (1998) Three-dimensional matrices of calcium polyphosphates support bone growth in vitro and in vivo. *J Mater Sci-Mater M* **9**(12), 743–748.

19. Dong J.A., Uemura T., Shirasaki Y. and Tateishi T. (2002) Promotion of bone formation using highly pure porous beta-TCP combined with bone marrow-derived osteoprogenitor cells. *Biomaterials* **23**(23), 4493–4502.
20. Ambrosio L., Guarino V., Sanginario V., Torricelli P., Fini M., Ginebra M.P., Planell J.A. and Giardino R. (2012) Injectable calcium phosphate based composites for skeletal bone treatments. *Biomed Mater* 29, **7**(2), 024113.
21. Deporter D.A., Komorin N., Howley T.P., Shig A., Ghent A., Hansel P. and Parisien K. (1988) Reconstituted bovine skin collagen enhances healing of bone wounds in the rat calvaria. *Calcified Tissue Int* **42**(5), 321–325.
22. Salgado A.J., Continho O.P. and Reis R.L. (2004) Bone tissue engineering: state of the art and future trends. *Macromol Biosci* **4**(8), 743–765.
23. Chen V.J., Smith L.A. and Ma P.X. (2006) Bone regeneration on computer-designed nano-fibrous scaffolds. *Biomaterials* **27,** 973–3979.
24. Seal B.L., Otero T.C. and Panitch A. (2001) Polymeric biomaterials for tissue and organ regeneration. *Mater Sci Eng: R: Rep* **34**, 147–230.
25. Guarino V., Causa F. and Ambrosio L. (2007) Porosity and mechanical properties relationship in PCL porous scaffolds. *J Appl Biomater Biomech* September–December, **5**(3), 149–157.
26. Salerno A., Oliviero M., Di Maio E., Iannace S. and Netti P.A. (2009) Design of porous polymeric scaffolds by gas foaming of heterogeneous blends. *J Mater Sci: Mater Med* **20**, 2043–2051.
27. Lantada A.D. and Morgado P.L. (2012) Rapid prototyping for biomedical engineering: current capabilities and challenges. *Annu Rev Biomed Eng* **14**, 73–96.
28. Causa F., Netti P.A., Ambrosio L., Ciapetti G., Baldini N., Pagani S., Martini D. and Giunti, A. (2006) Poly-ε-caprolactone/hydroxyapatite composites for bone regeneration: in vitro characterization and human osteoblast response. *J Biomed Mater Res* **76A**, 151.
29. Kim H.W., Knowles J.C. and Kim H.E. (2004) Hydroxyapatite/poly(e-caprolactone) composite coatings on hydroxyapatite porous bone scaffold for drug delivery. *Biomaterials* **25,** 1279–1287.
30. Hubbell J.A. (1995) Biomaterials in tissue engineering. *Biotechnology* **13,** 565–576.
31. Griffith L.G. and Naughton G. (2002). Tissue engineering – current challenges and expanding opportunities. *Science* **295**, 1009–1014.
32. Kim H.W., Knowles J.C. and Kim H.E. (2005) Hydroxyapatite and gelatin composite foams processed via novel freeze-drying and crosslinking for use as temporary hard tissue scaffolds. *J Biomed Mater Res* **72A**, 136–145.
33. Laurencin C T. and Freeman J.W. (2005) Ligament tissue engineering: an evolutionary materials science approach. *Biomaterials* **26,** 7530–7536.
34. Hench L.L. and Polak J.M. (2002). Third-generation biomedical materials. *Science* **295**, 1014–1017.
35. Guarino V., Causa F., Netti P.A., Ciapetti G., Pagani S., Martini D., Baldini N. and Ambrosio L. (2008) The role of hydroxyapatite as solid signal on performance of PCL porous scaffolds for bone tissue regeneration. *J Biomed Maters Res-Part B* **86B**, 548–557, DOI:10.1002/jbm.b.31055.
36. Guarino V., Scaglione S., Sandri M., Alvarez-Perez M.A., Tampieri A., Quarto R. and Ambrosio L. (2012) MgCHA particles dispersion in porous PCL scaffolds: *in vitro* mineralization and *in vivo* bone formation. *J Tissue Eng Reg Med* doi: 10.1002/term.1521.

37. Pitt C.G., Gratzel M.M. and Kimmel G.L. (1981) Aliphatic polyesters. 2. The degradation of poly(DL-lactide), poly(e-caprolactone) and their copolymers in vivo. *Biomaterials* **2**, 215–220.
38. Gabelnick H.L. (1983) Biodegradable implants: alternative approaches. In: Mishell DR, editor. *Advanced in Human Fertility and Reproductive Endocrinology, Vol. 2: Long Acting Steroid Contraception.* New York: Raven Press p. 149–173.
39. Salernitano E. and Migliaresi C. (2003) Composite materials for biomedical applications: a review. *J Appl Biomater Biomech* **1**, 3–18.
40. Betz M.W., Yeatts A.B., Richbourgh W.J., Caccamese J.F., Coletti D.P., Falco E.E. and Fisher J.P. (2010) Macroporous hydrogels upregulate osteogenic signal expression and promote bone regeneration. *Biomacromolecules* **11**, 1160–1168.
41. Liu Y., Wu G. and de Groot K. (2010) Biomimetic coatings for bone tissue engineering of critical-sized defects, *J R Soc Interface* **7**(5), S631-S647.
42. Bertinetti L., Drouet C., Combes C., Rey C., Tampieri A., Coluccia S. and Matra G. (2009) Surface characteristics of nanocrystalline apatites: effect of Mg surface enrichment on morphology, surface hydration species, and cationic environments. *Langmuir* **25**(10), 5647–5654.
43. Stayton P.S., Drobny G.P., Shaw W.J., Long J.R. and Gilbert M. (2003) Molecular recognition at the protein-hydroxyapatite interface. *Crit Rev Oral Biol Med* **14**(5), 370–376.
44. Sola A., Bellucci D., Raucci M.G., Zeppetelli S., Ambrosio L. and Cannillo V. (2012) Heat treatment of Na_2O-CaO-P_2O_5-SiO_2 bioactive glasses: densification processes and post sintering bioactivity. *J Biomed Mater ResPart A* **10A**(2), 305–322.
45. Hornez J.C., Chai F., Monchau F., Blanchemain N., Descamps M. and Hildebrand H.F. (2007) Biological and physico-chemical assessment of hydroxyapatite (HA) with different porosity. *Biomol Eng* **24**(5), 505–509.
46. Berry E.E. (1967) The structure and composition of some calcium deficient apatite. *J Inorg Nucl Chem* **29**(7), 317–327.
47. Eastoe J.E. (1961) The chemical composition of bone. In: Long C., editor, *Biochemists Handbook*. Princeton, NJ: Van Nostrand, pp. 715–720.
48. Hench L.L. (1991) Bioceramics: from concept to clinic. *J Am Ceram Soc* **74**(7), 487–510.
49. Ducheyne P. and Cuckler J.M. (1992) Bioactive ceramic prosthetic coatings. *Clin Orthop Relat Res* **276**, 102–114.
50. Wu C. and Chang J. (2006) A novel akermanite bioceramic preparation and characteristics. *J Biomater Appl* **21**(2), 119–129.
51. Hench L.L. Wheeler D.L. and Greenspan D.C. (1998) Molecular control of bioactivity in sol–gel glasses. *J Sol–Gel Sci Tech* **13**(1–3), 245–250
52. Dessì M., Raucci M.G., Zeppetelli S. and Ambrosio L. (2012) Design of injectable organic-inorganic hybrid for bone tissue repair. *J Biomed Mater Res A* August, **100**(8), 2063–2070.
53. Murugan R. and Ramakrishna S. (2005) Development of nanocomposites for bone grafting. *Compos Sci Technol* **65**, 2385–2406.
54. Vallet-Regí M. and González-Calbet J.M. (2004) Calcium phosphates as substitution of bone tissues. *Prog Solid State Chem* **32**(1–2), 1–31.
55. Viswanath B. and Ravishankar N. (2008) Controlled synthesis of plate-shaped hydroxyapatite and implications for the morphology of the apatite phase in bone. *Biomaterials* **29**(36), 4855–4863.

56. Vijayalakshmi U. and Rajeswari S. (2012) Influence of process parameters on the sol–gel synthesis of nano hydroxyapatite using various phosphorus precursors. *J Sol-Gel Sci Technol* **63**(1), 45–55.
57. Cao W. and Hench L.L. (1996) Bioactive materials. *Ceram Int* **22**, 493–507.
58. De Santis R., Catauro M., Di Silvio L., Manto L., Raucci M.G., Ambrosio L. and Nicolais L. (2007) Effects of polymer amount and processing conditions on the in vitro behaviour of hybrid titanium dioxide/polycaprolactone composites. *Biomaterials* **28**(18), 2801–2809.
59. Luciani G., Costantini A., Silvestri B., Tescione F., Branda F. and Pezzella A. (2008) Synthesis, structure and bioactivity of pHEMA/SiO_2 hybrids derived through in situ sol-gel process. *J Sol-Gel Sci Technol* **46**, 166–175.
60. Silvestri B., Luciani G., Costantini A., Tescione F., Branda F. and Pezzella A. (2009) In-situ sol-gel synthesis and characterization of bioactive pHEMA/SiO_2 blend hybrids. *J Biomed Mater Res – Part B Appl Biomater* **89**(2), 369–378.
61. Costantini A., Luciani G., Silvestri B., Tescione F. and Branda F. (2008) Bioactive poly(2-hydroxyethylmethacrylate)/silica gel hybrid nanocomposites prepared by sol-gel process. *J Biomed Materials Research – B Appl Biomater* **86**(1), 98–104.
62. Catauro M., Raucci M.G., De Marco D. and Ambrosio L. (2006) Release kinetics of ampicillin, characterization and bioactivity of TiO_2/PCL hybrid materials synthesized by sol-gel processing. *J Biomed Mater Res – Part A* **77**(2), 340–350.
63. Gupta R. and Kumar A. (2008) Bioactive materials for biomedical applications using sol–gel technology. *Biomed Mater* **3**, 03400.
64. Raucci M.G., Guarino V. and Ambrosio L. (2010) Hybrid composite scaffolds prepared by sol-gel method for bone regeneration. *Compos Sci Technol* **70**, 1861–1868.
65. Raucci M.G., D'Antò V., Guarino V., Sardella E., Zeppetelli S., Favia P. and Ambrosio L. (2010) Biomineralized porous composite scaffolds prepared by chemical synthesis for bone tissue regeneration. *Acta Biomater* **6**, 4090–4099.
66. Martin C., Winet H. and Bao J.Y. (1996) Acidity near eroding polylactide-polyglycolide in vitro and in vivo in rabbit tibial bone chamber. *Biomater* **17**, 2373–2380.
67. El-Ghannam A. (2005) Bone reconstruction: from bioceramics to tissue engineering. *Exp Rev Med Dev* **2**, 87–101.
68. Liu Q., de Wijn R.J., de Groot K. and van Blitterswijk A.C. (1998) Surface modification of nanoapatite by grafting organic polymer. *Biomater* **19**, 1067–1072.
69. Catauro M., Raucci M.G., Ausanio G. and Ambrosio L. (2007) Sol-gel synthesis, characterization and bioactivity of poly(ether-imide)/TiO_2 hybrid materials. *J Appl Biomat Biomech* **5**, 41–48.
70. Catauro M., Raucci M.G., de Gaetano F., Buri A., Marotta A. and Ambrosio L. (2004) Sol-gel synthesis, structure and bioactivity of polycaprolactone/ $CaO{\bullet}SiO_2$ hybrid material. *J Mater Sci Mater Med* **15**, 991–995.
71. Raucci M.G., Adesanya K., Di Silvio L., Catauro M. and Ambrosio L. (2010) The biocompatibility of silver-containing $Na_2O{\cdot}CaO{\cdot}2SiO_2$ glass prepared by sol–gel method. *J Biomed Mater Res B Appl Biomater* **92**, 102–110.
72. Landers R., Pfister A., Hubner U., John H., Schmelzeisen R. and Mulhaupt R. (2002) Fabrication of soft tissue engineering scaffolds by means of rapid prototyping techniques. *J Mater Sci* **37**, 3107–3116.

73. Melchels F.P.W., Feijena J., Grijpma D.W. (2010) A review on stereolithography and its applications in biomedical engineering. *Biomaterials* **31**, 6121–6130.
74. Wendel B., Rietzel D., Kuhnlein F., Feulner R., Hulder G. and Schmachtenberg E. (2008) Additive processing of polymers. *Macromol Mater Eng* **293**(10), 799–809.
75. Hieu L.C., Zlatov N., Vander Sloten J., Bohez E., Khanh L., Binh P.H., Oris P. and Toshev Y. (2005) Medical rapid prototyping applications and methods. *Assem Autom* **25**(4), 284–292.
76. Di Giacomo G.A.P., Cury P.R., de Araujo N.S, Sendyk W.R. and Sendyk C.R. (2005) Clinical application of stereolithographic surgical guides for implant placement: preliminary results. *J Periodontol* April 2005, **76**(4), 503–507, DOI 10.1902/jop.2005.76.4.50.
77. Ronca A., Ambrosio L. and Grijpma D.W. (2013) Preparation of designed poly(D,L-lactide)/nanosized hydroxyapatite composite structures by stereolithography. *Acta Biomater* **9,** 5989–5996.
78. Hollister S.J. (2005) Porous scaffold design for tissue engineering. *Nat Mater* **4**(7), 518–524.
79. Williams J.M., Adewunmi A., Schek R.M., Flanagan C.L., Krebsbach P.H., Feinberg S.E., Hollister S.J. and Das S. (2005) Bone tissue engineering using polycaprolactone scaffolds fabricated via selective laser sintering. *Biomaterials* **26**(23), 4817–4827.
80. Hutmacher D.W., Schantz T., Zein I., Ng K.W., Teoh S.H. and Tan K.C. (2001) Mechanical properties and cell cultural response of polycaprolactone scaffolds designed and fabricated via fused deposition modelling. *J Biomed Mater Res* **55**(2), 203–216.
81. Gloria A., Russo T., De Santis R. and Ambrosio L. (2009) 3D fiber deposition technique to make multifunctional and tailor-made scaffolds for tissue engineering applications. *J Appl Biomater Biomech* September–December, **7**(3), 141–152.
82. Giordano R.A., Wu B.M., Borland S.W., Cima L.G., Sachs E.M. and Cima M.J. (1996) Mechanical properties of dense polylactic acid structures fabricated by three dimensional printing. *J Biomat Sci-Polym E* **8**(1), 63–75.
83. Melchels F.P.W., Barradas A.M.C., van Blitterswijk C.A., de Boer J., Feijen J., Grijpma D.W. (2010) Effects of the architecture of tissue engineering scaffolds on cell seeding and culturing. *Acta Biomater* **6,** 4208–4217.
84. Bartolo P.J., Almeida H.A., Rezende R.A., Laoui T. and Bidanda B. (2008) Advanced processes to fabricate scaffolds for tissue engineering. In: Bidanda, B., and Bartolo, P.J., eds. *Virtual Prototyping and Bio Manufacturing in Medical Applications*. New York: Springer, pp. 149–170.
85. Yang S., Leong K.-F., Du Z. and Chua, C.K. (2002) The design of scaffolds for use in tissue engineering. Part II. Rapid prototyping techniques. *Tissue Eng* **8**, 1.
86. Kim K., Yeatts A., Dean D. and Fisher J.P. (2010) Stereolithographic bone scaffold design parameters: osteogenic differentiation and signal expression. *Tissue Eng: Part B* **16**(5).
87. Sun W., Starly B., Darling A. and Gomez C. (2004) Computer-aided tissue engineering: application to biomimetic modelling and design of tissue scaffolds. *Biotechn Appl Biochem* **39**(1), 49–58, February 2004.

88. Melchels F.P.W., Bertoldi K., Gabbrielli R., Velders A.H., Feijen J. and Grijpma D.W. (2010) Mathematically defined tissue engineering scaffold architectures prepared by stereolithography. *Biomaterials* **31,** 6909–6916.
89. Ronca A., Ambrosio L. and Grijpma D.W. (2012) Design of porous three-dimensional PDLLA/nano-hap composite scaffolds using stereolithography. *J App Biomater Funct Mater* **10**(3), 249–258.
90. Griffith M.L. and Halloran J.W. (1996) Freeform fabrication of ceramics via stereolithography. *J Am Ceram Soc* **79**(10), 2601–2608.
91. Mygind T., Stiehler M., Baatrup A., Li H., Zou X. Flyvbjerg A., Kassem M. and Bünger C. (2007) Mesenchymal stem cell ingrowth and differentiation on coralline hydroxyapatite scaffolds. *Biomaterials* **28**(6), 1036–1047.
92. Gauvina R., Chena Y.-C., Lee J.W., Somand P., Zorlutuna P., Nichol J.W., Baea H., Chend S. and Khademhosseini A. (2012) Microfabrication of complex porous tissue engineering scaffolds using 3D projection stereolithography. *Biomaterials* **33,** 3824–3834.
93. Nandakumar A., Barradas A., de Boer J., Moroni L., van Blitterswijk C. and Habibovic P. (2013) Combining technologies to create bioactive hybrid scaffolds for bone tissue engineering. *Biomatter* **3,** 1.
94. Chua C.K., Chou S.M. and Wong T.S. (1998) A study of the state-of-the-art rapid prototyping technologies. *Int J Adv Manuf Technol* **14,** 146–152.
95. Dhariwal B. and Boland H. E., (2004) Rapid prototyping of tissue-engineering constructs, using photopolymerizable hydrogels and stereolithography. *Tissue Eng* **10,** 1316–1322.
96. Lee K.W., Wang S., Fox B.C., Ritman E.L., Yaszemski M.J. and Lu L. (2007) Poly(propylene fumarate) bone tissue engineering scaffold fabrication using stereolithography: effects of resin formulations and laser parameters. *Biomacromolecules* **8,** 1077–1084.
97. Arcaute K., Mann B.K. and Wicker R.B. (2006) Stereolithography of three-dimensional bioactive poly(ethylene glycol) constructs with encapsulated cells. *Ann Biomed Eng* **34,** 1429–1441.
98. Melchels F.P.W., Feijen J. and Grijpma D.W. (2009) A poly(D,L-lactide) resin for the preparation of tissue engineering scaffolds by stereolithography. *Biomaterials* **30,** 3801–3809.
99. Cooke M.N., Fisher J.P., Dean D., Clare Rimnac and Mikos A.G. (2003) Use of stereolithography to manufacture critical-sized 3D biodegradable scaffolds for bone ingrowth. *J Biomed Mater Res Part B: App Biomater* **64B**(2), 65–69, 15 February 2003.
100. Heller C., Schwentenwein M., Russmueller G., Varga F., Stampfl J. and Liska R. (2009) Vinyl esters: low cytotoxicity monomers for the fabrication of biocompatible 3D scaffolds by lithography based additive manufacturing. *J Polym SciPart A: Polym Chem* **47**(24), 6941–6954.
101. Provin C. and Monneret S. (2002) Complex ceramic-polymer composite microparts made by microstereolithography. *IEEE Trans Electron Packag Manuf* **25**(1), 59–63.
102. Zhang X., Jiang X.N. and Sun C. (1999) Micro-stereolithography of polymeric and ceramic microstructures. *Sensor Actuat* **77,** 149–156.
103. Sharifia S., Blanquer S.B.G., van Kootena T.G. and Grijpma D.W. (2012) Biodegradable nanocomposite hydrogel structures with enhanced mechanical

properties prepared by photo-crosslinking solutions of poly(trimethylene carbonate)–poly(ethylene glycol)–poly(trimethylene carbonate) macromonomers and nanoclay particles. *Acta Biomater* **8,** 4233–4243.

104. Barry J.J., Evseev A.V., Markov M.A., Upton C.E., Scotchford C.A., Popov V.K. and Howdle, S.M. (2008) In vitro study of hydroxyapatite-based photocurable polymer composites prepared by laser stereolithography and supercritical fluid extraction. *Acta Biomater* **4,** 1603.
105. Bian W., Li D., Lian Q., Li X., Zhang W., Wang K. and Jin Z. (2012) Fabrication of a bio-inspired beta-Tricalcium phosphate/collagen scaffold based on ceramic stereolithography and gel casting for osteochondral tissue engineering. *Rapid Prototyping J* **18,** 1, 68–80.
106. Arcaute K., Mann B.K. and Wicker R.B. (2006) Stereolithography of three-dimensional bioactive poly(ethylene glycol) constructs with encapsulated cells. *Ann Biomed Eng* **34**(9), 1429–1441.
107. Lantada A.D. and Morgado P.L. (2012) Rapid prototyping for biomedical engineering: current capabilities and challenges. *Annu Rev Biomed Eng* **14,** 73–96.
108. Chua C.K., Leong K.F., Cheah C.M. and Chua S.W. (2003) Development of a tissue engineering scaffold structure library for rapid prototyping. Part 1: Investigation and classification. *Int J Adv Manuf Technol* **21**(4), 291–301
109. Pham Q.P., Sharma U. and Mikos A.G. (2006) Electrospinning of polymeric nanofibers for tissue engineering applications: A review. *Tissue Eng* **12**(5), 1197–1211.
110. Behonick D.J. and Werb Z. (2003) A bit of give and take: the relationship between the extracellular matrix and the developing chondrocyte. *Mech Dev* **120,** 1327–1336.
111. Li W.J., Laurencin C.T., Caterson E.J., Tuan R.S. and Ko F.K. (2002) Electrospun nanofibrous structure: a novel scaffold for tissue engineering. *J Biomed Mater Res* **60**(4), 613–621.
112. Smith L.A. and Ma P.X. (2004) Nano-fibrous scaffolds for tissue engineering. *Colloid Surface B* **39,** 125–131.
113. Hutmacher D.W. (2000) Scaffolds in tissue engineering bone and cartilage. *Biomaterials* **21,** 2529–2543.
114. Sharma B. and Elisseeff J.H. (2004) Engineering structurally organized cartilage and bone tissues. *Ann Biomed Eng* **32**(1), 148–159.
115. Rosso F., Marino G., Giordano A., Barbarisi M., Parmeggiani D. and Barbarisi A. (2005) Smart materials as scaffolds for tissue engineering. *J Cell Physiol* **203**(3), 465–470.
116. Deitzel J.M., Kleinmeyer J., Harris D. and Beck Tan N.C. (2001) The effect of processing variables on the morphology of electrospun nanofibers and textiles. *Polymer* **42,** 261–272.
117. Yarin A.L., Koombhongse S. and Reneker D.H. (2001) Bending instability in electrospinning of nanofibers. *J Appl Phys* **89**(5), 3018–3026.
118. Friess W. (1998) Collagen – biomaterial for drug delivery. *Eur J Pharm Biopharm* **45,** 113–136.
119. Bhardwaj N. and Kundu S.C. (2010) Electrospinning: A fascinating fiber fabrication technique. *Biotechnol Adv* **28,** 325–347.
120. Rho K.S., Jeong L., Lee G., Seo B.M., Park Y.J., Hong S.D., Roh S., Cho J.J., Park W.H. and Min B.M. (2006) Electrospinning of collagen nanofibers: effects on

the behavior of normal human keratinocytes and early-stage wound healing. *Biomaterials* **27**, 1452–1461.

121. Fertala A., Han W.B. and Ko F.K. (2001) Mapping critical sites in collagen II for rational design of gene-engineered proteins for cell-supporting materials. *J Biomed Mater Res* **57**, 48–58.
122. Venugopal J. and Ramakrishna S. (2005) Biocompatible nanofiber matrices for the engineering of a dermal substitute for skin regeneration. *Tissue Eng* **11**, 847–854.
123. Yoshimoto H., Shin Y.M., Terai H. and Vacanti J.P. (2003) A biodegradable nanofiber scaffold by electrospinning and its potential for bone tissue engineering. *Biomaterials* **24**, 2077–2082.
124. Chen J.L., Chu B. and Hsiao B.S. (2006) Mineralization of hydroxyapatite in electrospun nanofibrous poly(L-lactic acid) scaffolds. *J Biomed Mater Res* **79A**, 307–317.
125. Huang L., McMillan R.A., Apkarian R.P., Pourdeyhimi B., Conticello V.P. and Chaikof E.L. (2000) Generation of synthetic elastin-mimetic small diameter fibers and fiber networks. *Macromolecules* **33**, 2989–2997.
126. Nagapudi K., Brinkman W.T., Leisen J.E., Huang L., McMillan R.A., Apkarian R.P., Conticello V.P. and Chaikof E.L. (2002) Photomediated solid-state crosslinking of an elastin-mimetic recombinant protein polymer. *Macromolecules* **35**, 1730–1737.
127. Xu C.Y., Inai R., Kotaki M. and Ramakrishna S. (2004) Aligned biodegradable nanofibrous structure: a potential scaffolds for blood vessel engineering. *Biomaterials* **25**, 877–886.
128. Zong X., Ran S., Fang D., Hsiao B.S. and Chu B. (2003) Control of structure, morphology and property in electrospun poly(glycolide-*co*-lactide) non-woven membranes via post-draw treatments. *Polymer* **44**, 4959–4967.
129. Zong X.H., Bien H., Chung C.Y., Yin L.H., Fang D.F., Hsiao B.S., Chu B. and Entcheva E. (2005) Electrospun fine-textured scaffolds for heart tissue constructs. *Biomaterials* **26**, 5330–5338.
130. Yang F., Murugan R., Ramakrishna S., Wang X., Ma Y.X. and Wang S. (2004) Fabrication of nano-structured porous PLLA scaffold intended for nerve tissue engineering. *Biomaterials* **25**, 1891–1900.
131. Yang F., Xu C.Y., Kotaki M., Wang S. and Ramakrishna S. (2004) Characterization of neural stem cells on electrospun poly(L-lactic acid) nanofibrous scaffold. *J Biomat Sci-Polym E* **15**, 1483–1497.
132. Guarino V., Cirillo V., Taddei P., Alvarez-Perez M.A. and Ambrosio L. (2011) Tuning size scale and cristalliniy of PCL electrospun membranes via solvent permittivity to address hMSC response. *Macromol Biosci* **11**, 1694–1705 10.1002/mabi.201100204.
133. Matthews J.A., Wnek G.E., Simpson D.G. and Bowlin G.L. (2002) Electrospinning of collagen nanofibers. *Biomacromolecules* **3**, 232–238.
134. Agarwal S., Wendorff J.H. and Greiner A. (2008) Use of electrospinning technique for biomedical applications. *Polymer* **49**, 5603–5621.
135. Spasova M., Stoilova O., Manolova N., Rashkov I. and Altankov G. (2007) Preparation of PLLA/PEG nanofibers by electrospinning and potential applications. *J Bioact Compat Pol* **22**(1), 62–76.
136. Ramachandran K. and Gouma P.I. (2008) Electrospinning for bone tissue engineering. *Recent Patents Nanotechnol* **2**, 1–7.

137. Guarino V., Alvarez-Perez M.A., Cirillo V. and Ambrosio L. (2011) hMSC interaction with PCL and PCL/gelatin platforms: A comparative study on films and electrospun membranes. *J Bioact Comp Pol* **26**(2), 144–160.
138. Cai Q., Yang J., Bei J. and Wang S. (2002) A novel porous cells scaffold made of polylactide–dextran blend by combining phase-separation and particle-leaching techniques. *Biomaterials* **23**, 4483–4492.
139. Pavlov M.P., Mano J.F., Neves N.M. and Reis R.L. (2004) Fibers and 3D mesh scaffolds from biodegradable starch-based blends: production and characterization. *Macromol Biosci* **4**, 776–784.
140. Almany L. and Seliktar D. (2005) Biosynthetic hydrogel scaffolds made from fibrinogen and polyethylene glycol for 3D cell cultures. *Biomaterials* **26**, 2467–2477.
141. Wayne J.S., Mcdowell C.L., Shields K.J. and Tuan R.S. (2005) *In vivo* response of polylactic acid–alginate scaffolds and bone marrow-derived cells for cartilage tissue engineering. *Tissue Eng* **11**, 953–963.
142. Yoo H.S., Lee E.A., Yoon J.J. and Park T.G. (2005) Hyaluronic acid modified biodegradable scaffolds for cartilage tissue engineering. *Biomaterials* **26**, 1925–1933.
143. Alvarez Perez M.A., Guarino V., Cirillo V. and Ambrosio L. (2012) *In vitro* mineralization and bone osteogenesis in poly(ε-caprolactone)/gelatin nanofibers. *J Biomed Mater Res Part A* **100**, 3008–3019.
144. Di Martino A., Liverani L., Rainer A., Salvatore G., Trombetta M. and Denaro V. (2011) Electrospun scaffolds for bone tissue engineering. *Musculoskelet Surg* **95**, 69–80.
145. Kim H.W., Song J.H. and Kim H.E. (2006) Bioactive glass nanofiber–collagen nanocomposite as a novel bone regeneration matrix. *J Biomed Mat Res Part A* **79A**, 698–705.
146. Kim H.W., Lee H.H. and Chun G.S. (2008) Bioactivity and osteoblast responses of novel biomedical nanocomposites of bioactive glass nanofiber filled poly(lactic acid). *J Biomed Mat Res Part A* **85A**, 651–663.
147. Cirillo V., Guarino V. and Ambrosio L. (2013) Design of bioactive electrospun scaffolds for bone tissue engineering. *J Appl Biomater Funct Mater* **10**(3), 223–228.

6

3D bioceramic foams for bone tissue engineering

K. K. MALLICK and J. WINNETT, University of Warwick, UK

DOI: 10.1533/9780857099037.2.118

Abstract: In seeking to meet the current challenge of hard tissue augmentation mimicking native bone, clinically viable three-dimensional (3D) bioceramc foams can be incorporated as temporary scaffolds for the regeneration of new tissue. These biotemplates must fulfil a range of stringent and clinically demanding property specifications. In this chapter, we review the selection of biomaterials and the fabrication routes currently available to manufacture such foams for load bearing (*structural*) and non-load bearing (*non-structural*) applications. Although the processing approaches differ widely in the approach to optimizing the parametric properties, a well-designed methodology is considered a prerequisite to the success of the fabricated scaffolds in promoting new bone growth.

Key words: bone augmentation, tissue engineering, bioceramic foams, replication/reticulation, freeze casting, sol-gel, rapid prototyping (RP).

6.1 Introduction

The growing field of tissue engineering combines scientific and technical aspects from material science and engineering interfaced with biological sciences to generate structures that fulfil the clinical requirements in order to aid the natural reparation processes of the body. This is achieved through the substitution or restoration of diseased, damaged or malfunctioning tissues (Langer, 2009). Hard tissues provide requisite strength and structure to the body through bone, enamel and dentin. One of the most fascinating features of bone is its ability to regrow, remodel and adapt itself to its mechanical environment for small (<3 mm) defects. However, there are situations where bone is damaged beyond the limits of its self-healing ability and currently bone grafting is the optimal procedure to augment tissue regeneration. Although autologous grafts are still considered as the gold standard for such procedures, they are associated with problems such as donor site morbidity and the lack of available tissue (Salgado *et al.*, 2004; Mahendra and Maclean, 2007; Dawson and Oreffo, 2008). Alternatives such as auto- and xenografts are accompanied with further problems including the increased risk of disease transmission and reduced osteogenicity. Hence synthetic porous constructs are seen as the optimal candidate to replace natural grafts.

Such scaffolds are temporary structural platforms onto which cells of the appropriate lineage, morphology and dimensionality can attach, differentiate and proliferate. As such, there are many requirements that such structures must adhere to, including biocompatibility of the structure and any products of biodegradation, osteoconductivity, similarity of mechanical properties to the native tissue, three-dimensionality and interconnected porosity of appropriate pore sizes (Burg *et al.*, 2000; Dawson and Oreffo, 2008; Jones *et al.*, 2009). Of these, the pore size can have a significant impact on the biological functionality of the structure, and is one of the more readily variable features.

Different levels of porosity are required due to the impact they have on new bone growth: open macroporosity (100–1000 μm) allows for the infiltration of cells, nutrients and metabolites throughout the structure whilst microporosity provides cells with attachment sites (Hing, 2005; Karageorgiou and Kaplan, 2005). Furthermore, the high strength of bulk structures can lead to problems with stress shielding (Huiskes, 1988) and the introduction of porosity can significantly reduce strength. Therefore these constructs are generally suitable for non-structural applications but in some cases they can be used for structural augmentations where compressive strength increases with progressive ingrowth of the new bone.

A wide range of material classes can be used to produce bioscaffolds suitable for their use in bone tissue engineering, with metals, ceramics, polymers and composites having been investigated. Currently, the most popular materials are metals, in particular stainless steel and cobalt–chromium alloys, due to their high strength, ductility and wear resistance. Transition metals such as titanium and its alloys, such as Ti6Al4V, Ta, Zr and, more recently, Mg are increasing in popularity due to their impressive biocompatibility and mechanical similarities to native bone.

The main advantage of using bioceramics is their impressive biocompatibility and mechanical strength despite the fact that they are also brittle which can cause significant risk of fracture. Metal oxides such as Al_2O_3 and TiO_2 can be used in situations requiring high strength and bioinertness; glasses and glass ceramics have the ability to bond to bone through the formation of an apatitic layer upon contact with physiological media; and calcium phosphates (apatites) closely matched chemically to bone leading to significant bone growth whilst simultaneously being osteoclastically resorbed.

There are wide ranging techniques currently used to produce scaffold structures. These can be broadly divided into two types. Conventional techniques use the reaction of materials under certain conditions such as heat and/or pressure. In general, they are limited by the inability to specify the pore size, although in most cases the porosity is controllable. Rapid prototyping (RP) on the other hand utilizes a 'bottom-up' approach whereby building blocks are arranged in a predetermined manner using computer aided design (CAD) and manufacture and can be further subdivided based as powder, liquid or colloid. Although the pore size is readily definable the main issue for most of these techniques lies with the minimum resolution possible as well as significant costs associated with the machinery or capital overhead.

6.2 Biology of bone

Bone is an idealized composite biomaterial consisting of apatite crystals interspersed within a predominantly collagenous matrix (Rho *et al.,* 1998). It is a highly specialized tissue that provides an internal support network, a reservoir for minerals, protection for the internal organs and attachment sites for muscles (Weiner and Wagner, 1998). It has a hierarchical structure that can be divided into macro- (cancellous and cortical bone), micro- (10–500 μm; Haversian systems, osteons, trabeculae), sub-micron (1–10 μm, lamellae), nano- (1 nm–1 μm; fibrillar collagen and embedded mineral), and sub-nano (below 1 nm; mineral, collagen and non-collagenous proteins) regions (Weiner *et al.,* 1999; Rho *et al.,* 1998).

Cortical bone accounts for approximately 80% of the total mass of bone due to its relatively low porosity. This is because it is made up of osteons that provide a compact and orthotropic structure providing the high compressive strength with a Young's modulus of 17 GPa (Rho *et al.,* 1998). Trabecular bone has a much higher porosity between 50% and 90% depending on the location within the body. Although this porosity results in strength reduction up to a factor of ten, it however allows even distribution of loads and transmission of oxygen, cells, metabolites and nutrients throughout the structure. It is made up of a 3D, interconnected network of trabeculae that provides anisotropic mechanical behaviour (Wettergreen, 2008).

Bone undergoes a continuously occurring reparation mechanism whereby cylinders of old bone are removed and replaced by a layer of non-mineralized extracellular matrix that is eventually mineralized to integrate with the existing bone (Burr *et al.,* 1997). Bone that has been damaged is preferentially targeted. However, healthy bone is also remodelled so that, at any one time, 3–5% of the total bone is being remodelled (Davies, 2007). As bone is mechanosensitive, the mass architecture is remodelled in response to the mechanical loading placed upon it. Thus, any replacement must be of a suitable strength such that it does not fail and require costly and tricky corrections. Furthermore, the scaffold should not be too mechanically strong as this can cause stress shielding, a phenomenon whereby new bone does not need to be as strong as the bone it is replacing, leading to weakened bone (or bone loss) and implant loosening (Frost, 1987, 2003).

6.3 Biomaterials

Biomaterials can be broadly classified as metal, polymer, ceramic or composite (Planell *et al.*, 2009). Currently, the most prevalent implants for bone regenerative applications are those from metals and alloys, although ceramics such as apatites and Bioglass® are attracting attention due to their good biocompatibility and high compressive strength. Ceramics for use in biological situations, commonly known as bioceramics, fall into three categories: inert (non-resorbable), semi-inert (bioactive) or bioresorbable, as shown in Fig. 6.1.

6.3.1 Bioinert ceramics

Bioinert ceramics can be readily used in similar situations to metallic implants as they exhibit impressive hardness and wear characteristics, low friction and elicit no adverse immune response (Vallet-Regi, 2006; Planell *et al.*, 2009). It is also possible to generate some bioactivity due to a bony regrowth at the implant–tissue interface on porous structures. Oxides of some transition metals such as aluminium, titanium and zirconium are the most common inert ceramics.

Aluminium oxide (Al_2O_3, alumina) is an extremely hard polycrystalline material with low friction and wear characteristics as well as good biocompatibility (Costa *et al.*, 2007). It has a very high bulk strength which must be reduced through the introduction of porosity (Yoon *et al.*, 2008; Moon *et al.*, 2011) in order to minimize the risk of stress shielding. At the same time it is very brittle and often requires toughening via the incorporation of polymeric (Comas *et al.*, 2012) or composite (Costa *et al.*, 2008) coatings.

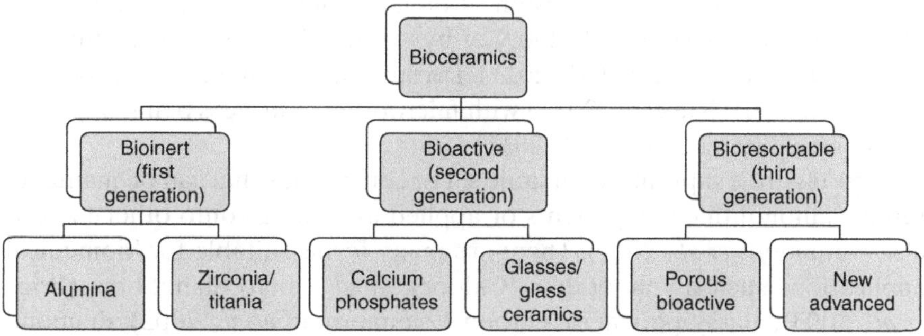

6.1 Types of bioceramics classed as bioinert, bioactive and bioresorbable.

It has been possible to initiate bone growth onto the surface of alumina structures through the use of naturally occurring biocompatible ions, micro- and nanoporosity (Karlsson *et al.*, 2003) and surface functionalization with organic products (Comas *et al.*, 2012). It has been suggested, however, that other ceramics and metals are more suited for bone scaffolding as a non-adherent fibrous membrane can form causing implant loosening (Andreiotelli *et al.*, 2009).

Zirconium oxide (ZrO_2, zirconia) has high mechanical strength and toughness, and good biocompatibility and chemical stability (Manicone *et al.*, 2007; Andreiotelli *et al.*, 2009). Although bioinert, it has been possible to grow osteoblasts directly onto the surface of porous structures with mechanical properties more suited to native bone (Andreiotelli *et al.*, 2009). Currently, the main concerns lie with the long-term performance of scaffolds and the cytotoxicity of zirconia powder (von Schewelov *et al.*, 2005). Zirconia undergoes phase changes from monoclinic to cubic and from cubic to tetragonal at high temperatures, both of which induce stress in the structure due to the volumetric expansion (Manicone *et al.*, 2007). This can be overcome through the stabilization with metallic oxides such as yttria to generate partially stabilized zirconia (PSZ), which also has enhanced flexural strength and fracture toughness (Rambo *et al.*, 2004). However, only small amounts of yttria can be used owing to radioactivity that can have a deleterious effect on surrounding tissues.

The impressive biocompatibility of Ti and its alloys is caused by the formation of a TiO_2 layer on their surface (Li, 1993; Fartash *et al.*, 1995) thus making these suitable biomaterials. TiO_2 possesses good mechanical, structural and chemical properties (Kokubo *et al.*, 2007). It occurs in three phases: rutile which is the most stable; anatase which elicits the strongest bodily reactions due to the absorption of OH^- and PO_4^{3-} ions; and brookite (Li, 1993; Tang *et al.*, 2004). Bony tissue bonds directly to the surface of the implant with a similar level of bioactivity as observed in titanium structures (Fartash *et al.*, 1995). This can be further enhanced through the use of coatings (Huang *et al.*, 2007), inclusion of bioactive ions (Park, 2011), nanotubular production (Xiao *et al.*, 2009) or by varying the structure and number of the –OH groups (Xie *et al.*, 2011). Porous scaffolds have been produced (Zhu *et al.*, 2007; Xie *et al.*, 2011) with microporous surfaces being of particular clinical interest (Xie *et al.*, 2011).

Carbons are a supremely versatile set of compounds that can be manufactured as monolithic components or applied as coatings onto other materials (Yamamoto *et al.*, 2011). There are many forms suitable for biomedical applications such as nanotubes (Chlopek *et al.*, 2006), nanofibres (Price *et al.*, 2003), glassy polymeric carbon (Zimmerman *et al.*, 2002), diamond like carbon (DLC) (Smart *et al.*, 2006) and graphene. However, the advantageous properties do not necessarily account for the extra cost and there are significant difficulties in producing macroporous scaffolds (Vinu *et al.*, 2005). Furthermore, there is conflicting data regarding the cytotoxicity of the different forms with adverse effects reported (Smart *et al.*, 2006).

6.3.2 Bioactive/bioresorbable ceramics

Recently, there has been a trend to replace second generation bioactive ceramics by third generation bioresorbable ceramics via the adaptation of existing materials or through the development of new materials to house biological agents. Both bioactive and bioresorbable ceramics can be produced as monolithic structures or applied to bioinert materials to initiate regrowth and enhance bonding between the implant and surrounding tissue.

Glass and glass ceramics

Since the discovery of bone bonding (Hench *et al.*, 1971), glasses and glass ceramics have been deemed suitable for use as bone scaffold materials. Of these silicon oxide based systems, it has been determined that Bioglass 45S5® has the optimal composition for bony regrowth. Bioactivity is caused by the chemical activity leading to the formation of bonds between the ceramic and both hard and soft tissues. Glass particles of 0.1–1 μm are formed via the nucleation and growth of uniform crystals through melt or sol-gel techniques (Vallet-Regi, 2006; Salinas and Vallet-Regi, 2007). Although melt techniques have been used to produce glasses for a range of applications, sol-gel techniques offer more exciting possibilities due to the potential to vary the porosity and surface properties of the scaffolds and/or to incorporate biological agents further enhancing the bioactivity (Salinas and Vallet-Regi, 2007). The brittle nature of glass is a major consideration when developing scaffolds, as is the difficulty of applying glasses to other materials, for example as coatings (Salinas and Vallet-Regi, 2007).

Silica based mesoporous materials

Mesoporous materials have an ordered, homogenous distribution of pores of different geometries and are suitable for use in a range of applications. They consist of a silica based matrix that has a large surface area and pore volume, with mesopores of 2–50 nm. The surface topography is similar to that of bioactive glasses and forms an apatitic layer *in vitro*. The range of porosities enhances bioactivity through the increased accessibility of cells. It is also possible to incorporate nanometric particles such as biological agents and controllably release these to accelerate regrowth (Trewyn *et al.*, 2004; Vallet-Regi, 2010; Garcia *et al.*, 2011).

Table 6.1 Main apatites used in biological applications

Apatites	Chemical formula	Ca/P ratio
Monocalcium phosphate monohydrate (MCPM)	$Ca(H_2PO_4)_2{\bullet}H_2O$	0.5
Monocalcium phosphate anhydrous (MCPA)	$Ca(H_2PO_4)_2$	0.5
Dicalcium phosphate dihydrate (Brushite) (DCPD)	$CaHPO_4{\bullet}2H_2O$	1
Dicalcium phosphate anhydrous (Monetite) (MCPA)	$CaHPO_4$	1
Octacalcium phosphate (OCP)	$Ca_8(H_2PO_4)_2(PO_4)_4{\bullet}5H_2O$	1.33
Tricalcium phosphate (TCP)	$Ca_3(PO4)_2$	1.5
Hydroxyapatite (HA)	$Ca_{10}(PO_4)_6(OH)_2$	1.67
Tetracalcium phosphate monoxide (Hilgenstockite) (TCPM)	$Ca_2(PO_4)_2O$	2

Calcium phosphates

Materials containing calcium and phosphate ions make up a family known as apatites, with eight main variations as shown in Table 6.1. These are the most closely matched materials to native bone, with hydroxyapatite (HA; Ca/P: 1.67) ($Ca_{10}(PO_4)_6(OH)_2$) the most crystallographically similar synthetic material available (Salinas and Vallet-Regi, 2007). Bone growth on apatites is generated via the ionic exchange between the implant and the surrounding host tissue.

HA resorbs slowly and encourages significant bony ingrowth (Vallet-Regi, 2001). It has a bulk Young's modulus of 80–110 GPa, which can be significantly reduced through the introduction of porosity to render the structure suitable for both structural and non-structural augmentations. Tricalcium phosphate (TCP) resorbs at a much greater rate and total resorption can occur before complete regrowth, hence it is predominantly used as a bone filler (Jang *et al.*, 2011). Biphasic combinations of β-TCP and HA, biphasic calcium phosphate (BCP) can have tailored resorption rates depending on the relative amount of each component and hence offer exciting possibilities (Macchetta *et al.*, 2009).

Calcium phosphates can be combined with inorganic salts to create a gel that can be injected into the appropriate location where it cures. Such apatitic bone cements maintain the biocompatibility, bioactivity and osteoconductivity of the apatite and hence some commercialization has been undertaken (Yuasa *et al.*, 2004). There can be problems with the safety, mechanical toughness of the cured product and curing time and hence research into these materials is still ongoing (Yuasa *et al.*, 2004).

6.4 Manufacturing techniques

There are two sets of techniques that can be used to produce porous scaffolds: conventional and rapid prototyping. Conventional techniques are based on the knowledge of how the material will react under certain conditions such as heat or pressure, while RP is based around CAD and manufacture. A comparison between the techniques outlined is given in Table 6.2.

6.4.1 Conventional techniques

Conventional techniques use material properties to generate porous structures and are currently the more prevalent due to the lower costs involved. Further enhancing the importance of this set of techniques is the wide range of porosities, pore strengths and mechanical properties that can be achieved.

Dry methods

Dry methods constitute the simplest manufacturing techniques and involve compressing powdered biomaterials. It is also possible to vary the porosity by changing the compression pressure (Salgado *et al.*, 2004). Although easy to carry out and with relatively low costs, there are limitations to the component geometry. Furthermore, there can be problems with the weak mechanical properties of the final structure, as well as with polymer springback (Singh *et al.*, 2006). By sintering the green bodies under pressure, the samples have increased strength. Thus hot isostatic pressing (Zhao *et al.*, 2006) and uniaxial hot pressing have been developed. However this does not overcome the issues with the fabrication of scaffolds of complex geometry (Bose *et al.*, 2009).

Slip casting

For most production methods, a biomaterial is mixed into slurry that is then used to produce the structure. This is achieved through the mixing of a powdered biomaterial with a liquid, dispersant and binder until the required viscosity is reached. For such techniques, the slurry rheology is an important parameter and must be optimized to ensure the desired characteristics of the construct are achieved. For slip casting, relatively viscous slurry is produced and cast into a porous mould. This absorbs the liquid component of

Table 6.2 Analysis of manufacturing techniques against specified scaffold requirements

Method	Pore size (μm)	Porosity (%)	Compressive strength (MPa)	Biological response	Reference
Bone	1–3500	10–90	2–23	N/a	(Rho *et al.*, 1998; Choi *et al.*, 1990)
Dry methods	N/a	N/a	0.11–0.23	N/a	(Koc *et al.*, 2004; Zhao *et al.*, 2006)
Phase separation/ freeze casting	8–200	30–75	2–36	Enhanced MC3T3 and ALP	(Fu *et al.*, 2008b; Fu *et al.*, 2008a; Macchetta *et al.*, 2009)
Replication techniques	Foam dependant (10–1000)	<95	0.3–8.4	Enhanced MG63, ALP and other gene groups	(Cunningham *et al.*, 2011; Kim *et al.*, 2011)
Gelcasting	0.1–1000	<90	2–32.6	Enhanced ALP and MG63	(Sepulveda *et al.*, 2000; Yang *et al.*, 2011; Li *et al.*, 2008)
Sol-gel techniques	0.037–1000	50–70	2–20	Increased osteogenic cell markers	(Arcos and Vallet-Regi 2010; Nieto *et al.*, 2009)
Solvent casting/ particulate leaching	Particle dependant (5–600)	50–90	0.3–9.3	Enhanced osteoblast growth	(Liu *et al.*, 2010; Le Ray *et al.*, 2010)
3DP	Feature >10 Layer >20	Tailored 30–80	0.2–45	bMSC decreased, ALP enhanced	(Leukers *et al.*, 2005a; Leukers *et al.*, 2005b)
SLS	Feature >10 Layer >76	Tailored 30–80	2–31	No change	(Shanjani *et al.*, 2010; Shanjani *et al.*, 2011)
SLA	Feature >1 Layer >1	Tailored 30–80	1.3–56	Comparable at best	(Jansen *et al.*, 2009; Melchels *et al.*, 2010)
Robocasting	Feature >30 Layer >30	Tailored 30–80	2–20	N/a	(Russias *et al.*, 2007; Miranda *et al.*, 2008)
Ink-jet printing	Feature >200 Layer >0.4	Tailored 30–80	N/a	N/a	(Lewis 2006; Lewis *et al.*, 2006)
FDM	Feature >100 Layer >100	Tailored 30–80	0.16–15	Enhanced	(Schantz *et al.*, 2005; Too *et al.*, 2002)

ALP: alkaline phosphatase; 3DP: three-dimensional printing; SLS: selective laser sintering; SLA: stereolithography; FDM: fused-deposition modelling.

the slurry, whilst the biomaterial adheres to the mould, before the sample is sintered after complete infiltration. It is also possible to incorporate porogens within the slurry to produce hierarchical structures; however this is a highly wasteful procedure (Koc *et al.,* 2004; Cachinho and Correia, 2007).

Solvent casting/particulate leaching

Solvent casting is one of the most popular techniques of producing porous scaffolds. Porogens are included into the slurry which, once sublimed, generate pores within the structure. As there is a wide range of porogens available, it is possible to generate pore sizes of 5–600 μm with a reasonable degree of control (Chuenjitkuntaworn *et al.,* 2010). One of the most important features of the porogen is that it must be non-cytotoxic in case of incomplete removal. To reduce the risk of partial sublimation, the structure must be less than 3 mm thick, and the structures also have weak mechanical properties (Mikos and Temenoff, 2000). Nevertheless, this method still attracts much research attention with work currently aimed at developing the optimal porogen with paraffin beads (Liu *et al.,* 2010) and sucrose (Le Ray *et al.,* 2010) having been suggested. Not only has this led to further variability in the porosity, but there is also a greater degree of control so that the mechanical strength and biological response can be maximized.

Phase separation/freeze casting

The freezing of a liquid leads to the growth of ice crystals as solid dendrites. The size and nature of these dendrites are related to the processing parameters during the freezing process, in particular the freezing rate and temperature, and close control over these allows for control over the final structure. Using a closely controlled system of heat and pressure, it is possible to bypass the liquid phase and sublime the solid ice as a gas leading to the generation of porous structures, with the process outlined in Fig. 6.2. This ensures that the structure generated during freezing is maintained (Fu *et al.,* 2008a, b; Mallick 2009). Freeze drying consists of three stages: freezing, primary drying and secondary drying. Primary drying, the sublimation stage, is the process of converting ice directly into a gas without melting. The product is held at a temperature lower than the eutectoid of the ice and controlled vacuum conditions are used to remove the vapour. During this process, some heat must be put into the system to maintain the constant temperature and avoid sublimation cooling, which can affect the process. It is imperative that only the required amount of heat is provided as too much can lead to sample collapse. Secondary drying is not so much of a feature in freeze casting of bioscaffolds as this removes any remaining liquid after sublimation, such as that within the material itself.

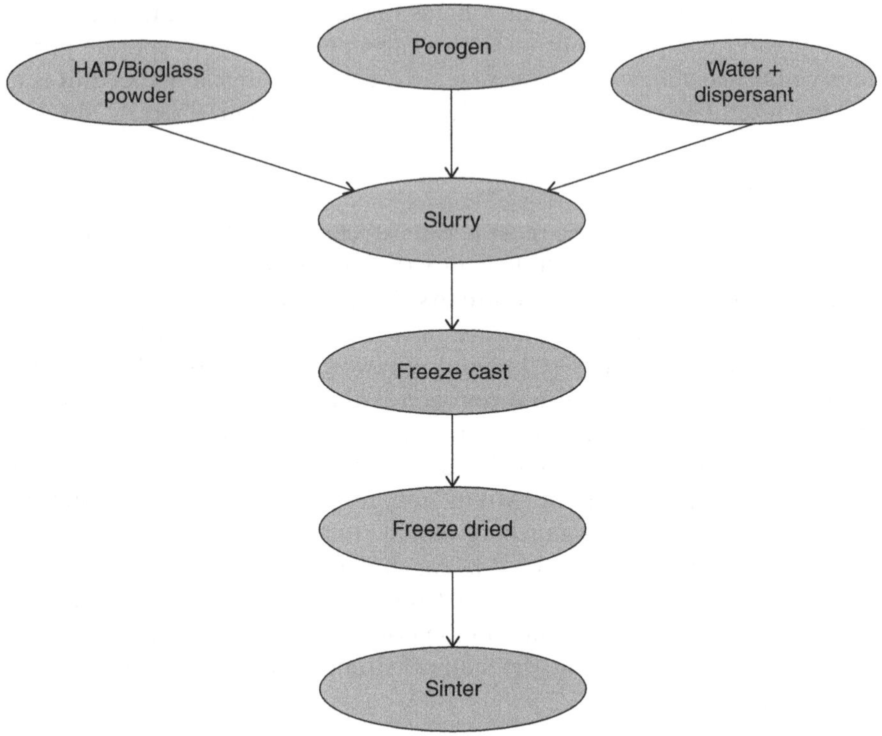

6.2 Outline of the freeze casting technique used to produce bioscaffolds.

There are many processing parameters affecting the final structure of the sample such as the solid loading of the slurry, freezing rate and temperature, and sublimation heat input and pressure. These can all be controlled, enabling the scaffold to have a relatively predictable structure (Fu *et al.*, 2008a, b). The main problems with this technique are the poor mechanical properties of the fabricated ceramics and limitations to the pore size (Fu *et al.*, 2008b), although current research aims to improve the potential range through the use of novel porogens. Combining porogens can lead to multiscale porosity and end-enhance the biological response (Burg *et al.*, 2000) though this can further complicate the drying procedures.

Replication techniques

There are many synthetic and naturally occurring porous structures which have a wide range of porosities and pore sizes suitable for tissue engineering purposes. These can therefore be used as a template from which the scaffold is based. The use of these templates is the simple theory behind

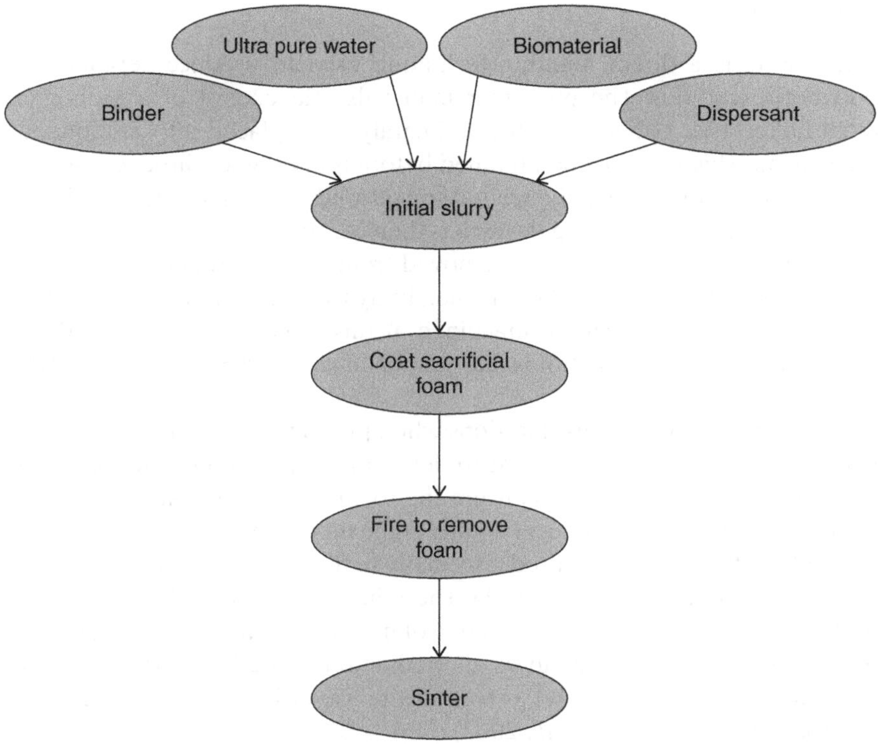

6.3 Foam replication technique used to produce bioscaffolds for tissue regenerative applications.

replication techniques that involve coating a sacrificial template with biomaterial slurry before pyrolysis of the template leave the desired construct (Cunningham *et al.*, 2011; Kim *et al.*, 2011; Mallick and Winnett, 2011), outlined in Fig. 6.3. The main limitation of this technique refers to the weak mechanical properties of the scaffold due to the generation of microcracks during template removal (Kim *et al.*, 2011). In general, foam reticulated structures have a compressive strength of between 0.3 and 2.5 MPa, although it is possible to generate composite structures relatively simply via dip coating, for example, due to the large pore sizes achievable thus improving the strength.

The wide range of templates that can be used include those with directionally aligned pores that generate structures more akin to natural bone and enhanced mechanical strength along the primary axis (Cunningham *et al.*, 2011). Furthermore, the introduction of porogens into the slurry generates constructs with both macro- and microporosity in a controllable manner such that biological performance can be enhanced (Mallick and Winnett, 2011; Mallick *et al.*, 2012).

Gel casting

Gel casting is a direct foaming technique used to produce ceramic and polymeric scaffolds. The precursor materials that consist of a monomer, cross linker, free radical initiator and catalyst are placed into an aqueous suspension. This slurry is then foamed before it undergoes direct consolidation where the binder is polymerized to consolidate the particles within the slurry. This process forms a gel which is then cast into an appropriate mould. After the gel has solidified, it is removed from the mould under controlled conditions and dried to produce a green body with such impressive mechanical properties that it can be machined at this stage. Finally, the binder is burnt out and the final scaffold sintered (Chen *et al.*, 2008; Yang *et al.*, 2011), as outlined in Fig. 6.4.

There are two main considerations when producing ceramic scaffolds via gel casting. Firstly, it is important to determine the optimal gelation speed which must be fast enough to ensure the foam does not collapse. Secondly, the gel rheology must enable all details of the mould to be filled and, as such, gels of high fluidity are required, especially in the production of highly complex shapes (Puertolas *et al.*, 2011). There has been a great deal of interest in this technique due to the wide range of materials that can be gel cast. This is currently aimed at the analysis and reduction of the inner stresses of the constructs (Chen *et al.*, 2008) as well as the use of low or completely non-cytotoxic monomers (Jia *et al.*, 2002).

Sol-gel techniques

Sol-gel techniques are a well-developed, robust and versatile set of methods for fabricating scaffolds (Arcos and Vallet-Regi, 2010). The process involves the formation of mineral phases as a result of the inorganic polymerization of soluble molecular precursors. For aerogel ceramics, these are known as aerogel precursors, whereas for glasses silicate precursors are used. Briefly, polymerization occurs after hydrolysis with connectivity spanning throughout the solvent leading to the generation of a wet gel that is strengthened using controlled thermal processes. This also removes the liquid by-product of polycondensation and other organic species such that only the scaffold remains (Hench and West, 1990; Avnir *et al.*, 2006) with the process as used for glasses outlined in Fig. 6.5.

One of the major advantages of these techniques is that there is no need for extreme temperatures and so organic materials can be incorporated into the scaffolds. By varying the precursor it is possible to tailor the porosity and, to an extent, the pore size (Avnir *et al.*, 2006; Vallet-Regi, 2010) whilst the method lends itself to the generation of thin films or coatings. The main limitations lie with the poor mechanical properties of the scaffolds, hence current research involving sol-gel processes is aimed at composite scaffold production (Puertolas *et al.*, 2011) as well as the inclusion of biological

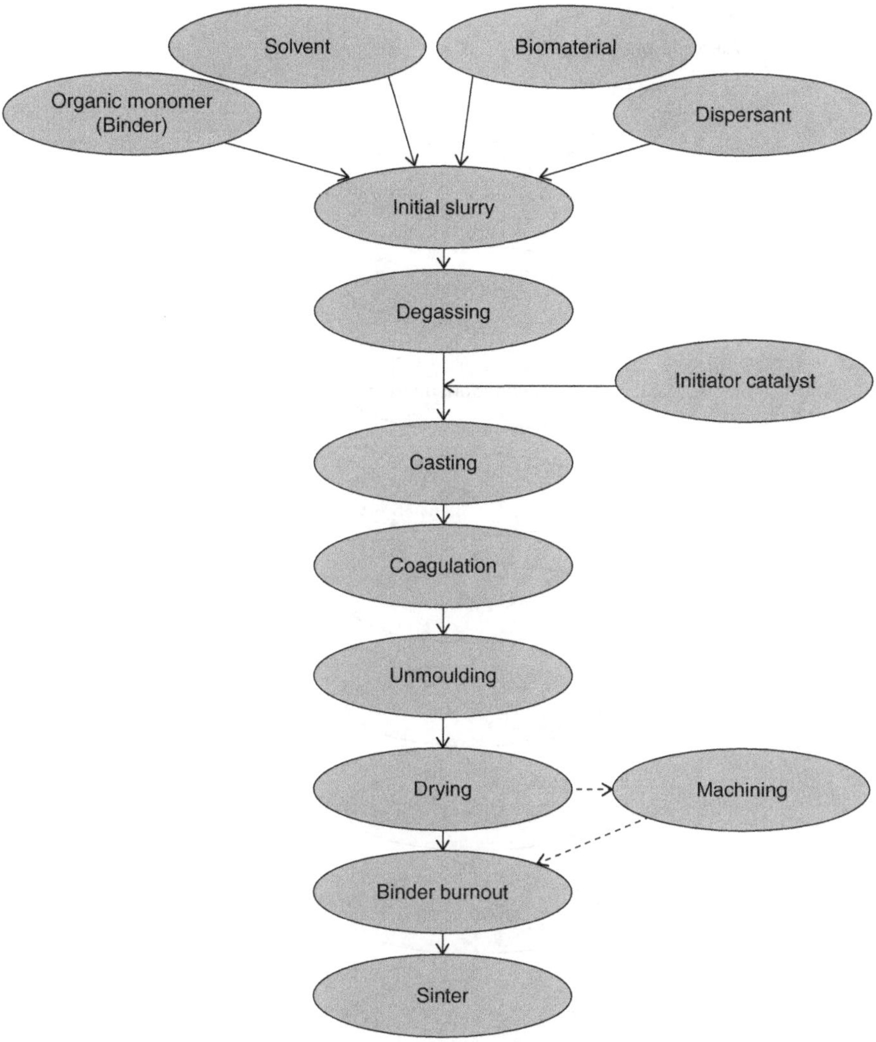

6.4 Flowchart for the gel casting technique as used to produce porous biomaterial constructs.

media to enhance the biological activity (Avnir *et al.,* 2006). Owing to the complexity of the procedure, coupled with the higher costs compared to other conventional techniques, sol-gel techniques are not used for large scale manufacture (Avnir *et al.,* 2006).

6.4.2 Rapid prototyping (RP) techniques

Rapid prototyping or solid freeform fabrication (SFF) techniques are based on the 'bottom-up' approach to manufacture. This involves starting

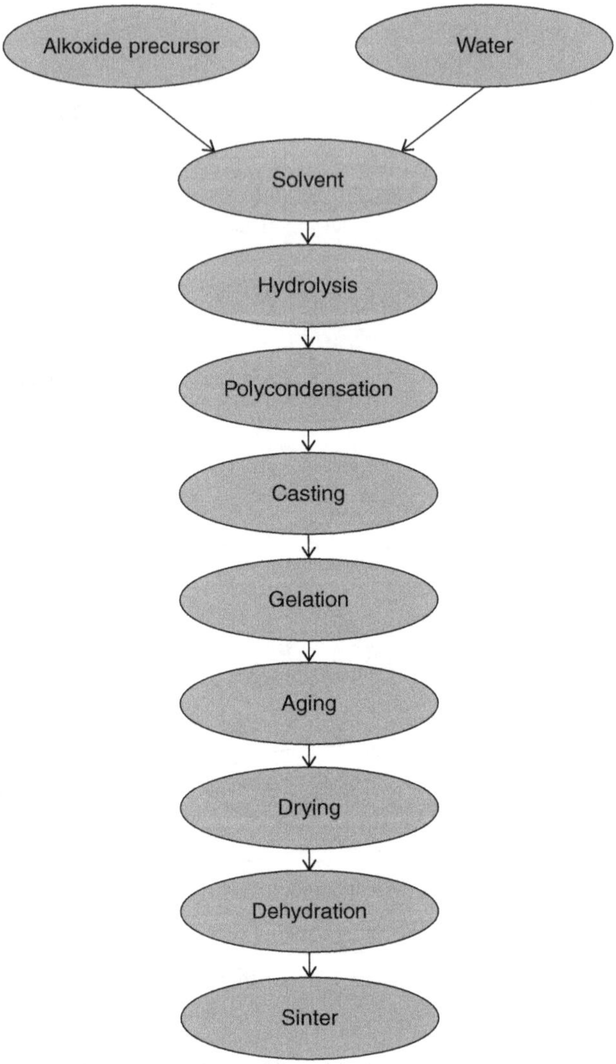

6.5 Sol-gel process as used to produce porous glasses suitable for bone regenerative purposes.

with small components or building units such as powders, liquids or colloids and arranging them in a predetermined manner using CAD and manufacture (Peltola *et al.*, 2008). Although it is possible to have a great deal of control over the porosity and pore size of the scaffolds, these techniques are associated with a much greater cost than conventional techniques and there are limitations to the resolution of the machinery.

Powder based techniques

Powder based techniques join thin layers of powder through means of an adhesive, as used in three-dimensional printing (3DP) (Will *et al.*, 2008; Wu *et al.*, 2011) or by melting as used in selective laser sintering (SLS) (Zhou *et al.*, 2008). A thin layer of powder is rolled onto the build table with individual particles then joined by either the localized spraying of an adhesive (3DP) or the selective melting through the use of a laser (SLS). Once each layer is complete the build table is lowered, a new layer of powder is deposited and the technique repeated until the desired geometry is obtained.

As the powder that has not been joined remains in place and is used as a base for the next layer of powder, there is no need for any support structures. However, this can also cause problems due to the entrapment of powders. Furthermore, 3DP generates fragile green bodies due to weak inter-particular bonds whilst SLS is highly inefficient (Shanjani *et al.*, 2010, 2011).

Liquid based techniques

Liquid based techniques operate on the principle of selectively drying a liquid biomaterial. As such, there is only one major liquid based SFF technique, commonly known as stereolithography (SLA), which combines materials with ultraviolet (UV) sensitivity (known as photopolymers) with the use of a laser. The build table is lowered further into a vat of the photopolymer with each iteration. Each layer is traced by the laser beam onto the surface, causing local solidification until the construct has been generated. The green body is then washed to remove any excess resin and finally cured in a UV oven (Melchels *et al.*, 2009, 2010).

It is a highly accurate technique and the maximum resolution of the finished structure can be as good as 1 μm. The main limitation with the widespread use of this technique, particularly with respect to ceramic foams, is the lack of suitable photosensitive biomaterials and the ability to liquidize them such that they can be selectively cured (Jansen *et al.*, 2010).

Colloidal ink based techniques

SFF techniques that require biomaterial slurries are known as colloidal ink based techniques and operate via the deposition of the slurry into the desired geometry. There are three major techniques within this subset. Robocasting requires the generation of a highly concentrated colloid that is capable of supporting its own weight, and depositing this in a predetermined manner. Thus a wide range of materials can be used to generate robocast structures, however the scaffolds are generally poor mechanically (Miranda *et al.*, 2008; Martinez-Vazquez *et al.*, 2010). Ink-jet casting deposits droplets of the colloid onto the build table where they solidify to produce constructs with a relatively high mechanical strength. Upon deposition, spreading of the colloid occurs which leads to the generation of long, thin pores that, although laminar, are not suitable for use as bone scaffolds (Lewis, 2006; Lewis *et al.,* 2006). To reduce the effect of this, 3D Bioplotting® has been developed whereby the droplets are instead dropped into a liquid medium where they solidify. Although heat can also be added to the ink to change the viscosity of the colloid, any temperature increase removes the possibility of incorporating biological agents into the scaffold (Leong *et al.,* 2003; Peltola *et al.,* 2008). Finally, fused-deposition modelling (FDM) uses a head-heated liquefier attached to a carriage to simultaneously heat and pump the filament material into place. This generates structures with mechanical anisotropy similar to natural bone without the risk of material encapsulation. However, a thermoplastic polymer must be used in this method and there is disproportionate lack of choice of bioactive, biocompatible biopolymers without cytotoxic properties. Furthermore, the use of heat also means that biological agents cannot be used which could otherwise enhance the biological activity (Too *et al.,* 2002; Kalita *et al.,* 2003).

6.5 Conclusions

Ceramic foams offer a great potential to overcome the limitations of existing grafting techniques of hard tissue regeneration. They can be classified depending on the effect they have when immersed in physiological media. Bioinert ceramics, such as titania and alumina, have no effect on the host location; bioactive ceramics, such as bulk structures from Bioglass® or HA, encourage bony regrowth on the surface; and bioresorbable ceramics, such as porous bioactive ceramics, become replaced by the native tissue over time. Generating porous constructs is highly important in maximizing repair through improved scaffold anchoring due to enhanced biological activity and thus a more even load distribution. Furthermore, such structures have mechanical properties more similar to those of natural bone and hence avoid potential problems with stress shielding. They can be produced from a wide number of techniques which can be divided into two categories. The first of these, conventional, utilizes the reactivity of the precursor constituents under certain heat or pressure, and is currently more prevalent due to the lower associated costs. Rapid prototyping, on the other hand, is based on the 'bottom-up' approach to manufacture, where building units such as powders, liquids or colloids are arranged in a predetermined manner using computer aided manufacture and design that allows significantly improved control over the final structure (construct) of the scaffold.

6.6 References

Andreiotelli, M., Wenz, H. J. and Kohal, R. J. (2009). Are ceramic implants a viable alternative to titanium implants? A systematic literature review. *Clinical Oral Implants Research,* **20**, 32–47. DOI 10.1111/j.1600-0501.2009.01785.x.

Arcos, D. and Vallet-Regi, M. (2010). Sol-gel silica-based biomaterials and bone tissue regeneration. *Acta Biomaterialia,* **6**, 2874–2888. DOI 10.1016/j.actbio.2010.02.012.

Avnir, D., Coradin, T., Lev, O. and Livage, J. (2006). Recent bio-applications of sol-gel materials. *Journal of Materials Chemistry,* **16**, 1013–1030. DOI 10.1039/B512706h.

Bose, S., Banerjee, A., Dasgupta, S. and Bandyopadhyay, A. (2009). Synthesis, processing, mechanical, and biological property characterization of hydroxyapatite whisker-reinforced hydroxyapatite composites. *Journal of the American Ceramic Society,* **92**, 323–330. DOI 10.1111/j.1551-2916.2008.02881.x.

Burg, K. J. L., Porter, S. and Kellam, J. F. (2000). Biomaterial developments for bone tissue engineering. *Biomaterials,* **21**, 2347–2359.

Burr, D. B., Forwood, M. R., Fyhrie, D. P., Martin, B., Schaffler, M. B. and Turner, C. H. (1997). Bone microdamage and skeletal fragility in osteoporotic and stress fractures. *Journal of Bone and Mineral Research,* **12**, 6–15.

Cachinho, S. C. P. and Correia, R. N. (2007). Titanium porous scaffolds from precursor powders: rheological optimization of TiH2 slurries. *Powder Technology,* **178**, 109–113. DOI 10.1016/j.powtec.2007.04.014.

Chen, B., Zhang, T., Zhang, J., Lin, Q. and Jiang, D. (2008). Microstructure and mechanical properties of hydroxyapatite obtained by gel-casting process. *Ceramics International,* **34**, 359–364. DOI 10.1016/j.ceramint.2006.10.021.

Chlopek, J., Czajkowska, B., Szaraniec, B., Frackowiak, E., Szostak, K. and Beguin, F. (2006). In vitro studies of carbon nanotubes biocompatibility. *Carbon,* **44**, 1106–1111. DOI 10.1016/j.carbon.2005.11.022.

Choi, K., Kuhn, J. L., Ciarelli, M. J. and Goldstein, S. A. (1990). The elastic-moduli of human subchondral, trabecular, and cortical bone tissue and the size-dependency of cortical bone modulus. *Journal of Biomechanics,* **23**, 1103–1113.

Chuenjitkuntaworn, B., Inrung, W., Damrongsri, D., Mekaapiruk, K., Supaphol, P. and Pavasant, P. (2010). Polycaprolactone/Hydroxyapatite composite scaffolds: Preparation, characterization, and in vitro and in vivo biological responses of human primary bone cells. *Journal of Biomedical Materials Research Part A,* **94A**, 241–251. DOI 10.1002/Jbm.A.32657.

Comas, H., Laporte, V., Borcard, F., Mieville, P., KRAUSS Juillerat, F., Caporini, M. A., Gonzenbach, U. T., Juillerat-Jeanneret, L. and Gerber-Lemaire, S. (2012). Surface functionalization of alumina ceramic foams with organic ligands. *ACS Appl Mater Interfaces,* **4**, 573–576.10.1021/am201638a.

Costa, H. D. S., Pereira, M. M., Andrade, G. I., Stancioli, E. F. B. and Mansur, H. S. (2007). Characterization of calcium phosphate coating and zinc incorporation on the porous alumina scaffolds. *Materials Research,* **10**, 27–29.

Costa, H. S., Mansur, A. A. P., Barbosa-Stancioli, E. F., Pereira, M. M. and Mansur, H. S. (2008). Morphological, mechanical, and biocompatibility characterization of macroporous alumina scaffolds coated with calcium phosphate/PVA. *Journal of Materials Science,* **43**, 510–524. DOI 10.1007/s10853-007-1849-6.

Cunningham, E., Dunne, N., Clarke, S., Choi, S. Y., Walker, G., Wilcox, R., Unger, R. E., Buchanan, F. and Kirkpatrick, C. J. (2011). Comparative characterisation of 3-D hydroxyapatite scaffolds developed via replication of synthetic polymer foams and natural marine sponges. *Tissue Science and Engineering,* S1.10.4172/2157-7552.S1-001.

Davies, J. E. (2007). Bone bonding at natural and biomaterial surfaces. *Biomaterials,* **28**, 5058–5067.10.1016/j.biomaterials.2007.07.049.

Dawson, J. I. and Oreffo, R. O. C. (2008). Bridging the regeneration gap: Stem cells, biomaterials and clinical translation in bone tissue engineering. *Archives of Biochemistry and Biophysics,* **473**, 124–131. DOI 10.1016/j.abb.2008.03.024.

Fartash, B., Liao, H., Li, J., Fouda, N. and Hermansson, L. (1995). Long-term evaluation of titania-based ceramics compared with commercially pure titanium in-vivo. *Journal of Materials Science-Materials in Medicine,* **6**, 451–454.

Frost, H. M. (1987). Bone mass and the mechanostat – a proposal. *Anatomical Record,* **219**, 1–9.

Frost, H. M. (2003). Bone's mechanostat: A 2003 update. *Anatomical Record Part a-Discoveries in Molecular Cellular and Evolutionary Biology,* **275A**, 1081–1101. DOI 10.1002/Ar.A.10119.

Fu, Q., Rahaman, M. N., Dogan, F. and Bal, B. S. (2008a). Freeze casting of porous hydroxyapatite scaffolds. I. Processing and general microstructure. *Journal of Biomedical Material Research B Applied Biomaterials,* **86**, 125–135.10.1002/jbm.b.30997.

Fu, Q., Rahaman, M. N., Dogan, F. and Bal, B. S. (2008b). Freeze casting of porous hydroxyapatite scaffolds. II. Sintering, microstructure, and mechanical behavior. *Journal of Biomedical Material Research B Applied Biomaterials,* **86**, 514–522.10.1002/jbm.b.31051.

Garcia, A., Izquierdo-Barba, I., Colilla, M., De Laorden, C. L. and Vallet-Regi, M. (2011). Preparation of 3-D scaffolds in the SiO_2–P_2O_5 system with tailored hierarchical meso-macroporosity. *Acta Biomaterialia,* **7**, 1265–1273. DOI 10.1016/j.actbio.2010.10.006.

Hench, L. L., Splinter, R. J., Allen, W. C. and Greenlee, T. K. (1971). Bonding mechanisms at the interface of ceramic prosthetic materials. *Journal of Biomedical Materials Research,* **5**, 117–141.10.1002/jbm.820050611.

Hench, L. L. and West, J. K. (1990). The sol-gel process. *Chemical Reviews,* **90**, 33–72.

Hing, K. A. (2005). Bioceramic bone graft substitutes: Influence of porosity and chemistry. *International Journal of Applied Ceramic technology,* **2**, 184–199.10.1111/j.1744-7402.2005.02020.x.

Huang, P., Xu, K. W. and Han, Y. (2007). Formation mechanism of biomedical apatite coatings on porous titania layer. *Journal of Materials Science-Materials in Medicine,* **18**, 457–463. DOI 10.1007/s10856-007-2005-1.

Huiskes, R. (1988). Stress patterns, failure modes and bone remodeling. *In:* Fitzgerald, R. H. and HIP SOCIETY (U.S.) (eds.) *Non-Cemented Total Hip Arthroplasty.* New York: Raven Press.

Jang, D. W., Kim, Y. H. and Lee, B. T. (2011). Microstructure control of TCP/TCP-(t-ZrO_2)/t-ZrO_2 composites for artificial cortical bone. *Materials Science and Engineering C-Materials for Biological Applications,* **31**, 1660–1666. DOI 10.1016/j.msec.2011.07.014.

Jansen, J., Boerakker, M. J., Heuts, J., Feijen, J. and Grijpma, D. W. (2010). Rapid photo-crosslinking of fumaric acid monoethyl ester-functionalized poly(trimethylene carbonate) oligomers for drug delivery applications. *Journal of Controlled Release,* **147**, 54–61.10.1016/j.jconrel.2010.06.031.

Jansen, J., Melchels, F. P. W., Grijpma, D. W. and Feijen, J. (2009). Fumaric acid monoethyl ester-functionalized poly(D,L-lactide)/N-vinyl-2-pyrrolidone resins for the preparation of tissue engineering scaffolds by stereolithography. *Biomacromolecules,* **10**, 214–220. DOI 10.1021/Bm801001r.

Jia, Y., Kanno, Y. and Xie, Z. P. (2002). New gel-casting process for alumina ceramics based on gelation of alginate. *Journal of the European Ceramic Society,* **22**, 1911–1916.

Jones, J. R., Atwood, R. C., Poologasundarampillai, G., Yue, S. and Lee, P. D. (2009). Quantifying the 3D macrostructure of tissue scaffolds. *Journal of Materials Science-Materials in Medicine,* **20**, 463–471. DOI 10.1007/s10856-008-3597-9.

Kalita, S. J., Bose, S., Hosick, H. L. and Bandyopadhyay, A. (2003). Development of controlled porosity polymer-ceramic composite scaffolds via fused deposition modeling. *Materials Science & Engineering C-Biomimetic and Supramolecular Systems,* **23**, 611–620. DOI 10.1016/S0928-4931(03)00052-3.

Karageorgiou, V. and Kaplan, D. (2005). Porosity of 3D biomaterial scaffolds and osteogenesis. *Biomaterials,* **26**, 5474–5491. DOI 10.1016/j.biomaterials.2005.02.002.

Karlsson, M., Palsgard, E., Wilshaw, P. R. and DI Silvio, L. (2003). Initial in vitro interaction of osteoblasts with nano-porous alumina. *Biomaterials,* **24**, 3039–3046. DOI 10.1016/S0142-9612(03)00146-7.

Kim, J., Lim, D., Kim, Y. H., Koh, Y. H., Lee, M. H., Han, I., Lee, S. J., Yoo, O. S., Kim, H. S. and Park, J. C. (2011). A comparative study of the physical and mechanical properties of porous hydroxyapatite scaffolds fabricated by solid freeform fabrication and polymer replication method. *International Journal of Precision Engineering and Manufacturing,* **12**, 695–701. DOI 10.1007/s12541-011-0090-z.

Koc, N., Timucin, M. and Korkusuz, F. (2004). Fabrication and characterization of porous tricalcium phosphate ceramics. *Ceramics International,* **30**, 205–211. DOI 10.1016/S0272-8842(03)00090-7.

Kokubo, T., Matsushita, T. and Takadama, H. (2007). Titania-based bioactive materials. *Journal of the European Ceramic Society,* **27**, 1553–1558. DOI 10.1016/j.jeurceramsoc.2006.04.015.

Langer, R. (2009). Perspectives and challenges in tissue engineering and regenerative medicine. *Advanced Materials,* **21**, 3235–3236. DOI 10.1002/adma.200902589.

LE Ray, A. M., Gautier, H., Bouler, J. M., Weiss, P. and Merle, C. (2010). A new technological procedure using sucrose as porogen compound to manufacture porous biphasic calcium phosphate ceramics of appropriate micro- and macrostructure. *Ceramics International,* **36**, 93–101. DOI 10.1016/j.ceramint.2009.07.001.

Leong, K. F., Cheah, C. M. and Chua, C. K. (2003). Solid freeform fabrication of three-dimensional scaffolds for engineering replacement tissues and organs. *Biomaterials,* **24**, 2363–2378.

Leukers, B., Gulkan, H., Irsen, S. H., Milz, S., Tille, C., Schieker, M. and Seitz, H. (2005a). Hydroxyapatite scaffolds for bone tissue engineering made by 3D printing. *Journal of Materials Science Materials in Medicine,* **16**, 1121–1124. DOI 10.1007/s10856-005-4716-5.

Leukers, B., Gulkan, H., Irsen, S. H., Milz, S., Tille, C., Seitz, H. and Schieker, M. (2005b). Biocompatibility of ceramic scaffolds for bone replacement made by 3D printing. *Materialwissenschaft Und Werkstofftechnik,* **36**, 781–787. DOI 10.1002/mawe.200500968.

Lewis, J. A. (2006). Direct ink writing of 3D functional materials. *Advanced Functional Materials,* **16**, 2193–2204. DOI 10.1002/adfm.200600434.

Lewis, J. A., Smay, J. E., Stuecker, J. and Cesarano, J. (2006). Direct ink writing of three-dimensional ceramic structures. *Journal of the American Ceramic Society,* **89**, 3599–3609. DOI 10.1111/j.1551-2916.2006.01382.x.

Li, J. (1993). Behavior of titanium and titania-based ceramics *in vitro* and *in vivo*. *Biomaterials,* **14**, 229–232.

Li, X., Li, D. C., Wang, L., Lu, B. H. and Wang, Z. (2008). Osteoblast cell response to beta-tricalcium phosphate scaffolds with controlled architecture in flow perfusion culture system. *Journal of Materials Science-Materials in Medicine,* **19**, 2691–2697. DOI 10.1007/s10856-008-3391-8.

Liu, Y. X., Kim, J. H., Young, D., Kim, S., Nishimoto, S. K. and Yang, Y. Z. (2010). Novel template-casting technique for fabricating beta-tricalcium phosphate scaffolds with high interconnectivity and mechanical strength and in vitro cell responses. *Journal of Biomedical Materials Research Part A,* **92A**, 997–1006. DOI 10.1002/Jbm.A.32443.

Macchetta, A., Turner, I. G. and Bowen, C. R. (2009). Fabrication of HA/TCP scaffolds with a graded and porous structure using a camphene-based freeze-casting method. *Acta Biomaterialia,* **5**, 1319–1327. DOI 10.1016/j.actbio.2008.11.009.

Mahendra, A. and Maclean, A. D. (2007). Available biological treatments for complex non-unions. *Injury-International Journal of the Care of the Injured,* **38**, Supplement 4, S7–S12. http://dx.doi.org/10.1016/S0020-1383(08)70004-4.

Mallick, K. K. (2009). Freeze casting of porous bioactive glass and bioceramics. *Journal of the American Ceramic Society,* **92**, S85–S94. DOI 10.1111/j.1551-2916.2008.02784.x.

Mallick, K. K. and Winnett, J. (2011). Porous biodegradable scaffolds for hard tissue engineering. In: *Biomaterials Science – Processing, Properties, and Applications*, Volume 228 (eds R. Narayan, A. Bandyopadhyay and S. Bose), John Wiley & Sons, Inc., Hoboken, NJ.

Mallick, K. K., Winnett, J., Van Grunsven, W., Lapworth, J. and Reilly, G. C. (2012). Three-dimensional porous bioscaffolds for bone tissue regeneration: Fabrication via adaptive foam reticulation and freeze casting techniques, characterization, and cell study. *Journal of Biomedical Materials Research Part A,* **100**, 2948–2959.10.1002/jbm.a.34238.

Manicone, P. F., Iommetti, P. R. and Raffaelli, L. (2007). An overview of zirconia ceramics: Basic properties and clinical applications. *Journal of Dentistry,* **35**, 819–826. DOI 10.1016/j.jdent.2007.07.008.

Martinez-Vazquez, F. J., Perera, F. H., Miranda, P., Pajares, A. and Guiberteau, F. (2010). Improving the compressive strength of bioceramic robocast scaffolds by polymer infiltration. *Acta Biomaterialia,* **6**, 4361–4368. DOI 10.1016/j.actbio.2010.05.024.

Melchels, F. P. W., Feijen, J. and Grijpma, D. W. (2009). A poly(D,L-lactide) resin for the preparation of tissue engineering scaffolds by stereolithography. *Biomaterials,* **30**, 3801–3809. DOI 10.1016/j.biomaterials.2009.03.055.

Melchels, F. P. W., Feijen, J. and Grijpma, D. W. (2010). A review on stereolithography and its applications in biomedical engineering. *Biomaterials,* **31**, 6121–6130. DOI 10.1016/j.biomaterials.2010.04.050.

Mikos, A. G. and Temenoff, J. S. (2000). Formation of highly porous biodegradable scaffolds for tissue engineering. *Electronic Journal of Biotechnology,* **3**, 1995–2000.

Miranda, P., Pajares, A., Saiz, E., Tomsia, A. P. and Guiberteau, F. (2008). Mechanical properties of calcium phosphate scaffolds fabricated by robocasting. *Journal of Biomedical Materials Research Part A,* **85A**, 218–227. DOI 10.1002/Jbm.A.31587.

Moon, Y. W., Shin, K. H., Koh, Y. H., Choi, W. Y. and Kim, H. E. (2011). Production of highly aligned porous alumina ceramics by extruding frozen alumina/camphene body. *Journal of the European Ceramic Society,* **31**, 1945–1950. DOI 10.1016/j.jeurceramsoc.2011.04.033.

Nieto, A., Areva, S., Wilson, T., Viitala, R. and Vallet-Regi, M. (2009). Cell viability in a wet silica gel. *Acta Biomaterialia,* **5**, 3478–3487. DOI 10.1016/j.actbio.2009.05.033.

Park, J. W. (2011). Increased bone apposition on a titanium oxide surface incorporating phosphate and strontium. *Clinical Oral Implants Research,* **22**, 230–234. DOI 10.1111/j.1600-0501.2010.01974.x.

Peltola, S. M., Melchels, F. P., Grijpma, D. W. and Kellomaki, M. (2008). A review of rapid prototyping techniques for tissue engineering purposes. *Annals of Medicine,* **40**, 268–280. DOI 10.1080/07853890701881788.

Planell, J. A., Best, S. M., Lacroix, D. and Merolli, A. (2009). *Bone Repair Biomaterials,* Cambridge, UK: Woodhead Publishing.

Price, R. L., Waid, M. C., Haberstroh, K. M. and Webster, T. J. (2003). Selective bone cell adhesion on formulations containing carbon nanofibers. *Biomaterials,* **24**, 1877–1887. DOI 10.1016/S0142-9612(02)00609-9.

Puertolas, J. A., Vadillo, J. L., Sanchez-Salcedo, S., Nieto, A., Gomez-Barrena, E. and Vallet-Regi, M. (2011). Compression behaviour of biphasic calcium phosphate and biphasic calcium phosphate-agarose scaffolds for bone regeneration. *Acta Biomaterialia,* **7**, 841–847. DOI 10.1016/j.actbio.2010.07.032.

Rambo, C. R., Cao, J. and Sieber, H. (2004). Preparation and properties of highly porous, bimorphic YSZ ceramics. *Materials Chemistry and Physics,* **87**, 345–352. DOI 10.1016/j.matchemphys.2004.05.031.

Rho, J. Y., Kuhn-Spearing, L. and Zioupos, P. (1998). Mechanical properties and the hierarchical structure of bone. *Medical Engineering & Physics,* **20**, 92–102.

Russias, J., Saiz, E., Deville, S., Gryn, K., Liu, G., Nalla, R. K. and Tomsia, A. P. (2007). Fabrication and in vitro characterization of three-dimensional organic/inorganic scaffolds by robocasting. *Journal of Biomedical Materials Research Part A,* **83A**, 434–445. DOI 10.1002/Jbm.A.31237.

Salgado, A. J., Coutinho, O. P. and Reis, R. L. (2004). Bone tissue engineering: state of the art and future trends. *Macromolecular Bioscience,* **4**, 743–765. DOI 10.1002/mabi.200400026.

Salinas, A. J. and Vallet-Regi, M. (2007). Evolution of ceramics with medical applications. *Zeitschrift Fur Anorganische Und Allgemeine Chemie,* **633**, 1762–1773. DOI 10.1002/zaac.200700278.

Schantz, J. T., Brandwood, A., Hutmacher, D. W., Khor, H. L. and Bittner, K. (2005). Osteogenic differentiation of mesenchymal progenitor cells in computer designed fibrin-polymer-ceramic scaffolds manufactured by fused deposition modeling. *Journal of Materials Science-Materials in Medicine,* **16**, 807–819.

Sepulveda, P., Binner, J. G. P., Rogero, S. O., Higa, O. Z. and Bressiani, J. C. (2000). Production of porous hydroxyapatite by the gel-casting of foams and cytotoxic evaluation. *Journal of Biomedical Materials Research,* **50**, 27–34.

Shanjani, Y., DE Croos, J. N. A., Pilliar, R. M., Kandel, R. A. and Toyserkani, E. (2010). Solid freeform fabrication and characterization of porous calcium polyphosphate structures for tissue engineering purposes. *Journal of Biomedical*

Materials Research Part B-Applied Biomaterials, **93B**, 510–519. DOI 10.1002/Jbm.B.31610.

Shanjani, Y., Hu, Y. X., Pilliar, R. M. and Toyserkani, E. (2011). Mechanical characteristics of solid-freeform-fabricated porous calcium polyphosphate structures with oriented stacked layers. *Acta Biomaterialia,* **7**, 1788–1796. DOI 10.1016/j.actbio.2010.12.017.

Singh, D., Lorenzo-Martin, M. D., Gutierrez-Mora, F., Routbort, J. L. and Case, E. D. (2006). Self-joining of zirconia/hydroxyapatite composites using plastic deformation process. *Acta Biomaterialia,* **2**, 669–675. DOI 10.1016/j.actbio.2006.06.004.

Smart, S. K., Cassady, A. I., Lu, G. Q. and Martin, D. J. (2006). The biocompatibility of carbon nanotubes. *Carbon,* **44**, 1034–1047. DOI 10.1016/j.carbon.2005.10.011.

Tang, G. X., Zhang, R. J., Yan, Y. N. and Zhu, Z. X. (2004). Preparation of porous anatase titania film. *Materials Letters,* **58**, 1857–1860. DOI 10.1016/j.matlet.2003.11.016.

Too, M. H., Leong, K. F., Chua, C. K., Du, Z. H., Yang, S. F., Cheah, C. M. and Ho, S. L. (2002). Investigation of 3D non-random porous structures by fused deposition modelling. *International Journal of Advanced Manufacturing Technology,* **19**, 217–223.

Trewyn, B. G., Whitman, C. M. and Lin, V. S. Y. (2004). Morphological control of room-temperature ionic liquid templated mesoporous silica nanoparticles for controlled release of antibacterial agents. *Nano Letters,* **4**, 2139–2143. DOI 10.1021/Nl048774r.

Vallet-Regi, M. (2001). Ceramics for medical applications. *Journal of the Chemical Society-Dalton Transactions*, 97–108. DOI 10.1039/B007852m.

Vallet-Regi, M. (2006). Revisiting ceramics for medical applications. *Dalton Transactions*, 5211–5220. DOI 10.1039/B610219k.

Vallet-Regi, M. (2010). Nanostructured mesoporous silica matrices in nanomedicine. *Journal of Internal Medicine,* **267**, 22–43. DOI 10.1111/j.1365-2796.2009.02190.x.

Vinu, A., Miyahara, M., Sivamurugan, V., Mori, T. and Ariga, K. (2005). Large pore cage type mesoporous carbon, carbon nanocage: a superior adsorbent for biomaterials. *Journal of Materials Chemistry,* **15**, 5122–5127. DOI 10.1039/B507456h.

VON Schewelov, T., Sanzén, L., Önsten, I., Carlsson, Å. and Besjakov, J. (2005). Total hip replacement with a zirconium oxide ceramic femoral head. *Journal of Bone & Joint Surgery, British Volume,* **87-B**, 1631–1635.10.1302/0301-620x.87b12.16873.

Weiner, S., Traub, W. and Wagner, H. D. (1999). Lamellar bone: structure-function relations. *Journal of Structural Biology,* **126**, 241–155.10.1006/jsbi.1999.4107.

Weiner, S. and Wagner, H. D. (1998). The material bone: Structure mechanical function relations. *Annual Review of Materials Science,* **28**, 271–298.

Wettergreen, M. (2008). *The Effect of Material Organization on the Structural Properties of Porous Architectures.* Doctor of Philosophy, Rice University.

Will, J., Melcher, R., Treul, C., Travitzky, N., Kneser, U., Polykandriotis, E., Horch, R. and Greil, P. (2008). Porous ceramic bone scaffolds for vascularized bone tissue regeneration. *Journal of Materials Science-Materials in Medicine,* **19**, 2781–2790. DOI 10.1007/s10856-007-3346-5.

Wu, C. T., Luo, Y. X., Cuniberti, G., Xiao, Y. and Gelinsky, M. (2011). Three-dimensional printing of hierarchical and tough mesoporous bioactive glass scaffolds with a controllable pore architecture, excellent mechanical strength and mineralization ability. *Acta Biomaterialia,* **7**, 2644–2650. DOI 10.1016/j.actbio.2011.03.009.

Xiao, X. L., Yang, L. X., Guo, M. L., Pan, C. F., Cai, Q. Y. and Yao, S. Z. (2009). Biocompatibility and in vitro antineoplastic drug-loaded trial of titania nanotubes prepared by anodic oxidation of a pure titanium. *Science in China Series B-Chemistry,* **52**, 2161–2165. DOI 10.1007/s11426-009-0240-7.

Xie, L., Liao, X. M., Yin, G. F., Huang, Z. B., Yan, D. H., Yao, Y. D., Liu, W. L., Chen, X. C. and Gu, J. W. (2011). Preparation, characterization, in vitro bioactivity, and osteoblast adhesion of multi-level porous titania layer on titanium by two-step anodization treatment. *Journal of Biomedical Materials Research Part A,* **98A**, 312–320. DOI 10.1002/Jbm.A.33120.

Yamamoto, O., Alvarez, K., Kashiwaya, Y. and Fukuda, M. (2011). Significant effect of a carbon layer coating on interfacial bond strength between bone and Ti implant. *Carbon,* **49**, 1588–1598. DOI 10.1016/j.carbon.2010.12.041.

Yang, J., Yu, J. and Huang, Y. (2011). Recent developments in gelcasting of ceramics. *Journal of the European Ceramic Society,* **31**, 2569–2591. DOI 10.1016/j.jeurceramsoc.2010.12.035.

Yoon, B. H., Choi, W. Y., Kim, H. E., Kim, J. H. and Koh, Y. H. (2008). Aligned porous alumina ceramics with high compressive strengths for bone tissue engineering. *Scripta Materialia,* **58**, 537–540. DOI 10.1016/j.scriptamat.2007.11.006.

Yuasa, T., Miyamoto, Y., Ishikawa, K., Takechi, M., Momota, Y., Tatehara, S. and Nagayama, M. (2004). Effects of apatite cements on proliferation and differentiation of human osteoblasts in vitro. *Biomaterials,* **25**, 1159–1166. DOI 10.1016/j.biomaterials.2003.08.003.

Zhao, J., Xiao, S. G., Lu, X., Wang, J. X. and Weng, J. (2006). A study on improving mechanical properties of porous HA tissue engineering scaffolds by hot isostatic pressing. *Biomedical Materials,* **1**, 188–192. DOI 10.1088/1748-6041/1/4/002.

Zhou, W. Y., Lee, S. H., Wang, M., Cheung, W. L. and Ip, W. Y. (2008). Selective laser sintering of porous tissue engineering scaffolds from poly(L)/carbonated hydroxyapatite nanocomposite microspheres. *Journal of Materials Science-Materials in Medicine,* **19**, 2535–2540. DOI 10.1007/s10856-007-3089-3.

Zhu, L., Ye, X., Tang, G., Zhao, N., Gong, Y., Zhao, Y., Zhao, J. and Zhang, X. (2007). Biomimetic coating of compound titania and hydroxyapatite on titanium. *Journal of Biomedical Materials Research Part A,* **83A**, 1165–1175. DOI 10.1002/Jbm.A.31401.

Zimmerman, R., Ila, D., Muntele, C., Rodrigues, M., Poker, D. B. and Hensley, D. (2002). Enhanced tissue adhesion by increased porosity and surface roughness of carbon based biomaterials. *Nuclear Instruments & Methods in Physics Research Section B-Beam Interactions with Materials and Atoms,* **191**, 825–829.

7

Titanium and NiTi foams for bone replacement

A. BANSIDDHI, Kasetsart University, Thailand and
D. C. DUNAND, Northwestern University, USA

DOI: 10.1533/9780857099037.2.142

Abstract: Extensive development in porous titanium and titanium alloys over the past 40 years has successfully opened up a wide range of biomedical applications and helped offer a view into the next generation of metal implants. This chapter will first provide basic understanding of the materials science underlying the use of titanium-based foams, including both titanium alloys and nickel titanium (NiTi) shape-memory alloys, as bone replacement materials. It will then review recent developments in fabrication techniques that allow these alloys to fulfill implant requirements, and also their current limitations, which lead to three generations of currently available foam structures and their properties in existing commercial products and research in progress. Finally, future trends for titanium-based foams are discussed.

Key words: porous titanium implants, nickel titanium (NiTi), powder metallurgy (PM), metal foam structure, biocompatible property.

7.1 Introduction

Titanium and its alloys are among the most common bone replacement materials and have been intensively used for many decades. Ongoing developments in both the materials and the associated fabrication techniques have allowed porous titanium alloys to fulfill the required structure and properties from a biomedical perspective. Yet, titanium foams are far from being fully optimized for bone implants. The purpose of this chapter is to provide a basic understanding of the use of titanium-based foams as bone replacement materials, to review recent developments in academic and industrial research for titanium alloy and nickel titanium (NiTi) foams, and to use these to forecast some possible future trends in research and product development.

The chapter begins with an introduction to the properties of titanium and its alloys, their advantageous structures and properties that contribute to dense and porous bone implants as well as their limitations that require compositional and structural optimization. It also discusses the significance of the shape recovery effect for NiTi bone replacements with a comparison between conventional Ti alloy and shape memory NiTi. The second section reviews several up-to-date process developments aimed at creating porous structures. Fabrication methods are discussed in the context of how they underpin and constrain the available pore characteristics and thus the mechanical and biological properties relative to natural bone. The final section attempts to anticipate future trends in the titanium-based bone replacement research and the next generation of Ti and NiTi foam products.

7.2 Titanium-based materials for replacing bones

Titanium is a Group IV transition metal that has enjoyed great interest as one of the most important biocompatible metals, thanks to a suite of suitable biological and biomechanical properties. From a biological perspective, commercially pure titanium (CP Ti) has moderate to low cytotoxicity and is non-carcinogenic (Keegan *et al.*, 2007). Titanium is considered biologically inert (Fujibayashi *et al.*, 2004) and has excellent resistance to corrosion, as its native titanium dioxide (TiO_2) surface forms spontaneously and reforms easily after breakage (Blackwood, 2003). This ensures low metal ion release from the surface, which motivated its use in many first-generation biomedical applications such as dental and orthopedic surgery (Elias *et al.*, 2008; Lautenschlager and Monaghan, 1993). In addition, CP Ti interferes less than denser steel and cobalt alloys with post-operative imaging methods such as magnetic resonance imaging (MRI) and computed tomography (CT) scanning, which are required to assess the position and health of the implant and the surrounding tissue (Knott *et al.*, 2010; Matsuura *et al.*, 2005). From a biomechanical perspective, titanium exhibits high compressive strength and low Young's modulus, comparable with the other biocompatible metals (Table 7.1), while being considerably lighter. Its low density also makes titanium an alternative for biodevices made from ceramics and polymers, where low toughness and strength have been persistent issues, respectively (Ryan *et al.*, 2006).

With respect to bone replacement materials, the stiffness of CP Ti is higher than that required for this application. In contrast to the aerospace and automobile applications for which titanium is often sought, bone implant applications rely on producing a certain range of mechanical properties, specifically matching those of bone, in order to avoid potentially severe implant failure. For example, a stiff titanium implant (with stiffness of 100 GPa) embedded in a less-stiff implant site (for example, cortical bone with stiffness of 20 GPa) can cause a mechanical mismatch or 'stress-shielding' effect during loading. This results in bone resorption that may significantly affect the long-term performance of the implant (Bobyn *et al.*, 1992). Further,

Table 7.1 Biological and mechanical properties of titanium and other biocompatible metals

Material	Density (g/cm^3)	Compressive strength (MPa)	Tensile strength (MPa)	Young's modulus (GPa)	Toughness ($MPa\ m^{1/2}$)
Titanium (CPTi)	4.4–4.5	590–1117	240–550	100–110	55–115
Ti64	–	–	895–930	112	–
Co-Cr alloys	8.3–9.2	450–1896	–	200–253	100
Stainless steel 316L	7.9–81	170–310	–	189–205	50–200
Cancellous bone	–	1.5–9.3	1.6–2.42	0.1–0.4	–
Cortical bone	–	205–219 Longitudinal 131–153 Transverse	172 Longitudinal 52 Transverse	18.2 Longitudinal 11.7 Transverse	–

Sources: Compiled from Krishna *et al.* (2007); Alvarez and Nakajima (2009); and Geetha *et al.* (2009)

pure titanium is susceptible to crack propagation and has poor tribological properties (Choubey *et al.*, 2004) due to its low plastic shear strength and the fairly weak mechanical protection induced by surface oxides, which limits service lifetime of titanium implants to only 10–15 years.

In order to overcome the drawbacks of pure titanium while retaining its advantages for bone replacement applications, many modifications have been made to the material composition leading to the introduction of new titanium and NiTi biomedical alloys.

7.2.1 Titanium and titanium alloys

Many metallic elements have been considered in trials for making new titanium alloys for bone replacement. The key criteria for selecting alloying elements are: (1) to correct the aforementioned disadvantages of titanium, specifically to lower the Young's modulus and (2) to maintain or even improve the biological environment presented to the host. The former goal was achieved through the use of beta (β) titanium alloys. The beta phase refers to one of the two allotropic forms of titanium having a body-centered cubic structure and is usually found at high temperature. The beta phase provides a lower modulus and better fatigue resistance compared to the martensitic alpha (α) phase, normally found in CP Ti at low temperature. To stabilize the beta phase at the low temperature characteristic of biomaterials, a decrease of the α to β transformation temperature (or beta transus temperature) is required and this can be accomplished by adding the so-called 'beta stabilizer' alloying elements including niobium

(Nb), molybdenum (Mo), tantalum (Ta), vanadium (V), chromium (Cr), manganese (Mn), cobalt (Co), nickel (Ni), and copper (Cu) (Long and Jack, 1998). Some alloying elements, such as zirconium (Zr) or tin (Sn), may be added to the alloys without affecting the stability of either phase and are considered neutral. Conversely, the presence of interstitials such as oxygen (O), nitrogen (N), carbon (C), and other metals such as aluminum (Al) and gallium (Ga), stabilizes the alpha phase (Abdel-Hady *et al.*, 2006). Although there is quite a variety of stabilized and neutral alloy chemistries that can be brought to bear towards mechanically optimal Ti alloys, in practice the biocompatibility of the alloying additions often becomes the dominant factor (Eisenbarth *et al.*, 2004).

For example, Ti-6Al-4V (commonly referred to as Ti64) with extra low interstitial content (the elements O, C, and N) exhibits excellent mechanical properties (i.e., high strength of 800–900 MPa while retaining a modulus near 100 GPa) due to its α+β microstructure. However, the passive oxide films of its additives (Al_2O_3 and VO_2) that form on the alloy surface are less stable than TiO_2, which can lead to corrosion and release of Al and V ions (Metikos-Hukovic *et al.*, 2003). Vanadium in its elemental and oxidized forms (i.e., V_2O_5) is cytotoxic and induces adverse tissue reactions, while Al is suspected to play a role in neurodegenerative disease (Eisenbarth *et al.*, 2004; Walker *et al.*, 1989). To date, Ti64 has still been the most commonly used titanium alloy for implant applications and has been continuously developed to overcome the biological concerns and maintain its advantageous mechanical properties. At the same time, various new α+β titanium alloys with comparable mechanical properties but more compatible alloying (V-free) additives like Ta, Nb, Mo, and Zr have been proposed (Niinomi *et al.*, 2002). Examples include Ti-6Al-7Nb, Ti-5Al-2.5Fe, Ti-5Al-3Mo-4Zr, and Ti-15(Sn,Zr)-4Nb-2Ta-0.2Pd.

Later generations of titanium implant materials have been developed with the following benefits: (1) higher beta phase content (including metastable β, near β and pure β) and (2) more biocompatible alloy additives (less Al, Fe, Mo). The common beta titanium alloys are Ti-13Nb-13Zr (Lopez *et al.*, 2001), Ti-Mo-Zr-Fe (TMZF) (Wang *et al.*, 1993), and the Ti-Nb-Zr-Ta (TNZT) series (Okazaki, 2001; Elias *et al.*, 2006), which provide lower modulus (50–80 GPa) and better fatigue properties than CP Ti due to strong wear resistance of their passive oxides (e.g., Nb_2O_5). Recently, low modulus beta titanium alloys with Cr, Mn, Sn, Fe, and Al additions were proposed (Ozaki *et al.*, 2004) as alternatives to those based on more expensive metals (Zr, Nb, Ta, and Mo). At this point, a modulus of bulk beta titanium alloys as low as 30 GPa can be achieved with the help of heat treating process control (Laheurte *et al.*, 2010). More information on recent titanium alloys can be found in the latest review by Niinomi *et al.* (2012). Yet, unlike CP Ti or Ti64, the alloys described above have not been launched as common commercial products as their process optimizations are still under investigation. It is noteworthy that suitable alloy chemistry and the selection of biocompatible elements alone do not determine the structure and properties of the final alloys which depend also on the ease of thermomechancial processing and surface treatment (Kokubo *et al.*, 2004). Therefore, intensive study of the structure–property–processing correlations of the alloys is recommended in order to develop future generation titanium alloy implants, before turning to applications. Some examples are provided in Section 7.3 for the case of porous structures.

7.2.2 Nickel titanium alloys

Near equiatomic NiTi alloys, with nickel content of 48–51 at%, have high corrosion resistance and excellent strength and also exhibit shape recovery behaviors, called the shape memory effect and the pseudoelastic effect (Otsuka and Ren, 2005). These shape recovery effects result from phase transformations between martensite (the low temperature phase) and austenite (the high temperature phase), which can be accomplished by changes in temperature or stress at biologically relevant temperatures and that result in predictable changes in strains (i.e., 8%). Many biomedical devices have exploited NiTi shape memory effects to promote beneficial shape changes at the surgical site that are activated by body heat or external heat sources. Examples include minimally invasive instruments, orthodontic wires, and cardiovascular self-expansion stents which make use of pseudoelasticity to restore their designed shape after implementation. The transformation temperatures define the operating temperatures of such shape recovery devices and are quite sensitive to composition (a change in Ni content of only 0.1 at%, for instance, alters the transformation temperature of NiTi by ~10°C). For this reason, careful monitoring of alloy composition is critical to ensure that the transitions occur within the required temperature range (e.g., at body temperature but not during storage and handling) (Elahinia *et al.*, 2012).

For bone implants, these phenomena promote unusually large recovery strain (>0.5%) and low Young's modulus (<40 GPa) for NiTi which are preferable to those of conventional implant alloys (i.e., stainless steel, Ti), and more closely match those of bone (recovery of more than 1% and Young's modulus of <20 GPa). Moreover, shape recovery of the implant after implementation possibly enhances bone implant self-fixation, without the additional fixing screws usually required. Regarding a concern about toxic Ni ion leakage, NiTi provides a protective Ti-based oxide layer which acts as an effective barrier to Ni ion release (Es-Souni and Fisher-Brandies, 2005). Similar to titanium, NiTi was tested *in vitro* and *in vivo* for biocompatibility and showed acceptable metal ion levels (Assad *et al.*, 2003; Shabalovskaya, 2002). Due to their biological and biomechanical advantages, NiTi alloys are becoming viable competitors to titanium alloys, especially in the bone implant field. Although near equiatomic NiTi alloys are the most well studied composition, biocompatible ternary NiTi alloys such as NiTiTa and NiTiNb (Bansiddhi *et al.*, 2011), and nickel-less Ti-based shape memory alloys (Biesiekierski, 2012) are also gaining interest.

7.3 Development of Ti-based foams for replacing bone

As development of new titanium alloys continues, a parallel approach to mitigating the mismatching stiffness between titanium and bone is underway through the use of 'porous structures' or 'foams'. These terms refer to solid materials containing voids or pores open to the surface (Gibson, 2005), similar to those inside cancellous bones. The benefits of porosity for the mechanical properties of materials, which justify the effort to create these porous Ti structures, are briefly described in this section. Because the limitations of Ti and its alloys constrain the manufacturing paths available for creating porosity, many novel solutions and potential future directions have been generated. The challenges, and the solutions available for conquering them, are also briefly reviewed in this section.

7.3.1 Principle and strategies to develop Ti-based foams for replacing bone

In order to mimic bone structure most closely, titanium alloy foams target highly interconnected, open cell structures, with pore volume fractions (i.e., porosities) of 30–80% and pore sizes ranging from 50–400 μm. While no clear consensus has emerged yet regarding the optimal pore size for bone ingrowth, this foam structure in general has been found to enhance permeability for body fluid and promote tissue ingrowth (Bansiddhi *et al.*, 2008). The open pore structure provides a suitable environment for osseointegration and osseoconduction, which improve bone implant fixation (Karageorgiou and Kaplan, 2005).

According to the Gibson and Ashby equations (Gibson and Ashby, 1997), validated for a wide range of foam materials (Equations [7.1] and [7.2]), higher porosity leads to lower strength and modulus.

$$\frac{\sigma}{\sigma_o} = C1\left(\frac{\rho}{\rho_o}\right)^{n1} \quad [7.1]$$

$$\frac{E}{E_o} = C2\left(\frac{\rho}{\rho_o}\right)^{n2} \quad [7.2]$$

where σ and σ_o are the yield strength of the foam and the dense metal, E and E_o are the stiffness or Young's moduli of the foam and the dense metal, and ρ and ρ_o are the density of the foam and the dense metal. $C1$ and $C2$ are constants representing the dependence of stiffness and strength on cell geometry, respectively, $n1$ and $n2$ are scaling exponents and vary with the type of material.

For example, a CP Ti foam with 50% porosity exhibits a modulus of 25 GPa, much closer to that of bone than the 100 GPa observed in full-density CP Ti. Even though these equations are only strictly valid for certain pore geometry at porosity above 70%, their predictive power is useful and they are routinely used to predict and understand the mechanical properties of a wide variety of foams. The effects of other foam characteristics such as pore size, pore shape, fenestration size (the size of the 'windows' that separate adjacent pores), strut size, connectivity, and pore roughness are not explicitly represented in these equations and are recommended for study to ensure a strongly predictive model. More information on foam properties and the use and limitations of these equations can be found in the work of Gibson and Ashby themselves (Gibson and Ashby, 1997).

7.3.2 Ti foams: advantages and constraints of Ti and Ti-based alloys

Despite the attractive properties Ti and Ti-based alloys exhibit constraints that complicate fabrication of high quality foamed – and even fully-densified – structures. For example, Ti and its alloys have very high melting points (1670°C for CP Ti), extreme chemical affinity with atmospheric gases including oxygen and nitrogen at high temperature, and high reactivity with most mold materials (Donachie, 2000). These properties make casting Ti alloys more difficult and less cost effective than casting aluminum foams. Oxygen is a very potent strengthener in titanium, but also reduces tensile ductility, leading to various grades of CP Ti grades reflecting oxygen content (Lautenschlager and Monaghan, 1993). Expensive high temperature and high vacuum production equipment is often needed to overcome the limitations in conventional Ti alloy casting.

Therefore, other approaches are more common for working with titanium-based materials, such as powder metallurgy (PM) (Singh *et al.*, 2010a). PM products are made by consolidating titanium powders under solid-state conditions, generally requiring lower temperatures and are subject to less stringent chemical reactivity restrictions. Since PM enables near-net shape production, machining and materials wastage can be minimized. Despite all these advantages, however, PM processing is still typically more expensive than casting approaches.

Forming foam structures by overcoming these processing restrictions without compromising the material properties is, as expected, even more difficult than processing the monolithic materials. The evolution of Ti foam architecture thus began with a solid Ti implant coated with a layer of porous material of Ti or other materials. The purpose of this design was to improve the interfacial bond between bone tissue and the outer surface of an implant thus enhancing bone implant fixation. Due to the abrupt and sizeable difference in density, and thus physical properties (e.g., fatigue strength), between the solid inner body and outer porous coating, fracture failure at the substrate/coating interface was often observed, and metal debris release became a concern (Cook and Thomas, 1991). The processing goal for this design was therefore to enhance the adhesion of the porous coating to the implant while maintaining the benefit in biological fixation. Although porous coatings helped to reduce the density of the implant, the overall strength and modulus of this implant design is still much closer to those of the bulk metal than to those of bone.

Fully porous implants therefore drew more attention because of their uniformly low density, full permeability for bone ingrowth, and strength and stiffness that are lower but sufficient to match those of bone. Diverse porous Ti structures under this implant category, processed using a variety of pore-forming strategies, have been shown to fulfill the basic structural requirements for bone implants, with better mechanical and biological compatibility than the porous coating. However, concerns about the high risk of crack initiation and corrosion due to the high surface area of porous implants have emerged, especially for load bearing implants. This has led to extensive studies on fully porous implants so as to understand the processing–structure–properties correlation of each foam structure and achieve an optimal balance of properties and price.

Though both implant designs are still in demand, some additional key requirements are noted. Firstly, a more effective implant design specification is needed that includes a gradient or hierarchical structure across the implant, with tailored and controllable pore sizes, shapes and distributions in each area. For example, one side of an implant with high porosity may promote high vascularization and direct osteogenesis whilst the other, denser, side may promote mechanical stability. This will allow the best fit implant structure for specific host sites, as the porous architectures of bone are themselves differentiated and graded. Secondly, customized complex implant contours are interesting for the purpose of precise adaptation to the region of implantation. Improving the flexibility and controllability of foaming techniques, without surrendering the resulting material properties, will be essential to meet this requirement.

Next, foam structures with a built-in biological and biomechanical enhancing mechanism are desirable. When implants are used in load bearing applications, rapid bone implant fixation is required in the early stage for fast rehabilitation, while resistance to fatigue as well as high toughness are desirable for increasing longevity and implant life. Surface modifications to adjust the pore roughness or bioactivity are examples of approaches used (Liu *et al.*, 2004).

Finally, foam structures with economically viable production methods will become very influential. Since the demand for bone implants has increased sharply in recent years, fabrication methods that provide high production rates and scalability at economical prices are crucial. To date, all of the above key areas have been explored but with varying depth and degree of understanding. These have been the underlying development goals of the currently available Ti-based foam products that are described in the following section.

7.4 Introduction to currently available Ti-based foams

A large selection of Ti-based foam structures has been proposed and developed for bone implant applications. Figure 7.1 presents three generations of foam structures, which represent three different strategies to produce Ti foam implants.

In early development (for convenience, these will be referred to as Generation I), foams were created using methods that relied solely on preserving the pre-existing porosity of Ti-based powder preforms

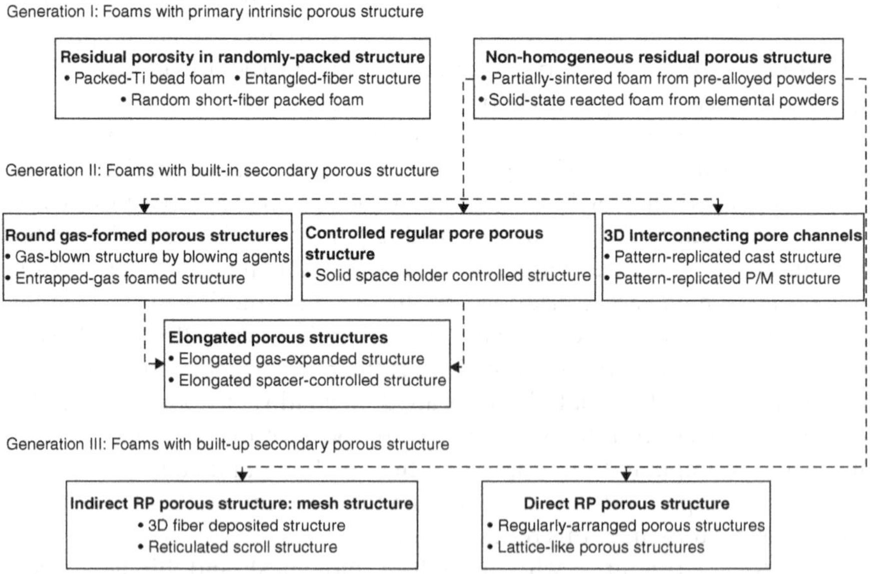

7.1 Three generations of conceptual Ti foam structures.

(unconsolidated powder compacts) during consolidation by sintering at elevated temperature, with or without an applied stress. Consequently, the primary strategy for optimizing structure in Generation I foams was to optimize consolidation parameters. Three-dimensional porosity with sizes of tens to a few hundreds of microns and a non-homogeneous distribution were usually observed, and somewhat near-net shaping of the product was achieved but not as a main feature.

Generation II foams introduced a secondary foam structure produced by entrapping extrinsic materials (in gaseous or solid form) in a controllable manner inside the foam. Macropores hundreds of microns in size were created into the consolidated material, in addition to the smaller primary intrinsic porosity preserved from Generation I as a feature of powder consolidation. Non-homogeneous pore distributions were still found but comparatively well-controlled and flexible pore architectures were obtained with less sensitivity to processing parameters. The aim in this generation then was to optimize the utilization of extrinsic additives without sacrificing the advantages of the base material. Similar to Generation I, approximately near-net shaping was achieved, but still relied on finishing operations.

Generation III foams were introduced not long after Generation II, with a shift based on rapid prototyping (RP) technology that combined tailor-made macroporosity several hundred microns in size with near-net shaping in a single procedure. At this time, foam structures within this generation are available with homogeneous pore structures following basic structural patterns.

Sections 7.5–7.7 review representative foam structures of Ti-based materials from each generation. Most of the structures are still under widespread investigation and some have been commercialized to end users. For each foam structure, the underlying processing strategies are provided not because it is essential that the reader understands them but rather to illuminate: (1) the key foam characteristics (e.g., pore size, shape, connectivity, and volume fraction) that can realistically be expected from each structure and (2) any fabrication-related features or limitations that should be expected (e.g., gradients in porosity, bulk vs surface-only porosity, external near-net shape). These in turn should help the reader to better understand the course of future research.

7.5 Generation I: foams with primary intrinsic porous structure

Pores are naturally present within a preform of solid powders. For example, mono-dispersed spherical powders can be randomly packed to provide density of up to 72%, that is, a porosity of at least 28%; these numbers will vary with changes in the shape and size distribution. Generation I foams exploit this fact to create two basic foam structures: residual porosity in randomly packed structures and non-homogeneous residual porous structures.

7.5.1 Residual porosity in randomly packed structure

A loose preform of solid Ti-based powders or wires is consolidated and lightly sintered under high vacuum or inert gas environment. The solid portions then become the foam which surrounds an interconnected pore channel structure which is an inverse structure to the solid. The porous architecture is therefore directly determined by the packing pattern of the solid, for instance, orderly packed powders, randomly packed chopped wires, and entangled wires.

Sintering a preform of spherical CP Ti powders with particle sizes ranging from 150–500 μm provides a solid with total porosity of 5–37% (Oh *et al.*, 2003). Since the pore shape is highly cusped and light sintering provided only partial bonding between powders, the porous structure was brittle and prone to crack propagation at low stresses, especially under fatigue conditions due to stress concentrations at powder necks. This foam structure with a porosity of 30% exhibits an elastic modulus in the range of cortical bone but higher porosity (more than 40%) is impractical due to insufficient strength.

To increase porosity while enhancing the strength, a transient liquid phase (Ti-Si) was introduced to substantially improve bonding between powders. A gradient in porosity of this structure was also demonstrated by sintering stacked layers of different powder sizes (Thieme *et al.*, 2001). Packed Ti bead structures were also sintered as coatings on the surfaces of monolithic Ti implants (Fig. 7.2a). Even though the porosity level is low in this approach, Ti bead-coated implants seemed applicable for tracheal prostheses in long-term *in vivo* testing (Schultz *et al.*, 2007). This type of foam structure is also available in the biomedical market in the form of ear, nose, throat, and hip implants.

In the case of random short fiber packing, an isotropic orientation of pore channels is more difficult to achieve due to the anisotropic shape of the solid phase. A maximum porosity of 50% is attainable at the pore size of 100–700 μm, due to fear of low bond strength at wire-contacting sites and possible collapse of the packing structure. However, Ti fiber mesh scaffolds have shown great success to promote long-term implant fixation, as reviewed elsewhere (Alvarez and Nakajima, 2009).

Recently, a CP Ti entangled-fiber structure was made by wire-twisting to form a bundle of twisted wires, followed by compaction in a steel mold and sintering. Either non-woven or quasi-ordered architectures can be achieved with this approach (He *et al.*, 2012). In the non-woven architecture, the pores have irregular sizes and shapes but the material is macroscopically homogeneous (Fig. 7.2b). High porosity up to 80% is attainable with this approach, and is determined by the compaction pressure. The average pore size in

(a) (b)

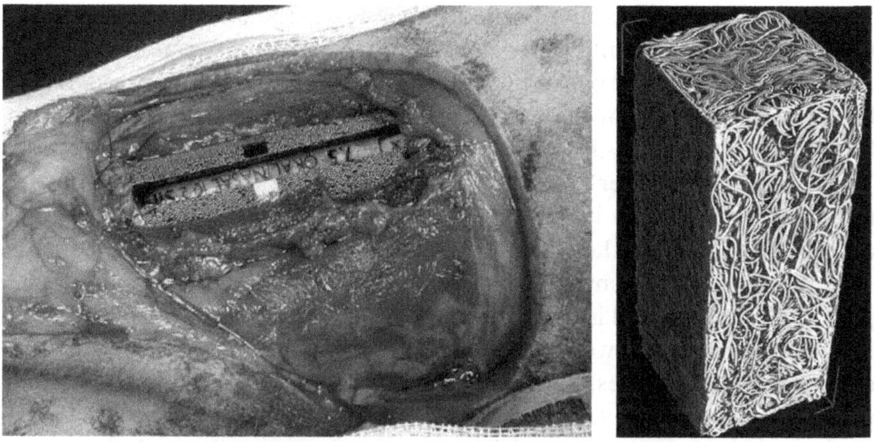

7.2 Foams with residual porosity in randomly packed structure; (a) Ti-bead foam structure used as prosthesis inserted and fixed in the sheep. (*Source*: Schultz *et al.*, 2007, with kind permission from Springer Science and Business Media.) and (b) CPTi entangled wires shown in the reconstructed 3D image of a rectangular prism-like sample (He *et al.*, 2012). (*Source*: Reprinted from *Journal of the Mechanical Behavior of Biomedical Materials*, 5, He *et al.* 'Porous titanium materials with entangled wire structure for load-bearing biomedical applications' pp. 16–31, Copyright (2011), with permission from Elsevier.)

wire entanglement depends on the porosity and wire diameter. For example, a wire diameter of 80 μm provides homogeneous porosity of 48% with a majority (70%) of pore sizes below 100 μm. As the porosity is increased to 82%, the proportion of pores under 100 μm drops to less than 25%. Due to the structural flexibility of continuous fiber structures, low modulus, and good toughness under tensile loading are possible. In addition, the tensile strain to failure is 15–40%, larger than that of the bone, and is due to local wires unwinding toward the tensile load. Despite these advantages, however, mismatch in mechanical properties of this structure is a concern as the tensile strength of the foam (48–108 MPa) with 40% porosity is appropriate for mimicking natural cortical bone (50–150 MPa (Wang *et al.*, 2010)), but the modulus is very low (0.04–1.2 GPa) and is comparable to cancellous bone.

7.5.2 Non-homogeneous residual porous structure

As described above, a porous structure is formed by the interstices within a preform of solid metal powders that are partially sintered. In this second type of structure, a smaller size of metal powders (typically < 60 μm) is used, easing the sintering process and often resulting in low porosity (<40%) and small pore sizes. Unlike the packed Ti bead structure, where the homogenous strut density is controlled directly by the starting material, struts in this foam are composed of packed powders. Two types of starting powders can be used: pre-alloyed powders and elemental powders. For example, Ti64 can be fabricated from pre-alloyed Ti64 powders versus a mixture of Ti, Al, and V powders. The former requires solid-state sintering to bond the powder at temperatures below its melting point, while the latter undergoes *in situ* alloying to synthesize the alloy at the same time as the structure is formed.

In the case of conventional sintering (CS), porosity results from incomplete densification and is controlled by sintering temperature and pressure. Partial sintering by lowering the processing temperature and/or pressure during consolidation results in an increase in porosity, interconnectivity, and pore size as well as a decrease in Young's modulus. However, pore shape tends to shift from a more rounded to a more irregular shape. As an example of this type of foam, a grade-4 CP Ti foam (40% porosity) formed at a low compacting pressure of 38.5 MPa and sintered at 1000–1100°C was found to best fit the stiffness of 20 GPa of cortical bone (Torres *et al.*, 2011). Higher porosity of up to 78% was possible, but with a risk of pore collapse (Wen *et al.*, 2001).

Other sophisticated processes such as hot isostatic pressing (HIP), hot pressing, spark plasma sintering (SPS), and laser sintering, can be used to sinter powder compacts with higher energy heating sources but usually resulting in a lower range of porosity. Laser sintering is very useful for surface modification as powders are partially melted and deposited (<20% porosity) on to a solid implant (Traini *et al.*, 2008). Similarly, plasma spraying was used to form rough and irregular surfaces with limited and non-interconnected porosity. Bulk structures with large pores (>100 μm) are impractical with these methods. To produce near-net shape products, metal powders can be mixed in a metal injection molding (MIM) process with polymeric binders that melt at low temperatures and carry the metal powders under pressure into an appropriately shaped mold, where they can be subjected to a high temperature sintering. Higher porosity is achieved with this method after the binder is removed but with a risk of high shrinkage and low strength.

In the case of solid-state reaction of elemental powders, a higher degree of porosity is formed by the Kirkendall effect as part of a rapid exothermic reaction (Fig. 7.3). For example, the self-propagating high temperature synthesis (SHS) is used to produce NiTi foams by ignition of a compressed elemental Ni and Ti powder mixture (Barrabes *et al.*, 2008). These foams have medium porosity of about 60% (of which about 72% of pores are open) and provide high strength, comparable modulus to bone, and a shape memory effect (Chu *et al.*, 2005). Since the exothermic reaction between Ni and Ti is

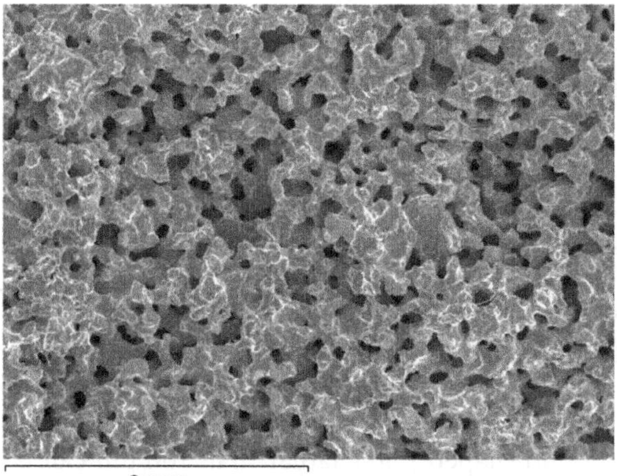

7.3 A reaction-based (SHS) NiTi foam structure (Barrabes *et al.*, 2008). (*Source*: Reprinted from *Materials Science & Engineering C-Biomimetic and Supramolecular Systems,* 28, Barrabes *et al.*, 'Mechanical properties of nickel-titanium foams for reconstructive orthopaedics', pp. 23–27, Copyright (2007), with permission from Elsevier.)

rapid and consolidation is complete within minutes, controlling the porous structure via controlling the process parameters and composition is difficult. However, the relatively lower sintering temperature, lower cost, and more versatile composition gained from elemental metallurgy is favorable. Using a similar concept, microwave (Tang *et al.*, 2011) and laser sintering (Shishkovsky *et al.*, 2008) can also be used to facilitate the production of NiTi foams with medium-to-low porosity in a short period of time, at low temperature and high heating rates. Furthermore, low-pressure hot pressing of Ti and Al-40 V powders followed by subsequent heat treatment at 1020°C can be used as an alternative to the expensive, high temperature and high pressure HIP process of pre-alloyed Ti64 powders (Ye *et al.*, 2010).

7.6 Generation II: foams with built-in secondary porous structure

A secondary porous structure in addition to the primary intrinsic porous structure can be made by introducing foreign additives (gas phases or solid phases) into the powder compact, before consolidation using the same routes as in Generation I foams. Before, during or after densification, these extrinsic phases are removed, and leave behind pores with their corresponding characteristics. Gas phases are present as bubbles inside a material compact, and as a result create rounded pores which may or may not be interconnected with each other and the surface. Solid phases can be prepared in various geometries so that, when used in a material compact and then removed, the pore structure replicates their shapes. Extensive studies in this foam generation allow a wide selection in porous titanium structures.

7.6.1 Round gas-formed porous structures

Secondary porous structures can be generated by gaseous phases via two routes: gaseous blowing agents (e.g., titanium hydride, TiH_2) and inert gases (e.g., argon). Blowing agents can be used to replace basic material powder (TiH_2 instead of Ti) or mixed into the powders before sintering or reactive synthesis. TiH_2 decomposes at low temperature (400°C) and releases hydrogen gas, which is then partially trapped within the compact. A porosity of 40% consisting of round-shaped pores larger than 100 μm can be formed in NiTi using such a method (Sadrnezhaad and Hosseini, 2009). One advantage of the rounded pores is reduced stress concentrators within the foam, which is preferable in avoiding mechanical failure and allows a strength and stiffness of 255 MPa and 20 GPa, respectively, to be achieved in foams with 40% porosity. This technique can also be used to form a porous surface layer on a solid implant, for example, by selectively laser melting a mixture of NiTi and TiH_2 powders on a NiTi core surface (Man *et al.*, 2005). However, the inherent limitations of using blowing agents are residual contamination and impurity phases as well as limited control over the size, shape, and distribution of the porosity. Process optimization and careful selection of blowing agents are therefore required.

Another approach to gas foaming is to isostatically press heated metal powders in a closed canister containing an inert gas. The pressurized gas during this HIP process is then trapped as individual pores within a densified metallic matrix and can subsequently expand into larger pores upon depressurization. The secondary pores are round and independent of the original powder preform (Fig. 7.4). Normally, porosity less than 20% can be produced in this way before the pressurized gas begins to escape through link-up of pores. Titanium and NiTi foams with porosities of 44% and 20%, respectively (resulting from entrapped gas of only 1%) can be achieved by performing the expansion under isothermal creep and superelastic conditions, respectively (Murray and Dunand, 2006; Oppenheimer and Dunand, 2005).

7.6.2 Controlled regular pore porous structure

Porous structures with determined pore size, shape, and volume fractions can be produced by using the solid space holder (or spacer) method. Space holders, usually in powder form, are uniformly mixed and arranged within the metal powder compact before being subjected to the PM densification

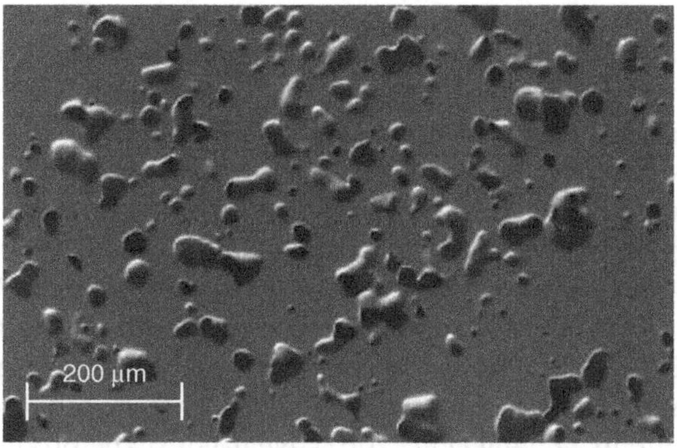

7.4 Scanning electron microscope (SEM) image of argon gas expansion-based foam structure containing 17% porosity after foaming for 170 h at 1200°C (Oppenheimer and Dunand, 2009). (*Source*: Reprinted from *Materials Science and Engineering A-Structural Materials Properties Microstructure and Processing,* 523, Oppenheimer *et al.*, 'Porous NiTi by creep expansion of argon-filled pores', pp. 70–76, Copyright (2012), with permission from Elsevier.)

processes described above (e.g., CS, HIP, SPS). An interconnected porous structure replicating the structure of the space holder material can then be created by removing the space holder after consolidation. The geometries and fraction of space holders mixed into the metal powders generally map directly to the pore geometries and porosity, respectively, within the consolidated metal part. The basic requirements for this method are that the space holders must be inert to the base materials, exhibit sufficient strength throughout the compacting process, and be easily removed after processing without residual contamination.

The most common space holders for producing Ti-based foams include inorganic materials such as ammonium bicarbonate (NH_4HCO_3) for NiTi (Wu *et al.*, 2007), CP Ti, TiNbSn (Nouri *et al.*, 2011), TiNbZr, and Ti-Mg composites (Wang *et al.*, 2011); sodium chloride (NaCl) for NiTi (Bansiddhi and Dunand, 2008); carbamide for CP Ti (Sharma *et al.*, 2011) and Ti64 (Bhattarai *et al.*, 2008); polymeric materials such as polymethyl methacrylate (PMMA) for CP Ti (Li *et al.*, 2012a); and metals, such as Mg for Ti and NiTi (Aydogmus and Bor, 2009; Esen and Bor, 2007), and Nb for NiTi (Bansiddhi and Dunand, 2011). Generally, spacer-defined pore sizes of 100–400 μm, accompanied by residual micropores of about 10 μm size, and a total porosity of 30–80%, can be achieved in this foam category. Since the pore geometry is directly derived from the geometry of the space holder, spherical, cubical, rectangular, acicular, and irregular pores can be created.

However, the overall pore structure and its mechanical properties are strongly influenced in each metal/space holder system by the compaction efficiency of the mixture, which depends on numerous other factors (Tuncer *et al.*, 2011). For example, increasing the size of the space holder may increase total porosity and strut thickness, but may also decrease strut density and mechanical properties due to poor compaction and resultant poor sintering. It also diminishes the number of spacer contacts, the size of the windows between pores and the specific surface area at a given relative density and these may affect bone ingrowth efficiency and load bearing functionality.

Despite the uniqueness of each material/spacer system, foams with porosity in the range 50–70% show comparable mechanical properties to bone, with strengths of 1–100 MPa and stiffnesses of 0.5–10 GPa. Gradient porosity can also be prepared by combining layers of material having different space holder sizes or contents (Zhang *et al.*, 2007). Near-net shape products with about 50% porosity have been proposed through use of MIM on Ti/binder/space holder mixtures, or gel casting of TiH_2/polymer gel/space holder into net shaped molds prior to sintering (Erk *et al.*, 2008; Kohl *et al.*, 2009). Recently, a NiTiCu foam was produced by liquid metal cast-replication of a SrF_2 space holder preform, another promising route to near-net shaping. In spite of the high reactivity and strong contamination tendencies of NiTi-based alloys, this method achieved about 60% porosity, with a superelastic effect and low stiffness (Young *et al.*, 2012). Figure 7.5 shows various pore structures generated by space holder methods. In an interesting extension of the traditional spacer approach, Ti and NiTi nanoparticles were used as self-templates: first stabilized as wet foams and emulsions, then dried and sintered. Submicron pore sizes and thin struts were produced throughout the bulk metallic foams (Studart *et al.*, 2012).

7.6.3 3D Interconnecting pore channels

Uniform 3D porous structures with high porosity and high pore interconnectivity can be pre-designed and produced by replication of a sacrificial polymeric pattern or sponge. In this method, a regular 3D wax pattern is generated using a rapid prototyping machine, embedded in refractory powders and removed by melting, thereby creating a mold whose cavity directly matches the original 3D wax pattern. Liquid titanium is then cast into the mold to produce an exact replica of the wax pattern. Foams with 60% porosity and large pore sizes can be made by this method as shown in Fig. 7.6a. However, the method suffers from the traditional drawbacks of titanium casting such as contamination, reaction with mold materials, and gradients in the cast microstructure (Lopez-Heredia *et al.*, 2008).

In a related method, a powder metallurgy approach was used to replicate a sponge structure by sintering a Ti-slurry-coated polymeric sponge. After sintering, the replicated titanium foam nearly resembled the network-like structure with hollow struts (Quadbeck *et al.*, 2011). In the case where TiH_2

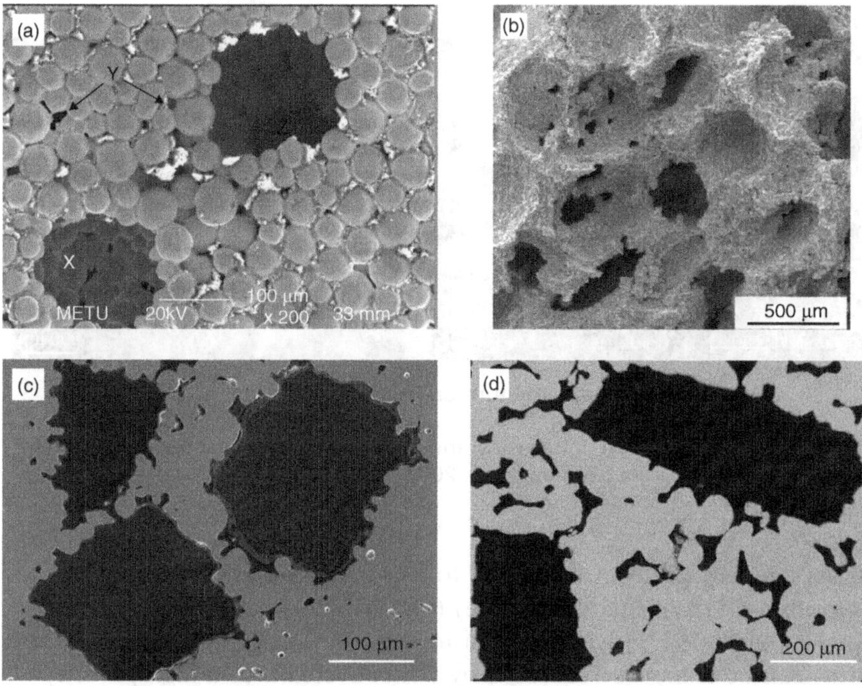

7.5 Pore structure of NiTi generated by different space holders; (a) Mg spheres (Asik *et al.*, 2012, with kind permission from Trans Tech Publications.), (b) rounded carbamide powders (Niu *et al.*, 2009). (*Source*: Reprinted from *Materials Science and Engineering A-Structural Materials Properties Microstructure and Processing,* 506, Niu *et al.*, 'Processing and properties of porous titanium using space holder technique', pp. 148–151, Copyright (2009), with permission from Elsevier.), (c) 'blocky' NaCl powders (Bansiddhi and Dunand, 2008). (*Source*: Reprinted from *Acta Biomaterialia,* 4, Bansiddhi and Dunand, 'Shape-memory NiTi foams produced by replication of NaCl space-holders', pp. 1996–2007, Copyright (2008), with permission from Elsevier Elsevier.), and (d) short Nb wires. (*Source*: Courtesy of Ampika Bansiddhi, Kasetsart University, 2012.)

was used instead of Ti, TiH_2 converted into Ti at high sintering temperature without contamination and provided a Ti wall with high porosity. A titanium sponge of 70–75% porosity provides compressive strength of 18–24 MPa (Cachinho and Correia, 2008; Lee *et al.*, 2009) and very low modulus. Interconnected high porosity (80%) foams with uniform porous structures can also be fabricated by chemical reversion. For example, Ti foams converted by electrochemical reduction of porous TiO_2 preforms built by a gel casting process show large pore sizes of 300 μm as well as micropores, a low modulus of 1 GPa, and a strength of 8 MPa (Singh *et al.*, 2010b). Another example is 3D interconnecting NiTi foams chemically converted from a

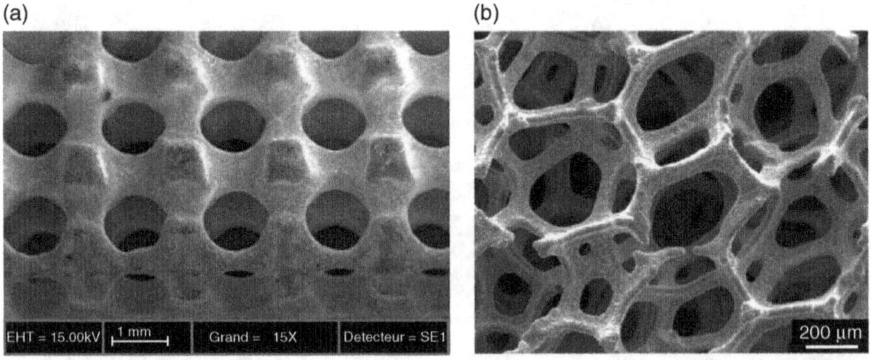

7.6 Replicated foam structure produced by different methods: (a) RP wax pattern + casting. (*Source*: Lopez-Heredia *et al.*, 2008, with kind permission from John Wiley and Sons.) (b) Chemical conversion of Ni foam. (*Source*: Gotman *et al.*, 2010, with kind permission from John Wiley and Sons.)

Ni sponge-like preform subjected to reactive interdiffusion with Ti ions (decomposed from titanium iodide) (Gotman, 2010). The high volumetric porosity (up to 93%) with a uniform pore size of 200–300 μm and dense struts can be achieved (Fig. 7.6b) by 1 h conversion of a high porosity Ni foam (i.e., 98%). A foam with 75% porosity produced by this method has a modulus of 3.5 GPa and relatively high yield strength of 40 MPa.

7.6.4 Elongated porous structures

Unidirectional narrow pore channels, or elongated pores, have been developed to mimic the pore architecture of bone, which accounts for its structural and mechanical anisotropy. The key processes used in these cases utilize gas expansion and space holder techniques. Anisotropic expansion of entrapped gas in densified bundles of solid titanium wires or solid titanium powders has been used (Spoerke *et al.*, 2007). High pressure deformation processes such as rolling, extruding, or uniaxial pressing have also been applied to elongate high pressure pores in a titanium compact, prior to or during its high temperature expansion (Davis *et al.*, 2001; Dunand, 2004). The only liquid-state route of this kind, the zone melting hydrogen process, was used to form elongated pores by drawing reacting gas along a molten NiTi ingot during solidification (Sugiyama *et al.*, 2007).

Using solid space holders instead of gas, a cylindrical preform made of an organic space holder coated with a mixture of titanium powders and binder was co-extruded to elongate the space holders and thus produce elongated pores after space holder removal (Dunand, 2004). Elongated ice space holders were created *in situ* inside an aqueous suspension of titanium powders

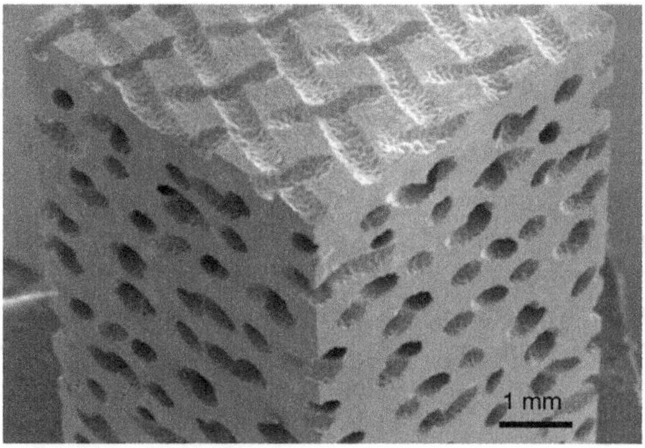

7.7 Elongated NiTi foam structure made by electrochemical dissolution of steel mesh (Neurohr and Dunand, 2011). (*Source*: Reprinted from *Acta Biomaterialia*, 7, Neurohr *et al.*, 'Shape-memory NiTi with two-dimensional networks of micro-channels', pp. 1862–1872, Copyright (2010), with permission from Elsevier.)

by using directional freezing. After sublimation of the ice and sintering of the metal, pores several millimeters in length and with 100 micron width were produced. A high compressive strength (40–60 MPa for 60% porosity) was shown, but with signs of embrittlement, possibly due to oxidation (Chino and Dunand, *et al.*, 2008). Camphene was cast using a uniaxial growth mode to form an elongated template, where suspended titanium powders in a camphene-based slurry migrated out of the growing solid to form a reverse aligned porous structure. Porosity as high as 69% was produced, with large pore sizes and a strength of 121 MPa (Yook *et al.*, 2012). Metal wires 100–500 μm in length can also be consolidated with Ti-based powders, followed by removal via sublimation (in the case of Mg wires in a Ti compact, Lee *et al.*, 2013) or electrochemical dissolution (in the case of steel wires in a Ti, Jorgensen *et al.* 2011, or NiTi compact, Neurohr and Dunand *et al.*, 2011). The latter processing route provides exact control over pore size, pore shape, and connectivity as shown in Fig. 7.7. Porous structures with hexagonal arrays or mesh arrays are possible by using woven wire space holders.

7.7 Generation III: foams with built-up secondary porous structure

Ti-based powders may be built up into a foam structure using solid free-form fabrication, which selectively positions the powders in 2D using RP methods, building layer-by-layer a 3D part without requiring a mold or sacrificial template. Generally, two porous structures are found within foams of this generation. The primary foam structure consists of residual interparticle porosity with 10–100 μm size, depending on the consolidation method, and is a shared feature with original Generation I foams. The secondary foam structure consists of porosity created directly in a precisely controlled manner and corresponding to the pre-designed model. Powder consolidation can be done after or during layer synthesis, referred to as 'indirect' and 'direct' metal depositions, respectively. Either way, accurate, customized porous structures with well-defined pore size and connectivity were produced, rather than most porous structures with a pore size range obtained in Generation II foams. Functionally graded porosity and external near-net shaping are intrinsic features of this foam generation, as computer-aided design (CAD) models may be used to assist RP formation, sometimes coupled with 3D micro-computed tomography (μCT) or MRI data of the target prototype.

In the view of the manufacturer, these RP methods display advantages when compared to traditional processes in terms of providing patient-specific internal and external structures with better repeatability, more regularly controlled porosity and pore shapes, sufficient pore interconnectivity and thin wall structures, based on pre-imaging of the patient bone. In addition, they avoid operational problems such as labor intensive processes, expensive post processing, incomplete removal of residual chemicals, use of volatile or hazardous porogenic materials, etc. However, manufacturing challenges remain to be overcome, including high capital costs, low output, difficulty in optimizing the free-flow morphology of the powder, low dimensional accuracy and shrinkage when liquid phases are used, powder surface finishes, and removal of powders left in pores. Nonetheless, this foam generation is quite promising for individual, customized implant design and fabrication.

7.7.1 Indirect RP porous structure: mesh structure

A slurry of 45 micron Ti64 powder was pushed through a nozzle to form fibers, which later were deposited and aligned layer by layer into a 3D structure, followed by drying and sintering (Li *et al.*, 2007). Completely interconnected pores with large sizes in the range of 100–1000 μm and porosities around 40–75% were produced in this 3D fiber deposition (3DP). Pore characteristics (i.e., pore size, pore connectivity) were directly influenced by a combination of fiber diameter, fiber spacing, and fiber orientation. As shown in Fig. 7.8, a meshed structure with porosity of 68% was created by depositing 400 μm fibers with a fiber spacing of 800 μm and a 0/45 lay down pattern. The interconnecting pore sizes are in the same range as the individual pore size. By reducing fiber spacing from 800 to 200 μm with a lay down angle of 0/90, a gradient pore size of 160–660 μm was observed

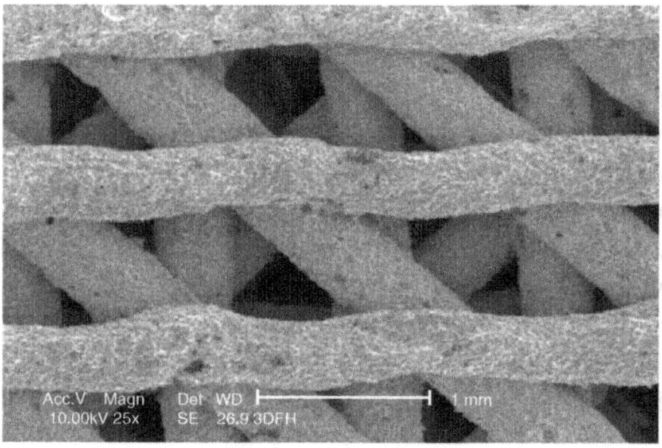

7.8 Fiber deposition-based foam structure produced with fiber spacing of 800 μm and 0/45 lay down pattern (Li *et al.*, 2007). (*Source*: Reprinted from *Biomaterials*, 28, Li *et al.*, 'Bone ingrowth in porous titanium implants produced by 3D fiber deposition', pp. 2810–2820, Copyright (2007), with permission from Elsevier.)

with 53% porosity. *In vivo* testing in a goat lumbar spine model showed a positive effect on bone ingrowth as porosity and pore size increased. This method provides the capability to mimic pore characteristics at a specific implantation site; however, lengthy slurry preparation times and shrinkage of the porous structure during drying and sintering remain challenges.

Another example of an indirect RP structure is the reticulated titanium scroll. As opposed to the usual layer-by-layer built-up structure, a 3D microarchitecture was created by directly printing TiH_2 ink filaments into pliable planar mesh arrays, rolling the layers into a cylinder shape (scroll), and then partial vacuum sintering the scroll to reduce TiH_2 to titanium. The complex ingredients of TiH_2 ink and efficient direct ink writing (DIW) system seem to be keys to success in producing a hierarchical pore size distribution in this manner. However, a more complex folding pattern of the arrays is required for a more complex 3D outer shape (Hong *et al.*, 2011).

7.7.2 Direct RP porous structure

High-energy sources may be used to directly bond Ti-based powders during a deposition process without the need for post sintering. Lasers are commonly used to form the structure of Ti-based foams and NiTi foams, for example, in the so-called laser engineering net shaping (LENS™) and selective laser melting (SLM) methods. An electron beam source was recently applied in an electron beam melting (EBM) process. External shape and internal porous structures are more highly defined than those from indirect fabrication processes; however, more expensive laser or electron gun equipment and inert environments are required. Many foam structures can be obtained from direct RP methods and categorized into two groups: regularly arranged porous structures and lattice-like porous structures.

Regularly arranged RP porous structures can be prepared by using LENS™ equipped with a focused Nd-YAG laser beam. A designed array of powders is deposited into the molten metal pool generated by the laser beam on a surface. The deposited powders solidify rapidly and strongly bond to the layer below. By scanning the laser in a predefined pattern, a regular structure can be built up until a 3D part is formed. Pore characteristics are determined by a combination of processing parameters such as laser power, scan speed, powder feed rate, distance between laser tracks, and upward increment.

Commercially pure Ti foams and equiatomic NiTi alloy foams were demonstrated with powder sizes of 50–150 μm. Rough surfaces and pore sizes up to 800 μm can be found in CP Ti foams, with a majority smaller than 500 μm as shown in Fig. 7.9a (Xue *et al.*, 2007). Net shape, functional hip stems with up to 20% porosity fabricated by LENS™ are displayed in Fig. 7.9b (Krishna *et al.*, 2007). A medium-low porosity (20–50%) with less than two thirds of the porosity being open is observed, perhaps due to liquid phase sintering. By adjusting the processing parameters to minimize liquid formation, porosity up to 70% was demonstrated but with very low modulus. Also, loosening of powders in weakly bonded areas as well as fatigue of the foams could become concerns in load bearing applications.

Commercially pure Ti and Ti64 foams having porosity of only 35–42% and 23–32%, respectively, were said to have suitable moduli for human cortical bone replacement. Similarly, NiTi foams having 12–42% porosity exhibit the desired modulus and up to 4% recovery strain (Bandyopadhyay *et al.*, 2009). *In vitro* human osteoblast cell (HOB) testing showed improvement in cell adhesion and proliferation within the critical pore size of 200 μm, and *in vivo* testing in male Sprague-Dawley rats showed fast rates of tissue generation and integration (Bandyopadhyay *et al.*, 2010).

Many lattice-like porous structures were produced by SLM and EBM methods (Fig. 7.10). A precise laser/electron beam with a small spot size (tens of microns in diameter) was used to selectively melt powders (25–60 μm) in the cross section of a powder bed at the focusing level on to a metal platform. After each successive layer is melted in this way, the platform is lowered and a new powder layer is filled in on top, ready for another selective melting patterned to match the corresponding 2D slice of the foam design or model. When all successive layers are done, a 3D implant with a patient-specific external structure and highly defined internal porous structure with spatial replication is produced.

Several hundred microns pore size and powdery surface roughness are usually observed in this foam structure. In SLM, pore overhanging (due to

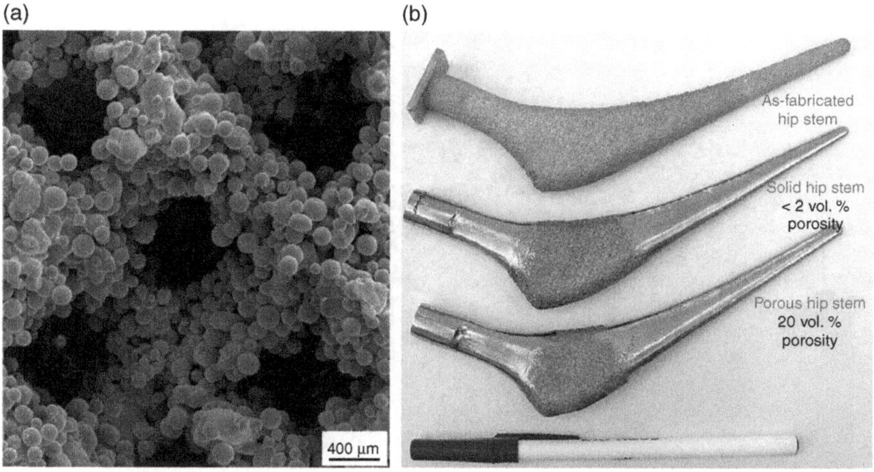

7.9 Regularly arranged foam structure produced by direct RP (LENS™); (a) porous Ti with 58% porosity containing 800 μm pores (Xue *et al.*, 2007). (*Source*: Reprinted from *Acta Biomaterialia,* 3, Xue *et al.*, 'Processing and biocompatibility evaluation of laser processed porous titanium', pp. 1007–1018, Copyright (2007), with permission from Elsevier.) and (b) net shaped, functional porous CP Ti hip stem products with up to 20% porosity (Krishna *et al.*, 2007). (*Source*: Reprinted from *Acta Biomaterialia,* 3, Krishna *et al.*, 'Low stiffness porous Ti structures for load-bearing implants', pp. 997–1006, Copyright (2007), with permission from Elsevier.)

melting) creates errors in pore diameter relative to the design, which limits creation of pore sizes below about 450 μm (Warnke *et al.*, 2009). Micro-CT scans as well as finite element analysis (FEA) are utilized to measure, predict, and adjust the mismatch between the designed and as-product structures; however, only porous structures with identical arrays throughout the structure can be controlled (Van Bael *et al.*, 2011). The benefit of using electron beam instead of laser is a greater energy density thus shorter building time, and better control of strut density and size. Complete melting of powder with no evidence of poor interlayer bonding was found with uneven surface roughness that was proposed to favor cell surface curvature and corrugation.

Due to the ability to produce tailor-made foam structures using these advanced methods, many attempts have been made to pre-design 3D lattice structures that are able to provide predetermined mechanical properties and highly open porosity for biological purposes, including cubic mesh arrays, honeycomb-like structures, rhombic dodecahedra, diamond, and hatched structures. These lattice-like structures allow predictable mechanical behavior which can be manipulated by adjusting structural parameters such as strut size, length, and spacing. However, as the structural parameters

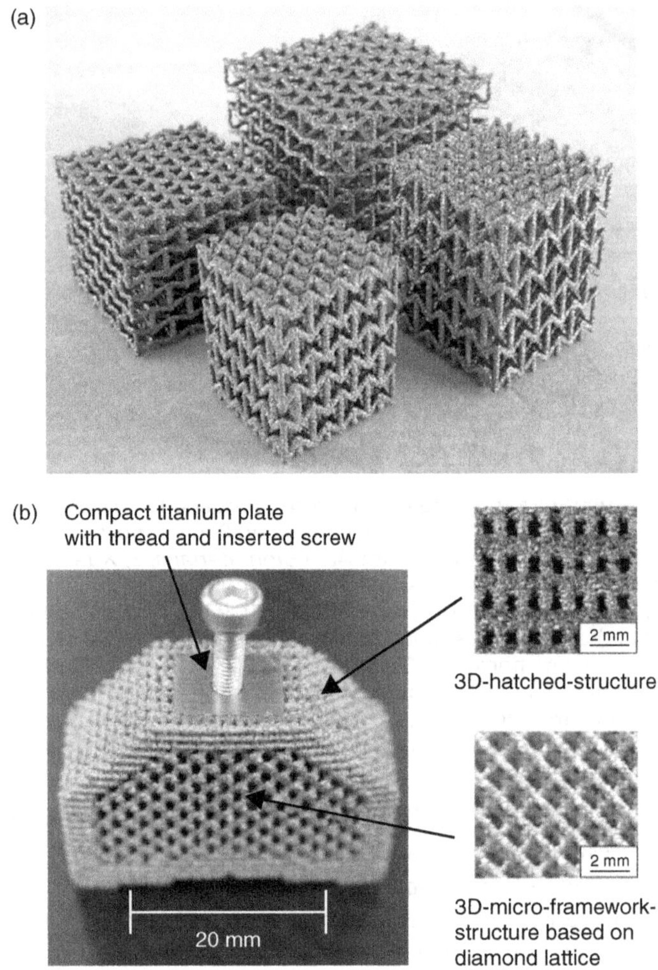

7.10 Lattice-like foam structures produced by direct RP; (a) Ti64 re-entrant lattice auxetic structure fabricated by EBM (Yang *et al.*, 2012). (*Source*: Reprinted from *Acta Materialia*, 60, Yang *et al.*, 'Compressive properties of Ti-6Al-4V auxetic mesh structures made by electron beam melting', pp. 3370–3379, Copyright (2012), with permission from Elsevier.) and (b) Ti64 hybrid structures used in the interbody fusion cage consisting of two different titanium structures and an integrated compact titanium plate with thread for the fixation of implantation instruments (symbolized by a screw). (*Source*: Heinl *et al.*, 2007, with kind permission from John Wiley and Sons.)

also directly control pore characteristics (pore size, porosity) a balance must be struck between biological and mechanical needs.

For instance, 15 mm Ti64 mesh array cubes provide porosities of 50–70%, increased by reducing the number of struts and the strut size (Parthasarathy *et al.*, 2010). A large range of stiffness and strength (0.6–2.9 GPa and 7–163 MPa) allows a wide range of applications. For example, thicker-strut foams (lower porosity) can be used in load bearing regions (i.e., mandible cortical bone and dental abutment reconstruction regions) while thinner struts can be used for replacing trabecular bone and in non-load bearing applications (i.e., cranial replacement). Similarly, Li *et al.* produced cylinders of Ti64 honeycomb-like structure with regular 500 μm pore size and 66% porosity exhibiting mechanical behavior in the same range, for example, strength of 116 MPa and stiffness of 2.5 GPa (Li *et al.*, 2009).

Twelve-facet rhombic dodecahedron cubes with porosity ranging from 60–92% provided a final effective modulus ranging from 23 to 78 MPa, suitable for trabecular bone (Harrysson *et al.*, 2008). Due to uniform deformation through the entire regular structure, the compressive fatigue strength of a Ti64 rhombic dodecahedron array of high porosity 60–85% was found to be greater than the predicted value (Li *et al.*, 2012b). Auxetic (negative Poisson's ratio) Ti64 mesh structures were designed with FEA and produced with 13–30% density. Compressive strengths of 13–76 MPa as well as moduli of 0.5–50 GPa were achieved with modification of strut length being the key controlling parameter (Yang *et al.*, 2012). Heinl *et al.* (2008) designed a diamond lattice structure with 81% porosity and 1.23 mm pore size, and a hatched lattice structure with 61% porosity and 450 micron pore size. The mechanical properties of the former were comparable to trabecular bone while those of the latter fell between trabecular and cortical bone. Three-dimensional micro-framework, functionally graded foams consisting of an inner diamond-like structure and outer frame hatched structure were proposed as a model interbody fusion cage. Anisotropic properties were found, with higher stiffness and strength when loading parallel to the layer-by-layer building direction (Heinl *et al.*, 2008).

7.8 Outlook to next generation Ti-based foams

From the first to the third generation of Ti-based foams, several foam structures have been proposed and fabricated, and many found to fulfill most or all of the structural requirements for replacing bone. The inherent mechanical and biological properties of these foams are well understood and within the acceptable ranges. The trend for future foam structure development does not therefore need to completely rethink the nature of mechanical and biological compatibility, but rather continue along the same trajectory which led to Generation III; that is, to produce foams with highly defined internal porous structures and net shape external structures that can be matched with specific host implantation sites. In addition, the most attractive future foam structures must incorporate a few extra efficiency-related improvements.

First, from a metallurgical perspective, new alloy compositions (in particular new beta titanium alloys) that provide lower modulus will enhance property matching between the implant material and the host bone. Foaming of such materials with thermomechanical processing to control texture will achieve even lower modulus while preserving high strength and fatigue endurance. In the case of NiTi-based alloys, optimization of superelasticity in foams by adjusting compositions and heat treatments can enable very low stiffnesses. Foams of titanium alloys with a self-tunable modulus as well as low modulus Ni-free superelastic and shape memory titanium alloys will also become interesting choices for creating new implant features.

Second, from a structural perspective, more complex hierarchical porous structures are required in order to make a true tailor-made foam implant. Fabrication techniques that present the most *flexibility* and *reliability* in the control of porous structure and external structure will promote individualized implant design. This will open a pathway to a wider range of potential applications since current Ti foams are focused on main body component replacements such as spinal fusion devices and hip implants. For long-term reliability, control of internal structural features such as microporosity in metal struts, surface roughness, and fenestration sizes, is essential to reduce stress concentration sites where fatigue failure, wear, and stress corrosion occur (Niionomi, 2008). Careful design of external structural features such as minimizing implant junctions where titanium–titanium or titanium–dissimilar materials interact can reduce fretting and generation of metallic debris. Developing NiTi implants with self-interlocking mechanisms based on the shape recovery effect is another option for preventing such debris.

Third, from a mechanical perspective, foam structures with *predictability* and *reproducibility* of in-use properties are desirable. This requires not only better process control during fabrication, but also sophisticated modeling and characterization techniques capable of establishing the actual mechanical and biological, as well as biomechanical, behavior of the foam products. Computational techniques such as μCT, numerical modeling (Shen *et al.*, 2006), or engineering finite element (Shen and Brinson, 2007) have been successful in extrapolating and anticipating the mechanical/biomechanical properties of foams. As these methods continue to mature, they will provide even more attractive alternatives for lengthy experimental runs. Even so, modeling and *ex situ* testing must eventually be solidly correlated to fatigue and fracture toughness measurements using patient-specific implants under highly realistic *in vivo* environments, and this will become a must-do procedure for new materials and designs. It should be kept in mind that a balance between mechanical properties and biological performance will always be necessary.

Finally, from a biological perspective, built-in bioactive interfaces on implants will advance implant *biofunctionality* and *long-term stability* by enhancing bone ingrowth and apposition. Since Ti is biotolerant and NiTi is biocompatible but not bioactive, it is necessary to either combine them with other more bioactive biomaterials, or to perform surface modifications. Plasma spraying or sol-gel coating of hydroxyapatite (Cachinho and Correia, 2008), surface-induced mineralization coating of calcium phosphate (Wheeler *et al.*, 1997), hydroxyapatite/titania hybrid coating layers (Lee *et al.*, 2009), bioactive titania nanotubes or nanotitanate coatings (Fan *et al.*, 2011; Wu *et al.*, 2008), resorbable mineral bone cement (Glorius *et al.*, 2011), and chemical modification such as oxidizing heat treatments and ion implantation have been used to provide bioactive surfaces, surface roughness, and wear resistance (Bansiddhi and Dunand, 2008; Shabalovskaya *et al.*, 2008). Other fruitful approaches are hybrid materials such as Ti foam-bioactive nanofiber hybrids (Sargeant *et al.*, 2008a, b), organoapatite layer-deposited Ti foams (Spoerke *et al.*, 2005) and impregnated bioactive silica-polymer nanocomposites (Singh *et al.*, 2010b). *In situ* surface modification or hybrid material production with fewer processing steps would be very attractive.

Up to this point, all of the above areas have been explored, but with different depth and degrees of understanding. It is noteworthy that since the lead time from research laboratory-to-industry is in the neighborhood of 10 years, it is likely that several foam structures reviewed in the previous section will be launched in the near future. However, development of these important and promising materials will not stop there, and any Ti-based foam candidates which successfully integrate the above features should be quickly embraced and studied.

7.9 Future trends

Research-scale titanium-based foams have reached the early stages of maturity. The basic requirements for bone implants have driven development of three generations of titanium-based foams, as reviewed in Sections 7.5–7.7 of this chapter, and allow some visibility into the next generation of titanium foam products, as discussed in Section 7.8. The overall trends incline toward optimization of the current processing technologies, and at the same time innovation of novel foam products that are more economically viable and provide tailor-made, long-term implants at an industrial scale.

There are several key areas needing improvement. First is the relationship between the structural features and properties of foams. Each manufacturing process produces unique structural features leading to unique mechanical constraints. Besides the standard structural indices such as pore size, shape, and volume fraction, more subtle factors such as pore fenestration, surface roughness, orientation (if non-spherical), and micro pores present in the foam walls must be considered, as well as the chemistry of the alloy (in particular oxygen content for Ti and NiTi, and Ni/Ti ratio for NiTi) and its heat treatment. Additional mechanical testing, including fatigue and fracture (in particular within the saline environment of the body) is also needed. Suitable modeling (e.g., finite element method) and characterization (e.g., tomography) techniques are necessary to allow accurate prediction of structure–property relationships and define the acceptable limits of each foam structure, and thus its acceptable processing methods. In addition, *in situ* evaluation of the physicochemical properties and mechanical responses of foams under simulated physiological or *in vivo* environments is needed in order to assess osteoinductive potential.

The second key area is design and selection of materials offering greater variety and efficiency for bone implants. New titanium alloys with low modulus but high fatigue resistance, Ni-free titanium alloys, or NiTi alloys with improved shape recovery properties (i.e., shape memory, superelasticity) in porous form, will bring the technology one step closer to mimicking natural bone. A proper selection of NiTi composition should also allow biomechanical self-fixation (i.e., self-expansion) of the implant during deployment in a bone cavity site. Designed gradients in porosity will also offer a range of functionality in an implant, for example, promoting bone implant fixation at a highly porous interface while still providing a fatigue-resistant implant core. Also, pore surface modification approaches (either by surface treatments or a surface composite with other biomaterials) can further enhance biointerface-mediated cell responses and thus prolong implant life.

Finally, successful deployment of porous Ti or NiTi implants requires technologies that allow economic mass production for standard implants, while preserving flexible manufacturing options for tailor-made implants, a market with high and steeply rising demand. In general, minimizing energy usage and maximizing productivity are the keys to leveraging the current lab-scale processes into industrial-scale processes. For tailor-made designs, process controllability in outer shape, porosity, and pore dimensions as well as strut density are quite desirable. Many automated methods based on pre-designed structures (e.g., using computer-aided design and manufacturing (CAD/CAM) and RP technology) were developed in the latest generation of titanium-based foams, yet with unfavorable trade-off in high cost.

A successful integration of these three research trends may fulfill the requirements for the next generation of titanium-based bone implants.

7.10 Sources of further information and advice

The following publications are recommended as useful literature reviews on:

(i) Titanium and NiTi alloys

Geetha, M., Singh, A. K., Asokamani, R. and Gogia, A. K. (2009), 'Ti based biomaterials, the ultimate choice for orthopaedic implants – A review', *Progress in Materials Science,* **54,** 397–425.

Niinomi, M., Nakai, M. and Hieda, J. (2012), 'Development of new metallic alloys for biomedical applications', *Acta Biomaterialia,* **8,** 3888–3903.

Otsuka, K. and Ren, X. (2005), 'Physical metallurgy of Ti-Ni-based shape memory alloys', *Progress in Materials Science,* **50,** 511–678.

Rack, H. J. and Qazi, J. I. (2006), 'Titanium alloys for biomedical applications', *Materials Science & Engineering C-Biomimetic and Supramolecular Systems,* **26,** 1269–1277.

(ii) Metal foams

Banhart, J. (2001), 'Manufacture, characterisation and application of cellular metals and metal foams', *Progress in Materials Science,* **46,** 559-U3.

Gibson, L. J. (2005), 'Biomechanics of cellular solids', *Journal of Biomechanics,* **38,** 377–399.

Lefebvre, L. P., Banhart, J. and Dunand, D. C. (2008), 'Porous metals and metallic foams: Current status and recent developments', *Advanced Engineering Materials,* **10,** 775–787.

(iii) Titanium and NiTi foams

Bansiddhi, A., Sargeant, T. D., Stupp, S. I. and Dunand, D. C. (2008), 'Porous NiTi for bone implants: A review', *Acta Biomaterialia,* **4,** 773–782.

Dunand, D. C. (2004), 'Processing of titanium foams', *Advanced Engineering Materials,* **6,** 369–376.

Elahinia, M. H., Hashemi, M., Tabesh, M. and Bhaduri, S. B. (2012), 'Manufacturing and processing of NiTi implants: A review', *Progress in Materials Science,* **57,** 911–946.

Ryan, G., Pandit, A. and Apatsidis, D. P. (2006), 'Fabrication methods of porous metals for use in orthopaedic applications', *Biomaterials,* **27,** 2651–2670.

Shabalovskaya, S. A. (2002), 'Surface, corrosion and biocompatibility aspects of Nitinol as an implant material', *Bio-Medical Materials and Engineering,* **12,** 69–109.

Singh, R., Lee, P. D., Dashwood, R. J. and Lindley, T. C. (2010), 'Titanium foams for biomedical applications: a review', *Materials Technology,* **25,** 127–136.

Wen, C. E., Xiong, J. Y., Li, Y. C. and Hodgson, P. D. (2010), 'Porous shape memory alloy scaffolds for biomedical applications: a review', *Physica Scripta,* T139.

7.11 References

Abdel-Hady, M., Hinoshita, K. and Morinaga, M. (2006) 'General approach to phase stability and elastic properties of beta-type Ti-alloys using electronic parameters', *Scripta Materialia,* **55**(5), 477–480.

Alvarez, K. and Nakajima, H. (2009) 'Metallic scaffolds for bone regeneration', *Materials,* **2**(3), 790–832.

Aşık, E. E., Nakaş, G. I. and Bor, A. S. (2012) 'The processing and fatigue characterization of TiNi and Ti-6Al-4V porous material', *Key Engineering Materials.* **493–494**, 930–935.

Assad, M., Chernyshov, A. V., Jarzem, P., Leroux, M. A., Coillard, C., Charette, S. and Rivard, C. H. (2003) 'Porous titanium-nickel for intervertebral fusion in a sheep model: Part 2. Surface analysis and nickel release assessment', *Journal of Biomedical Materials Research Part B-Applied Biomaterials,* **64B**(2), 121–129.

Aydogmus, T. and Bor, S. (2009) 'Processing of porous TiNi alloys using magnesium as space holder', *Journal of Alloys and Compounds,* **478**(1–2), 705–710.

Bandyopadhyay, A., Espana, F., Balla, V. K., Bose, S., Ohgami, Y. and Davies, N. M. (2010) 'Influence of porosity on mechanical properties and *in vivo* response of Ti6Al4V implants', *Acta Biomaterialia,* **6**(4), 1640–1648.

Bandyopadhyay, A., Krishna, B. V., Xue, W. C. and Bose, S. (2009) 'Application of laser engineered net shaping (LENS) to manufacture porous and functionally graded structures for load bearing implants', *Journal of Materials Science-Materials in Medicine,* **20**, 29–34.

Banhart, J. (2001) 'Manufacture, characterisation and application of cellular metals and metal foams', *Progress in Materials Science,* **46**(6), 559–632.

Bansiddhi, A. and Dunand, D. C. (2008) 'Shape-memory NiTi foams produced by replication of NaCl space-holders', *Acta Biomaterialia,* **4**(6), 1996–2007.

Bansiddhi, A. and Dunand, D. C. (2011) 'Processing of NiTi Foams by transient liquid phase sintering', *Journal of Materials Engineering and Performance,* **20**(4–5), 511–516.

Bansiddhi, A., Sargeant, T. D., Stupp, S. I. and Dunand, D. C. (2008) 'Porous NiTi for bone implants: A review', *Acta Biomaterialia,* **4**(4), 773–782.

Barrabes, A., Sevilla, P., Planell, J. A. and Gil, F. J. (2008) 'Mechanical properties of nickel-titanium foams for reconstructive orthopaedics', *Materials Science and Engineering C-Biomimetic and Supramolecular Systems,* **28**(1), 23–27.

Bhattarai, S. R., Khalil, K. A. R., Dewidar, M., Hwang, P. H., Yi, H. K. and Kim, H. Y. (2008) 'Novel production method and in-vitro cell compatibility of porous Ti-6Al-4V alloy disk for hard tissue engineering', *Journal of Biomedical Materials Research Part A,* **86A**(2), 289–299.

Biesiekierski, A., Wang, J., Gepreel, M. A. H. and Wen, C. (2012) 'A new look at biomedical Ti-based shape memory alloys', *Acta Biomaterialia,* **8**(5), 1661–1669.

Blackwood, D. J. (2003) 'Biomaterials: Past successes and future problems', *Corrosion Reviews,* **21**(2–3), 97–124.

Bobyn, J., Mortimer, E. S., Glassman, A. H., Engh, C. A., Miller, J. E. and Brooks, C. E. (1992) 'Producing and avoiding stress shielding: Laboratory and clinical observations of noncemented total hip arthroplasty', *Clinical Orthopaedic Related Research,* **274**, 79–96.

Cachinho, S. C. P. and Correia, R. N. (2008) 'Titanium scaffolds for osteointegration: Mechanical, *in vitro* and corrosion behaviour', *Journal of Materials Science-Materials in Medicine,* **19**(1), 451–457.

Chino, Y. and Dunand, D. C. (2008) 'Directionally freeze-cast titanium foam with aligned, elongated pores', *Acta Materialia,* **56**(1), 105–113.

Choubey, A., Basu, B. and Balasubramaniam, R. (2004) 'Tribological behaviour of Ti-based alloys in simulated body fluid solution at fretting contacts', *Materials Science and Engineering A-Structural Materials Properties Microstructure and Processing,* **379**(1–2), 234–239.

Chu, C. L., Chung, C. Y., Lin, P. H. and Wang, S. D. (2005) 'Fabrication and properties of porous NiTi shape memory alloys for heavy load-bearing medical applications', *Journal of Materials Processing Technology,* **169**(1), 103–107.

Cook, S. D. and Thomas, K. A. (1991) 'Fatigue failure of non cemented porous coated implants. A retrieval study'. *The Journal of Bone and Joint Surgery,* **73B**, 20–24.

Davis, N. G., Teisen, J., Schuh, C. and Dunand, D. C. (2001) 'Solid-state foaming of titanium by superplastic expansion of argon-filled pores', *Journal of Materials Research,* **16**(5), 1508–1519.

Donachie, M. J. (2000) *Titanium: A Technical Guide.* 2nd Edition. Materials Park, Ohio. ASM International.

Dunand, D. C. (2004) 'Processing of titanium foams', *Advanced Engineering Materials,* **6**(6), 369–376.

Eisenbarth, E., Velten, D., Muller, M., Thull, R. and Breme, J. (2004) 'Biocompatibility of beta-stabilizing elements of titanium alloys', *Biomaterials,* **25**(26), 5705–5713.

Elahinia, M. H., Hashemi, M., Tabesh, M. and Bhaduri, S. B. (2012) 'Manufacturing and processing of NiTi implants: A review', *Progress in Materials Science,* **57**(5), 911–946.

Elias, C. N., Lima, J. H. C., Valiev, R. and Meyers, M. A. (2008) 'Biomedical applications of titanium and its alloys', *Journal of Materials,* **60**(3), 46–49.

Elias, L. M., Schneider, S. G., Schneider, S., Silva, H. M. and Malvisi, E. (2006) 'Microstructural and mechanical characterization of biomedical Ti-Nb-Zr(-Ta) alloys', *Materials Science and Engineering A-Structural Materials Properties Microstructure and Processing,* **432**(1–2), 108–112.

Erk, K. A., Dunand, D. C. and Shull, K. R. (2008) 'Titanium with controllable pore fractions by thermo reversible gel casting of TiH_2', *Acta Materialia,* **56**(18), 5147–5157.

Es-Souni, M. and Fischer-Brandies, H. (2005) 'Assessing the biocompatibility of NiTi shape memory alloys used for medical applications', *Analytical and Bioanalytical Chemistry,* **381**(3), 557–567.

Esen, Z. and Bor, S. (2007) 'Processing of titanium foams using magnesium spacer particles', *Scripta Materialia,* **56**(5), 341–344.

Fan, X. P., Feng, B., Weng, J., Wang, J. X. and Lu, X. (2011) 'Processing and properties of porous titanium with high porosity coated by bioactive titania nanotubes', *Materials Letters,* **65**(19–20), 2899–2901.

Fujibayashi, S., Neo, M., Kim, H. M., Kokubo, T. and Nakamura, T. (2004) 'Osteoinduction of porous bioactive titanium metal', *Biomaterials,* **25**(3), 443–450.

Geetha, M., Singh, A. K., Asokamani, R. and Gogia, A. K. (2009) 'Ti based biomaterials, the ultimate choice for orthopaedic implants – A review', *Progress in Materials Science,* **54**(3), 397–425.

Gibson, L. J. (2005) 'Biomechanics of cellular solids', *Journal of Biomechanics,* **38**(3), 377–399.

Glorius, S., Nies, B., Farack, J., Quadbeck, P., Hauser, R., Standke, G. and Stephani, G. (2011) 'Metal foam – bone cement composites: Mechanical and biological properties and perspectives for bone implant design', *Advanced Engineering Materials,* **13**(11), 1019–1023.

Gotman, I. (2010) 'Fabrication of load-bearing NiTi scaffolds for bone ingrowth by Ni foam conversion', *Advanced Engineering Materials,* **12**(7), B320–B325.

Harrysson, O. L. A., Cansizoglu, O., Marcellin-Little, D. J., Cormier, D. R. and West, H. A. (2008) 'Direct metal fabrication of titanium implants with tailored materials and mechanical properties using electron beam melting technology', *Materials Science and Engineering C-Biomimetic and Supramolecular Systems,* **28**(3), 366–373.

He, G., Liu, P. and Tan, Q. B. (2012) 'Porous titanium materials with entangled wire structure for load-bearing biomedical applications', *Journal of the Mechanical Behavior of Biomedical Materials*, **5**(1), 16–31.

Heinl, P., Muller, L., Korner, C., Singer, R. F. and Muller, F. A. (2008) 'Cellular Ti-6Al-4V structures with interconnected macro porosity for bone implants fabricated by selective electron beam melting', *Acta Biomaterialia,* **4**(5), 1536–1544.

Heinl, P., Rottmair, A., Korner, C. and Singer, R. F. (2007) 'Cellular titanium by selective electron beam melting', *Advanced Engineering Materials,* **9**(5), 360–364.

Hong, E., Ahn, B. Y., Shoji, D., Lewis, J. A. and Dunand, D. C. (2011) 'Microstructure and mechanical properties of reticulated titanium scrolls', *Advanced Engineering Materials,* **13**(12), 1122–1127.

Karageorgiou, V. and Kaplan, D. (2005) 'Porosity of 3D biomaterial scaffolds and osteogenesis', *Biomaterials,* **26**(27), 5474–5491.

Keegan, G. M., Learmonth, I.D. and Case, C.P. (2007) 'Orthopaedic metals and their potential toxicity in the arthroplasty patient, a review of knowledge and future strategies', *The Journal of Bone and Joint Surgery*, **89B**(5), 567–573.

Knott, P. T., Mardjetko, S. M., Kim, R. H., Cotter, T. M., Dunn, M. M., Patel, S. T. and Tager, D. S. (2010) 'A comparison of magnetic and radiographic imaging artifact after using three types of metal rods: Stainless steel, titanium, and vitallium', *Spine Journal,* **10**(9), 789–794.

Kohl, M., Habijan, T., Bram, M., Buchkremer, H. P., Stover, D. and Koller, M. (2009) 'Powder metallurgical near-net-shape fabrication of porous NiTi shape memory alloys for use as long-term implants by the combination of the metal injection molding process with the space-holder technique', *Advanced Engineering Materials,* **11**(12), 959–968.

Kokubo, T., Kim, H. M., Kawashita, M. and Nakamura, T. (2004) 'Bioactive metals: Preparation and properties', *Journal of Materials Science-Materials in Medicine,* **15**(2), 99–107.

Krishna, B. V., Bose, S. and Bandyopadhyay, A. (2007) 'Low stiffness porous Ti structures for load-bearing implants', *Acta Biomaterialia*, **3**(6), 997–1006.

Laheurte, P., Prima, F., Eberhardt, A., Gloriant, T., Wary, M. and Patoor, E. (2010) 'Mechanical properties of low modulus beta titanium alloys designed from the electronic approach', *Journal of the Mechanical Behavior of Biomedical Materials,* **3**(8), 565–573.

Lautenschlager, E. P. and Monaghan P. (1993) 'Titanium and titanium alloys as dental materials', *International Dental Journal*, **43**(3), 245–253.

Lee, J. H., Kim, H. E. and Koh, Y. H. (2009) 'Highly porous titanium (Ti) scaffolds with bioactive microporous hydroxyapatite/TiO_2 hybrid coating layer', *Materials Letters,* **63**(23), 1995–1998.

Lee, J. W., Lee, J. S., Kim, M. G. and Hyun, S. K. (2013) 'Fabrication of porous titanium with directional pores for biomedical applications', *Materials Transactions,* **54**(2), 137–142.

Lefebvre, L. P., Banhart, J. and Dunand, D. C. (2008) 'Porous metals and metallic foams: Current status and recent developments', *Advanced Engineering Materials,* **10**(9), 775–787.

Li, B. Q., Yan, F. and Lu, X. (2012a) 'Effect of microstructure on the tensile property of porous Ti produced by powder metallurgy technique', *Materials Science and Engineering A-Structural Materials Properties Microstructure and Processing,* **534**, 43–52.

Li, J. P., Habibovic, P., van den Doel, M., Wilson, C. E., de Wijn, J. R., van Blitterswijk, C. A. and de Groot, K. (2007) 'Bone ingrowth in porous titanium implants produced by 3D fiber deposition', *Biomaterials,* **28**(18), 2810–2820.

Li, S. J., Murr, L. E., Cheng, X. Y., Zhang, Z. B., Hao, Y. L., Yang, R. and Wicker, R. B. (2012b) 'Compression fatigue behavior of Ti-6Al-4V mesh arrays fabricated by electron beam melting', *Acta Materialia,* **60**(3), 793–802.

Li, X., Wang, C. T., Zhang, W. G. and Li, Y. C. (2009) 'Fabrication and characterization of porous Ti6Al4V parts for biomedical applications using electron beam melting process', *Materials Letters,* **63**(3–4), 403–405.

Liu, X. Y., Chu, P. K. and Ding, C. X. (2004) 'Surface modification of titanium, titanium alloys, and related materials for biomedical applications', *Materials Science and Engineering R-Reports*, **47**(3–4), 49–121.

Long, M. and Jack, H. J. (1998) 'Titanium alloys in total joint replacements – a materials science perspective', *Biomaterials*, **19**(18), 1621–1639.

Lopez, M. F., Gutierrez, A. and Jimenez, J. A. (2001) 'Surface characterization of new non-toxic titanium alloys for use as biomaterials', *Surface Science,* **482**, 300–305.

Lopez-Heredia, M. A., Goyenvalle, E., Aguado, E., Pilet, P., Leroux, C., Dorget, M. and Layrolle, P. (2008) 'Bone growth in rapid prototyped porous titanium implants', *Journal of Biomedical Materials Research Part A,* **85A**(3), 664–673.

Man, H. C., Zhang, S., Cheng, F. T. and Guo, X. (2005) 'Laser fabrication of porous surface layer on NiTi shape memory alloy', *Materials Science and Engineering A-Structural Materials Properties Microstructure and Processing,* **404**(1–2), 173–178.

Matsuura, H., Inoue, T., Ogasawara, K., Sasaki, M., Konno, H., Kuzu, Y. and Ogawa, A. (2005) 'Quantitative analysis of magnetic resonance imaging susceptibility artifacts caused by neurosurgical biomaterials: Comparison of 0.5, 1.5. and 3.0 Tesla magnetic fields', *Neurologia Medico-Chirurgica,* **45**(8), 395–398.

Metikos-Hukovic, M., Kwokal, A. and Piljac, J. (2003) 'The influence of niobium and vanadium on passivity of titanium-based implants in physiological solution', *Biomaterials,* **24**(21), 3765–3775.

Murray, N. G. D. and Dunand, D. C. (2006) 'Effect of initial preform porosity on solid-state foaming of titanium', *Journal of Materials Research,* **21**(5), 1175–1188.

Neurohr, A. J. and Dunand, D. C. (2011) 'Shape-memory NiTi with two-dimensional networks of micro-channels', *Acta Biomaterialia,* **7**(4), 1862–1872.

Niinomi, M. (2002) 'Recent metallic materials for biomedical applications', *Metallurgical and Materials Transactions A-Physical Metallurgy and Materials Science,* **33**(3), 477–486.

Niinomi, M., Nakai, M. and Hieda, J. (2012) 'Development of new metallic alloys for biomedical applications', *Acta Biomaterialia,* **8**(11), 3888–3903.

Niu, W. J., Bai, C. G., Qiu, G. B. and Wang, Q. (2009) 'Processing and properties of porous titanium using space holder technique', *Materials Science and Engineering A-Structural Materials Properties Microstructure and Processing,* **506**(1–2), 148–151.

Nouri, A., Hodgson, P. D. and Wen, C. (2011) 'Effect of ball-milling time on the structural characteristics of biomedical porous Ti-Sn-Nb alloy', *Materials Science and Engineering C-Materials for Biological Applications,* **31**(5), 921–928.

Oh, I. H., Nomura, N., Masahashi, N. and Hanada, S. (2003) 'Mechanical properties of porous titanium compacts prepared by powder sintering', *Scripta Materialia,* **49**(12), 1197–1202.

Okazaki, Y. (2001) 'A new Ti-15Zr-4Nb-4Ta alloy for medical applications', *Current Opinion in Solid State and Materials Science,* **5**(1), 45–53.

Oppenheimer, S. M. and Dunand, D. C. (2009) 'Porous NiTi by creep expansion of argon-filled pores', *Materials Science and Engineering A-Structural Materials Properties Microstructure and Processing,* **523**(1–2), 70–76.

Otsuka, K. and Ren, X. (2005) 'Physical metallurgy of Ti-Ni-based shape memory alloys', *Progress in Materials Science,* **50**(5), 511–678.

Ozaki, T., Matsumoto, H., Watanabe, S. and Hanada, S. (2004) 'Beta Ti alloys with low Young's modulus', *Materials Transactions,* **45**(8), 2776–2779.

Parthasarathy, J., Starly, B., Raman, S. and Christensen, A. (2010) 'Mechanical evaluation of porous titanium (Ti6Al4V) structures with electron beam melting (EBM)', *Journal of the Mechanical Behavior of Biomedical Materials,* **3**(3), 249–259.

Quadbeck, P., Kummel, K., Hauser, R., Standke, G., Adler, J., Stephani, G. and Kieback, B. (2011) 'Structural and material design of open-cell powder metallurgical foams', *Advanced Engineering Materials,* **13**(11), 1024–1030.

Rack, H. J. and Qazi, J. I. (2006) 'Titanium alloys for biomedical applications', *Materials Science and Engineering C-Biomimetic and Supramolecular Systems,* **26**(8), 1269–1277.

Ryan, G., Pandit, A. and Apatsidis, D. P. (2006) 'Fabrication methods of porous metals for use in orthopaedic applications', *Biomaterials,* **27**(13), 2651–2670.

Sadrnezhaad, S. K. and Hosseini, S. A. (2009) 'Fabrication of porous NiTi-shape memory alloy objects by partially hydrided titanium powder for biomedical applications', *Materials and Design,* **30**(10), 4483–4487.

Sargeant, T. D., Guler, M. O., Oppenheimer, S. M., Mata, A., Satcher, R. L., Dunand, D. C. and Stupp, S. I. (2008a) 'Hybrid bone implants: Self-assembly of peptide amphiphile nanofibers within porous titanium', *Biomaterials,* **29**(2), 161–171.

Sargeant, T. D., Oppenheimer, S. M., Dunand, D. C. and Stupp, S. I. (2008b) 'Titanium foam-bioactive nanofiber hybrids for bone regeneration', *Journal of Tissue Engineering and Regenerative Medicine*, **2**(8), 455–462.

Schultz, P., Vautier, D., Charpiot, A., Lavalle, P. and Debry, C. (2007) 'Development of tracheal prostheses made of porous titanium: A study on sheep', *European Archives of Oto-Rhino-Laryngology,* **264**(4), 433–438.

Shabalovskaya, S. A. (2002) 'Surface, corrosion and biocompatibility aspects of Nitinol as an implant material', *Bio-Medical Materials and Engineering,* **12**(1), 69–109.

Sharma, M., Gupta, G. K., Modi, O. P., Prasad, B. K. and Gupta, A. K. (2011) 'Titanium foam through powder metallurgy route using acicular urea particles as space holder', *Materials Letters,* **65**(21–22), 3199–3201.

Shen, H. and Brinson, L. C. (2007) 'Finite element modeling of porous titanium', *International Journal of Solids and Structures,* **44**(1), 320–335.

Shen, H., Oppenheimer, S. M., Dunand, D. C. and Brinson, L. C. (2006) 'Numerical modeling of pore size and distribution in foamed titanium', *Mechanics of Materials,* **38**(8–10), 933–944.

Shishkovsky, I. V., Volova, L. T., Kuznetsov, M. V., Morozov, Y. G. and Parkin, I. P. (2008) 'Porous biocompatible implants and tissue scaffolds synthesized by selective laser sintering from Ti and NiTi', *Journal of Materials Chemistry,* **18**(12), 1309–1317.

Singh, R., Lee, P. D., Dashwood, R. J. and Lindley, T. C. (2010a) 'Titanium foams for biomedical applications: A review', *Materials Technology,* **25**(3–4), 127–136.

Singh, R., Lee, P. D., Jones, J. R., Poologasundarampillai, G., Post, T., Lindley, T. C. and Dashwood, R. J. (2010b). 'Hierarchically structured titanium foams for tissue scaffold applications', *Acta Biomaterialia,* **6**(12), 4596–4604.

Spoerke, E. D., Murray, N. G., Li, H. L., Brinson, L. C., Dunand, D. C. and Stupp, S. I. (2005) 'A bioactive titanium foam scaffold for bone repair', *Acta Biomaterialia,* **1**(5), 523–533.

Studart, A. R., Nelson, A., Iwanovsky, B., Kotyrba, M., Kundig, A. A., Dalla Torre, F. H. and Loffler, J. F. (2012) 'Metallic foams from nanoparticle-stabilized wet foams and emulsions', *Journal of Materials Chemistry,* **22**(3), 820–823.

Sugiyama, M., Hyun, S. K., Tane, M. and Nakajima, H. (2007) 'Fabrication of lotus-type porous NiTi shape memory alloys using the continuous zone melting method and tensile property', *High Temperature Materials and Processes*, **26**(4), 297–301.

Tang, C. Y., Zhang, L. N., Wong, C. T., Chan, K. C. and Yue, T. M. (2011) 'Fabrication and characteristics of porous NiTi shape memory alloy synthesized by microwave sintering', *Materials Science and Engineering A-Structural Materials Properties Microstructure and Processing,* **528**(18), 6006–6011.

Thieme, M., Wieters, K. P., Bergner, F., Scharnweber, D., Worch, H., Ndop, J. and Grill, W. (2001) 'Titanium powder sintering for preparation of a porous functionally graded material destined for orthopaedic implants', *Journal of Materials Science-Materials in Medicine,* **12**(3), 225–231.

Torres, Y., Pavon, J. J., Nieto, I. and Rodriguez, J. A. (2011) 'Conventional powder metallurgy process and characterization of porous titanium for biomedical applications', *Metallurgical and Materials Transactions B-Process Metallurgy and Materials Processing Science,* **42**(4), 891–900.

Traini, T., Mangano, C., Sammons, R. L., Mangano, F., Macchi, A. and Piattelli, A. (2008) 'Direct laser metal sintering as a new approach to fabrication of an isoelastic functionally graded material for manufacture of porous titanium dental implants', *Dental Materials,* **24**(11), 1525–1533.

Tuncer, N., Arslan, G., Maire, E. and Salvo, L. (2011) 'Investigation of spacer size effect on architecture and mechanical properties of porous titanium', *Materials Science and Engineering A-Structural Materials Properties Microstructure and Processing,* **530**, 633–642.

Van Bael, S., Kerckhofs, G., Moesen, M., Pyka, G., Schrooten, J. and Kruth, J. P. (2011) 'Micro-CT-based improvement of geometrical and mechanical controllability of selective laser melted Ti6Al4V porous structures', *Materials Science and Engineering A-Structural Materials Properties Microstructure and Processing,* **528**(24), 7423–7431.

Walker, P. R., LeBlanc, J. and Sikorska, M. (1989) 'Effects of aluminum and other cations on the structure of brain and liver chromatin', *Biochemistry*, **28**, 3911–3915.

Wang, K., Gustavson, L. and Dumbleton, J. (1993). 'The characterization of Ti-12Mo-6Zr-2Fe. A new biocompatibility titanium alloy developed for surgical implants'. *Beta titanium in the 1990s,* The Mineral, Metals and Materials Society, Warrendale, Pennsylvania, 2697–2704.

Wang, X., Nyman, J. S., Domg, X., Leng, H. and Reyes, M. (2010), 'Fundamental biomechanics in bone tissue engineering', *Synthesis Lectures on Tissue Engineering,* **2**(1), 1–225.

Wang, Y. Q., Tao, J., Zhang, J. L. and Wang, T. (2011) 'Effects of addition of NH_4HCO_3 on pore characteristics and compressive properties of porous Ti-10%Mg composites', *Transactions of Nonferrous Metals Society of China,* **21**(5), 1074–1079.

Warnke, P. H., Douglas, T., Wollny, P., Sherry, E., Steiner, M., Galonska, S. and Sivananthan, S. (2009) 'Rapid prototyping: Porous titanium alloy scaffolds produced by selective laser melting for bone tissue engineering', *Tissue Engineering Part C-Methods,* **15**(2), 115–124.

Wen, C. E., Mabuchi, M., Yamada, Y., Shimojima, K., Chino, Y. and Asahina, T. (2001) 'Processing of biocompatible porous Ti and Mg', *Scripta Materialia,* **45**(10), 1147–1153.

Wheeler, D. L., Campbell, A. A., Graff, G. L. and Miller, G. J. (1997) 'Histological and biomechanical evaluation of calcium phosphate coatings applied through surface-induced mineralization to porous titanium implants', *Journal of Biomedical Materials Research,* **34**, 539–543.

Wu, S. L., Chung, C. Y., Liu, X. M., Chu, P. K., Ho, J. P. Y., Chu, C. L. and Luk, K. D. K. (2007) 'Pore formation mechanism and characterization of porous NiTi shape memory alloys synthesized by capsule-free hot isostatic pressing', *Acta Materialia,* **55**(10), 3437–3451.

Wu, S. L., Liu, X. M., Hu, T., Chu, P. K., Ho, J. P. Y., Chan, Y. L. and Luk, K. D. K. (2008) 'A biomimetic hierarchical scaffold: Natural growth of nanotitanates on three-dimensional microporous Ti-based metals', *Nano Letters,* **8**(11), 3803–3808.

Xue, W., Krishna, B. V., Bandyopadhyay, A. and Bose, S. (2007) 'Processing and biocompatibility evaluation of laser processed porous titanium', *Acta Biomaterialia,* **3**(6), 1007–1018.

Yang, L., Harrysson, O., West, H. and Cormier, D. (2012). 'Compressive properties of Ti-6Al-4V auxetic mesh structures made by electron beam melting', *Acta Materialia,* **60**(8), 3370–3379.

Ye, B., Matsen, M. R. and Dunand, D. C. (2010) 'Enhanced densification of Ti-6Al-4V powders by transformation-mismatch plasticity', *Acta Materialia,* **58**(11), 3851–3859.

Yook, S. W., Jung, H. D., Park, C. H., Shin, K. H., Koh, Y. H., Estrin, Y. and Kim, H. E. (2012) 'Reverse freeze casting: A new method for fabricating highly porous titanium scaffolds, with aligned large pores', *Acta Biomaterialia,* **8**(6), 2401–2410.

Young, M. L., DeFouw, J. D., Frenzel, J. and Dunand, D. C. (2012) 'Cast-replicated NiTiCu foams with superelastic properties', *Metallurgical and Materials Transactions A-Physical Metallurgy and Materials Science,* **43A**(8), 2939–2944.

Zhang, Y. P., Li, D. S. and Zhang, X. P. (2007) 'Gradient porosity and large pore size NiTi shape memory alloys', *Scripta Materialia,* **57**(11), 1020–1023.

8
Bioceramics for skeletal bone regeneration

G. C. WANG, Z. F. LU and H. ZREIQAT, The University of Sydney, Australia

DOI: 10.1533/9780857099037.2.180

Abstract: In this chapter, the physical, mechanical and chemical properties of calcium phosphate (Ca-P) and calcium silicate (Ca-Si) based bioceramics, and their effects on their biological performances, are reviewed and discussed. While Ca-P and Ca-Si ceramics have many advantages, their mechanical properties and bioactivity remain suboptimal, impairing their wide translation into the clinic. This chapter discusses the approaches taken to improve the mechanical properties of these bioceramics which include: topographical modifications of bioceramics at the micron and nanoscale levels; development of composite bioceramics; polymer-coated bioceramics; and the introduction of bioactive trace elements, biomolecules and growth factors into these bioceramics.

Key words: bioceramic, bone regeneration, mechanical property, osteoconductivity, bioactive trace elements.

8.1 Introduction

Compared to metal and polymer materials, ceramics are inorganic, non-metallic solids, usually brittle, hard and strong in compression, but weak in shearing and tension. A vast number of ceramics have been developed for applications in various fields; however, only few of them, such as zirconia (ZrO_2), alumina (Al_2O_3), hydroxyapatite ($Ca_5(PO_4)_3(OH)$) (HAp), are currently in clinical use. Bioceramics for use in bone tissue applications need to meet the following criteria: (1) non-toxic (2) non-carcinogenic (3) non-allergic (4) non-inflammatory (5) biocompatible and (6) bio-functional (Park and Bronzino, 2003). Bioceramics are classified, depending on their interaction with bone tissues, into three families: bioinert, bioactive and biodegradable/bioresorbable ceramics. Bioinert ceramics are usually encapsulated with a thick fibrous tissue after their *in vivo* implantation, leading to possible aseptic loosening of the implants (Park and Bronzino, 2003). Typical bioinert ceramics including zirconia, alumina and silicon nitride based ceramics are chemically stable; bioactive ceramics, on the other hand, are capable of forming a chemical bond with the surrounding bone (Park and Bronzino, 2003). Bioresorbable ceramics are degradable upon their *in vivo* implantation and can be replaced by endogenous tissue. Among bioresorbable ceramics, bioglasses, calcium phosphate (Ca-P) and calcium silicate (Ca-Si) based bioceramics and glass-ceramics are the most studied for skeletal bone regeneration applications.

A biomaterial used in bone regeneration should act as a temporary scaffold within the tissue defect for cell adhesion, proliferation and migration and be able to degrade at a rate that matches bone formation. A three-dimensional (3D) porous structure with highly interconnected struts can serve as a reservoir of water, nutrients, cytokines and growth factors, all of which are necessary for appropriate cell and tissue ingrowth (Arvidson *et al.*, 2011; Yang *et al.*, 2001, 2002.). In addition, it can also act as a temporary matrix for cell proliferation and extracellular matrix (ECM) deposition with consequent bone ingrowth until the new bony tissue is totally regenerated. The key challenge in developing biomaterials in the form of scaffolds for the repair and regeneration of large bone defects in load-bearing applications is to strike the right balance between material properties, implant architecture and osteoconductivity to satisfy functional and regenerative requirements.

This chapter will focus on the bioresorbable Ca-P and Ca-Si based ceramics. Material properties including crystal structure, chemical stability (degradation), mechanical properties and their effects on the biological performance of these bioceramics will be comprehensively discussed.

A large number of techniques have been developed to fabricate ceramic scaffolds with porous structures. However, different methods result in the production of scaffolds with totally different structural features including porosity, pore size, thickness and micro-/nanostructure of the struts, all of which are of great importance to the mechanical properties of the scaffolds (Burg *et al.*, 2000; Gauthier *et al.*, 1998; Gerhardt and Boccaccini, 2010; Mallick *et al.*, 2012; Otsuki *et al.*, 2006). Therefore, it is challenging to compare the mechanical properties of ceramic scaffolds with different structural features. Consequently, in this chapter the introduction and comparison of the mechanical properties will be based on dense ceramics with similar density, instead of the porous ceramic scaffolds.

8.2 Calcium phosphate (Ca-P) based bioactive ceramics for bone regeneration

Ca-P bioceramics have a unique characteristic for bone substitution and regeneration due to their compositional resemblance to bone mineral. Around 60~70 wt% of human bone mass is made of a Ca-P mineral with a chemical composition similar to HAp (Zhang *et al.*, 2011). Therefore, synthetic HAp and related Ca-P compounds such as α- and β-tricalcium phosphate (TCP) have been extensively used for tooth and bone replacement and regeneration. This section will focus on these selected Ca-P based synthetic biomaterials which have been tested and used for skeletal regeneration.

Ca-P compounds are abundant in nature with different Ca/P ratios, chemical composition and/or crystal structures. Detailed information on the chemical composition and crystal structures of those Ca-P compounds which appear in human calcified tissue is given in Table 8.1. Although these Ca-P compounds are present in human hard tissues, only a few, including dicalcium phosphate dehydrate (DCPD), dicalcium phosphate anhydrate(DCPA), α-tricalcium phosphate (α-TCP), β-tricalcium phosphate (β-TCP) and HAp, as summarized in Table 8.2, are used in biomedicine and biomedical field. The most extensively studied are α-TCP, β-TCP and HAp ceramics which have been widely used in the orthopaedic and dental applications. All of these – α-TCP, β-TCP and HAp – can be converted from one form into the other under specific thermodynamic conditions. For example, HAp can thermally decompose into TCP upon losing H_2O molecules. The decomposition temperatures vary significantly depending on the fabrication method of the HAp and atmospheric conditions (dry or wet), as well as the physiochemical properties (Ca/P ratio, type and amount of inclusions, particle size and shapes) (Cihlar *et al.*, 1999; LeGeros and LeGeros 1993; Liao *et al.*, 1999; Locardi *et al.*, 1993; Ravaglioli and Krajewski, 1992). β-TCP is a stable phase at low temperatures (below 1125°C) whereas α-TCP is stable only at higher temperatures (1125–1430°C) (Kwon *et al.*, 2003; Safronova *et al.*, 2011). Therefore, β-TCP will transform into α-TCP when the sintering temperature exceeds the phase transformation temperature. The reversible phase transformation was also reported by some researchers (Liao *et al.*, 1999). In biphasic Ca-P (HAp/TCP) ceramics sintered at temperatures higher than the β to α transformation temperature (Liao *et al.*, 1999), it was observed that β-TCP transformed to Ca-deficient HAp (CDHA) or oxy-hydroxyapatite (OHAp) after sintering, which was thought to be caused by the rehydration reaction in the course of cooling processes (Kong *et al.*, 2008). α-TCP can also transfer to CDHA by hydrolysis reaction with water, a cementitious-type reaction endowing self-setting and hardening properties upon α-TCP (Carrodeguas and De Aza, 2011; Chow, 2000; Li *et al.*, 1997). Extensive studies have been carried out to study the potential use of α-TCP, β-TCP and HAp as bone cement, as a carrier for drug delivery, and for compositing with polymers or other inorganic materials. Since this chapter deals with ceramics for the purpose of skeletal bone regeneration, only those Ca-P based ceramics which have been widely tested and/or used for this application will be discussed. Other applications of Ca-P compounds in the form of surface coatings (Narayanan *et al.*, 2008; Wang and Zreiqat,

Table 8.1 Chemical and crystallographic data of calcium phosphate

Ca/P	Compound	Formula	Space group	Solubility at 25°C, –log (K_{sp})
0.5	Monocalcium phosphate monohydrate (MCPM)	$Ca(H_2PO_4)_2.H_2O$	Triclinic $P\bar{1}$	1.14
0.5	Monocalcium phosphate anhydrate (MCPA)	$Ca(H_2PO_4)_2$	Triclinic $P\bar{1}$	1.14
1.0	Dicalcium phosphate dehydrate (DCPD, 'brushite')	$CaHPO_4.2H_2O$	Monoclinic *Ia*	6.59
1.0	Dicalcium phosphate anhydrate (DCPA, 'monetite')	$CaHPO_4$	Triclinic $P\bar{1}$	6.90
1.33	Octacalcium phosphate (OCP)	$Ca_8(HPO_4)_2(PO_4)_4).5H_2O$	Triclinic $P\bar{1}$	96.6
1.5	α-tricalcium phosphate (α-TCP)	α-$Ca_3(PO_4)_2$	Monoclinic $P2_1/a$	25.5
1.5	β-tricalcium phosphate (β-TCP)	β-$Ca_3(PO_4)_2$	Rhombohedral *R*3c*H*	28.9
1.2–2.2	Amorphous calcium phosphate (ACP)	$Ca_x(PO_4)_y.nH_2O$	N/A	Cannot be measured precisely
1.5–1.67	Calcium-deficient hydroxyapatite (CDHAp)	$Ca_{10-x(}HPO_4)_x(PO_4)_{6-x}(OH)_{2-x}$ $(0 < x < 1)$	N/A	≈85.1
1.67	Hydroxyapatite (HAp)	$Ca_{10}(PO_4)_6(OH)_2$	Monoclinic $P2_1/b$ or hexagonal $P6_3/m$	116.8
2.0	Tetracalcium phosphate (TTCP)	$Ca_4(PO_4)_2O$	Monoclinic $P2_1$	38–44

Source: Dorozhkin and Epple 2002, with permission from Elsevier.

Table 8.2 Calcium phosphate used in biomedicine and biomedical applications

Ca-P compounds	Applications in biomedicine and biomedical fields
DCPD	Used in calcium phosphate cement in bone repair and replacement (Panzavolta *et al.*, 2011; Driessens *et al.*, 1998)
DCPA	Used in calcium phosphate cement in bone repair and replacement (Chow and Takagi, 2001; Hsu *et al.*, 2008)
ACP	Used in calcium phosphate cements (Dorozhkin 2010; Combes and Rey, 2010); used as coatings on metallic implants (Maxian *et al.*, 1993; Zhao *et al.*, 2011, Garcia *et al.*, 1998); used as drug delivery carriers (Ito *et al.*, 2012; He *et al.*, 2011)
β-TCP, α-TCP, HAp	Used in calcium phosphate bone cement (Huan and Chang, 2009; Kwon *et al.*, 1997; Costantino *et al.*, 2000; Saito *et al.*, 2002; Kveton and Coelho, 2004); used as bone graft in form of blocks, granules or porous ceramics; used as biphasic calcium phosphate (BCP); used as coatings on metal implants (Narayanan *et al.*, 2008); used as composite scaffolds with polymer or other inorganic materials; carrier for drug release (Nyan *et al.*, 2009)

2010), drug delivery carriers (Carrodeguas and De Aza, 2011; Chow, 2000; Ginebra *et al.*, 2012), bone cement (Carrodeguas and De Aza, 2011; Chow 2000) and composites with polymers (Barinov, 2010; Boccaccini *et al.*, 2010; Neumann and Epple, 2006) will not be discussed in detail here. Readers can refer to the references cited for further information.

8.3 Properties of Ca-P bioceramics: degradability, bioactivity and mechanical properties

Degradability, bioactivity and mechanical properties of Ca-P ceramics are strongly correlated and of great importance for the biomedical application. To produce a Ca-P bioceramic with the optimized biological performance, efforts are needed to delicately adjust these properties to obtain a balance.

8.3.1 Degradability and bioactivity

Degradability is one of the prerequisite properties for materials used for bone regeneration. The ideal material needs to degrade at a rate matching the development of new tissue, so that by the time the injury site has been totally regenerated the scaffold is completely resorbed (Langer and Vacanti, 1993; Jones and Hench, 2003). Factors influencing the degradation of the biomaterials can be divided into two categories: (1) purely physico-chemical (ionic crystal dissolution) and (2) cellular (and enzymatic) activity at implantation sites (activated resorption). Both physico-chemical dissolution and activated resorption are influenced by the inherent physical and chemistry properties and crystallographic properties of the ceramic materials. Physical factors include form (granules, particulates or bulks), pore size, porosity and structural features (determining the surface area directly contacting the fluid). Chemical factors include chemical composition, stoichiometry or non-stoichiometry and ionic substitution (ion types and concentration). Crystallographic factors include crystal structures, crystallinity, grain size and crystal defects/imperfections. The degradation of biomaterials can also be influenced by the external conditions including the forces that can cause wear or abrasion of ceramic surfaces (Rey *et al.*, 2011), ionic strength and pH values of the surrounding fluid environments.

At physiological pH values (about 7.4), the *in vitro* solubility of Ca-P compounds tested in simulated body fluid (SBF) or other buffer solutions (Driessens *et al.*, 1992; Hyakuna *et al.*, 1990; Klein *et al.*, 1983; Kohri *et al.*, 1993), being expressed as the amount of Ca^{2+} and orthophosphate (PO_4^{3-}) ions in solution at the equilibrium, is known to decrease in the order of α-TCP > β-TCP >> HAp (Rey *et al.*, 2011). Notably, the degradation of HAp was found to be strongly dependent on its crystallinity and stoichiometry (Rey *et al.*, 2011). Amorphous HAp has a faster degradation rate than crystallized HAp (Nagano *et al.*, 1996) whose degradation rate declines as crystallinity increases (Xue *et al.*, 2004b). HAp is far less degradable than calcium-deficient hydroxyapatite (CDHAp) (Ducheyne *et al.*, 1993). Therefore, well-crystallized stoichiometric HAp is considered to be the most stable Ca-P compound. Dissolution of Ca-P compounds is accompanied by other chemical reactions (Driessens *et al.*, 1992; Hyakuna *et al.*, 1990) such as precipitation of Ca and P and hydrolysis of TCP, making the degradation of Ca-P ceramics a complex process. In contrast to *in vitro* degradation, biological factors are involved in the *in vivo* degradation of Ca-P ceramics including pH variation due to cell-mediated factors; osteolysis due to the action of osteoclasts; infection and diseases; degree of bone contact; type of bone; and the general health of the patient (Lu *et al.*, 2002). Cells involved in the *in vivo* degradation of bioceramics may be osteoclasts, multinucleated giant cells, monocytes and macrophages directly available in bone marrow (Frayssinet *et al.*, 1993; Heymann *et al.*, 1999, 2001; Kimakhe *et al.*, 1998; Koerten and van der Meulen, 1999).

The differences in the stability and degradation of these Ca-P compounds result in different bioactivity (referred to here as the *in vitro* bone-like apatite formation ability); osteoconductivity (the ability of biomaterials to serve as a template or scaffold for bone cells and permit bone growth on its surface or down into porous structures (Albrektsson and Johansson, 2001; LeGeros and LeGeros, 2006)); and osteoinductivity (the ability of biomaterials to form bone when implanted in non-bony sites, associated with osteogenesis (Albrektsson and Johansson, 2001; LeGeros and LeGeros, 2006)). The bioactivity of Ca-P compounds is dependent on the dissolution/precipitation in the SBF (Daculsi *et al.*, 1990). The release of Ca and P ions will increase the supersaturation of the fluid with respect to bone-like apatite. Radin and co-workers (Radin and Ducheyne, 1993) compared the *in vitro* apatite-forming ability of different Ca-P ceramics and reported that this ability increases in the following order: poorly-crystallized HAp < well-crystallized HAp < α-TCP < β-TCP. Further, both HAp and TCP have been proved to be osteoconductive (Ghosh *et al.*, 2008; Yuan *et al.*, 2001, 2002) *in vivo*. The degradation of Ca-P based materials is critical for the apatite formation whereby high dissolution rate may (1) compromise the mechanical strength of the ceramics, especially those with highly porous structures (scaffold) before the appropriate amount of ECM and new bone are formed, (2) lead to high local concentration of dissolved ingredients which may be harmful to the cells and (3) induce an acidic microenvironment (caused by the dissolution products) which may dissolve and demineralize the mineralized bone matrix (Yuan *et al.*, 2001). Therefore, the degradation of the Ca-P ceramics or Ca-P composites should be well controlled to an optimal rate. Biphasic calcium phosphate (BCP) bioceramics, e.g. HAp/β-TCP is a class of composite bioceramic known for its excellent osteoconductivity (Arinzeh *et al.*, 2005; Bodde *et al.*, 2007; Bouler *et al.*, 1996) and, in some cases, even osteoinductivity (Bodde *et al.*, 2007). BCP has a degradation rate that falls between those of HAp and TCP, being more degradable than HAp and less degradable than TCP. BCP ceramics can be obtained either by sintering a mixture of TCP and HAp powders or by sintering CDHAp above the temperature at which HAp transforms into TCP. Importantly, the degradation rate of BCP ceramics can be tailored by adjusting the ratios of HAp/TCP (Kohri *et al.*, 1993). The osteoconductivity and osteoinductivity of BCP have been previously confirmed by a number of *in vivo* (Daculsi *et al.*, 2003; Lee *et al.*, 2001; Lobo and Arinzeh, 2010) and *in vitro* studies (Radin and Ducheyne, 1993; Sous *et al.*, 1998; Toquet *et al.*, 1999). Currently, BCP ceramics have been recommended for use as an alternative or additive to autogeneous bone for orthopaedic and dental applications. These are available not only in the form of scaffolds but also as particulates, blocks and injectable paste, with or without polymer carriers (LeGeros *et al.*, 2003).

8.3.2 Mechanical properties

Although Ca-P ceramics have been widely used in the biomedical field, their poor mechanical properties, especially brittleness, limit their widespread utilization, especially for load-bearing applications (Lavernia and Schoenung, 1991). The mechanical properties of Ca-P based ceramics are

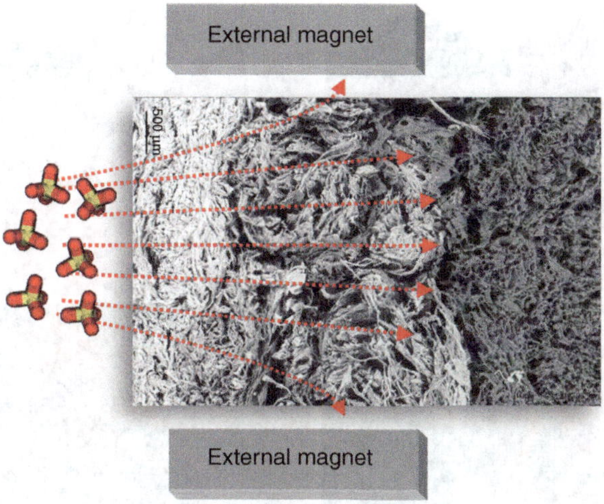

Plate I (Chapter 1) Schematic drawing of the functioning of a superparamagnetic osteochondral scaffold.

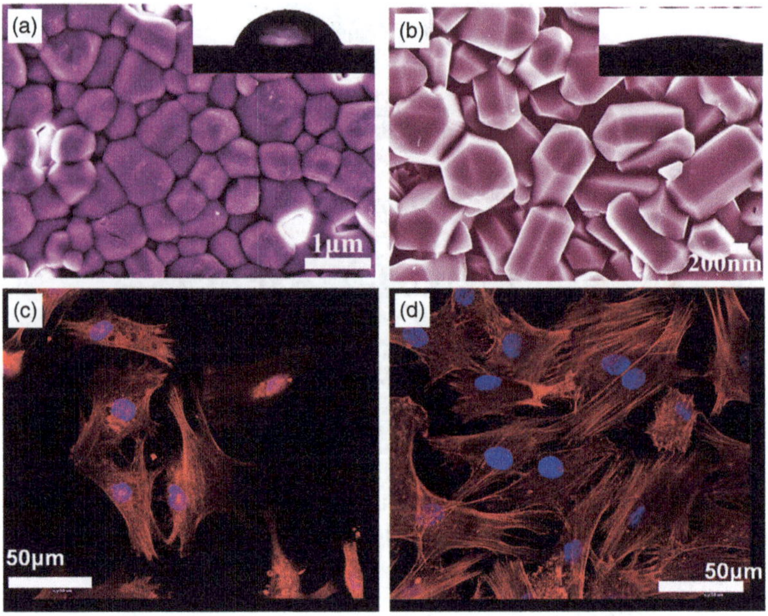

Plate II (Chapter 8) Surface grains of BCP ceramics were converted into HAp single crystals after hydrothermal treatment in sodium chloride (NaCl) solution in an autoclave sterilizer. (a) Scanning electron microscopy (SEM) image of as-sintered ceramics, (b) SEM image of ceramics with HAp single crystals, (c) Confocal laser scanning microscope (CLSM) image of HOBs cultured on as-sintered ceramics for 5 hours and (d) CLSM image of HOBs cultured on ceramics with HAp single crystals. (Cytoskeleton of HOBs were stained with rhodamine phalloidin (Red) and cellular nucleus were stained with DAPI).

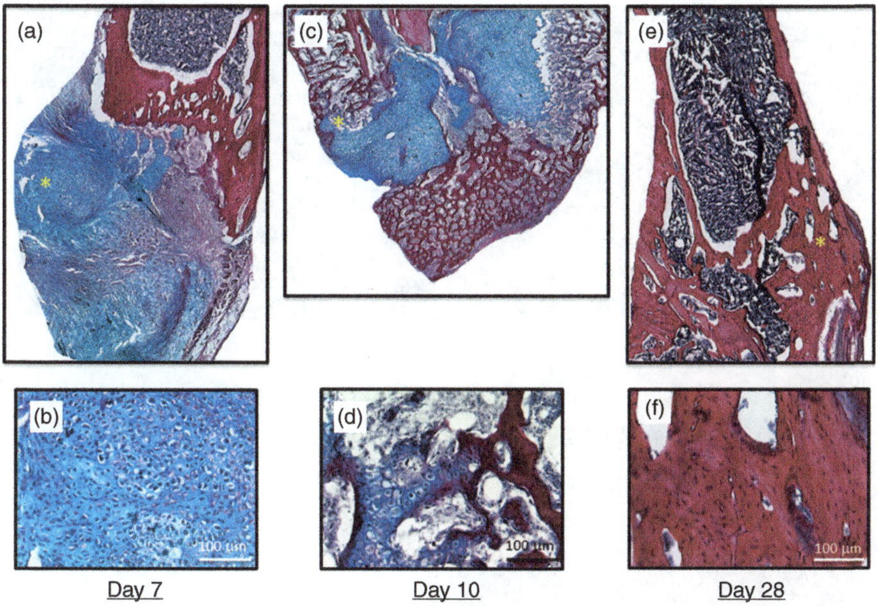

Plate III (Chapter 9) Endochondral Bone Repair During Fracture Healing: Hall's and Brunt's Quadruple Staining (bone = red, cartilage = blue) of non-stabilized murine tibia fracture after 7 (a, b), 10 (c, d) or 28 (e, f) days of healing. At 7 days the fracture callus is composed largely of cartilage (b), which then begins to transition to bone after 10 days of healing (d, and see Bahney *et al.* JBMR 2013), and is completely bone by 28 days post fracture. *Denotes location of high magnification image.

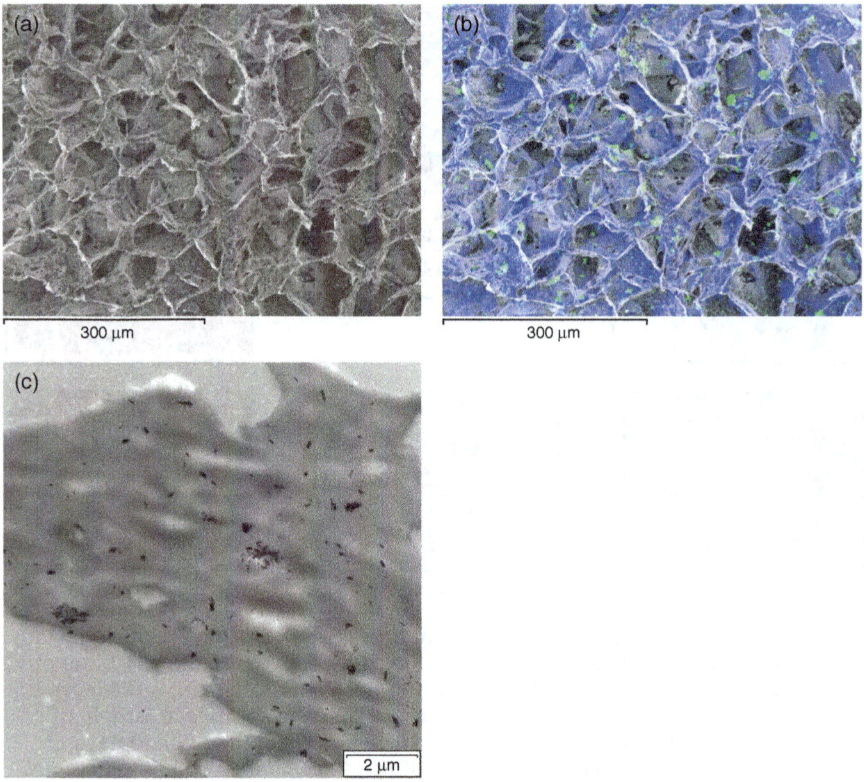

Plate IV (Chapter 11) nHAP/PCL scaffold (7.5% w/w) fabricated by TIPS. (a) SEM image, (b) SEM/EDX image with Ca (in green) and C (in blue) maps overlaid, (c) TEM image of a section of the scaffold displaying embedded nHAP crystals (scale bar 2 μm)

governed by their phase composition, grain size and structures (macro, micro and nanoscale) (Lavernia and Schoenung, 1991). Viswanath *et al.* (2008) reported that the fracture toughness of β-TCP single crystals is higher than that of HAp single crystals due to its extensive crack bridging phenomenon. Compared to HAp, it is suggested that TCP is quasi-brittle in nature. Metsger *et al.* (1999) reported that the compressive strength, Young's modulus and fracture toughness of dense TCP ceramics (315 MPa, 21 GPa and 2.34 J cm^{-3}, respectively) are superior to those of HAp ceramics (70 MPa, 9.2 GPa and 0.36 J cm^{-3}, respectively). This advantage is maintained even when they are in a form of porous scaffolds with 50% total volume porosity (Metsger *et al.*, 1999). Due to the differences in the mechanical properties of HAp and TCP, the proportion of TCP and HAp is of great significance to the mechanical properties of the BCP ceramics. There is an optimum content of β-TCP in the BCP matrix at which the BCP ceramics have the best mechanical properties (Metsger *et al.*, 1999; Raynaud *et al.*, 2002.). A high amount of β-TCP was found to be detrimental to the densification and the mechanical properties of the BCP ceramics. In contrast, a small amount of β-TCP in the HAp ceramic enhanced the overall mechanical properties of the BCP ceramic. For HAp ceramics, optimization of the sintering conditions is critical since phase transformation from HAp to TCP has a detrimental effect on the mechanical properties of HAp ceramics as it causes the decrease in the ceramics density (Gross and Rodriguez-Lorenzo, 2004; Kothapalli *et al.*, 2004; Muralithran and Ramesh, 2000). Below the decomposition temperature of HAp ($T_{HAp\text{-}TCP}$), the density of the HAp ceramics increases with the increase in sintering temperatures, resulting in the enhancement of the flexural strength. After the sintering temperature reaches $T_{HAp\text{-}TCP}$, HAp will dehydrate and decompose to β-TCP resulting in decreased mechanical strength (Gross and Rodriguez-Lorenzo, 2004; Kothapalli *et al.*, 2004; Muralithran and Ramesh, 2000). Further increase in the sintering temperature causes the phase transformation between α/β forms of TCP leading to more detrimental effects on the mechanical properties (Raynaud *et al.*, 2002). As alluded to earlier, β-TCP is a stable phase at lower temperature whilst α-TCP is stable at higher temperature. β-TCP will transform into α-TCP accompanied with the volume expansion (Enderle *et al.*, 2005) if the temperature is higher than their phase transformation temperatures ($T_{\alpha\text{-}\beta}$). This expansion will cause microcracks and prevent further densification of the ceramic bodies, thus reducing the mechanical strength (Enderle *et al.*, 2005; Ryu *et al.*, 2002.).

High porosity combined with interconnected pore structure and large surface area are key elements for the ceramic scaffolds to allow uniform bone ingrowth, efficient vascularization and diffusion of nutrients throughout the porous structure. The degree of porosity plays a significant role in determining the mechanical properties of the Ca-P porous ceramic scaffolds

(Le Huec *et al.*, 1995; Metsger *et al.*, 1999; Woodard *et al.*, 2007). In general, the mechanical properties of ceramics will decline as the porosity increases (Metsger *et al.*, 1999).

8.4 Enhancement of bioactivity and mechanical properties of Ca-P bioceramics

Although the bioactivity and osteoconductivity of Ca-P ceramics have been widely demonstrated, their biological performance in some occasion remains suboptimal. In addition, due to the inherent brittleness, most of Ca-P ceramics exhibit low fracture toughness, limiting their clinic application. Therefore, it is of great necessity to enhance their bioactivity and overcome the brittleness problem.

8.4.1 Enhancement of bioactivity

Various methods have been developed to enhance Ca-P ceramics' osteoconductivity and/or osteoinductivity including ionic substitutions by bioactive trace elements and incorporation of bioactive molecules (growth factors and biomolecules).

Trace elements such as zinc (Zn), strontium (Sr) and silicon (Si) are essential in the human body with significant effects on bone formation and have been shown to accelerate fracture healing. Their beneficial effects are summarized below.

Zn is an important essential trace element in the human body with significant effects on bone formation (Sandstead, 2000). At the cellular level, Zn plays a significant role in enhancing osteoblast proliferation (MacDonald, 2000), increasing the alkaline phosphatase (ALP) activity and DNA content in bone tissues (Yamaguchi and Miwa, 1991; Yamaguchi *et al.*, 1987) as well as selectively inhibiting osteoclast functions (Kishi and Yamaguchi, 1994).

Si is an essential trace element for metabolic processes associated with the development of bone and connective tissues (Carlisle, 1970). It plays an important role during the early stage of bone formation and the calcified process by increasing the mRNA expression of key osteoblastic genes, such as type I collagen (Arumugam *et al.*, 2004; Reffitt *et al.*, 2003).

Sr is another beneficial trace element for bone formation and it has been introduced gradually in the treatment for osteoporosis. Sr enhances osteoprogenitor cell replication and collagen synthesis in rat osteoblast cell culture (Canalis *et al.*, 1996). *In vivo* study (rat model) showed that Sr can stimulate bone formation and reduce bone resorption while preserving bone mineralization thus leading to increased bone vertebral mass.

Si-, Sr- and Zn-substituted Ca-P ceramics have been widely studied over the past decade. Their biological superiority over pure TCP (Botelho *et al.*,

Table 8.3 Examples of using trace elements to modify the currently used biomaterials

Trace elements	Encouraging results
Si	More rapid remodelling of bone surrounding the Si-HAp granules compared to pure HAp; superior osteoconductive properties to pure HAp granules *in vivo* (Porter *et al.*, 2004c)
	Si-TCP and Si-HAp showed increased osteogenesis of osteoblast-like cells with corresponding increased new bone matrix formation *in vitro* (Botelho *et al.*, 2006)
Zn	Zn-HAp had an obviously accelerated and enhanced osteogenesis *in vivo* compared to HAp group (Calasans-Maia *et al.*, 2008)
	Enhanced osteoconductivity, increased proliferation rate, higher alkaline phosphatase (ALP) activity and increased collagen secretion were found for the cells cultured on Zn-calcium phosphate cements (Pina *et al.*, 2010)
Sr	Sr-HAp coated implants showed improved osseointegration compared to HAp coated implants (Lim *et al.*, 2010)
	Sr-HAp stimulated osteoblast activity and exerted inhibitory effects on osteoclast proliferation (Capuccini *et al.*, 2009)

2006) and HAp (Botelho *et al.*, 2006; Calasans-Maia *et al.*, 2008; Capuccini *et al.*, 2009; Lim *et al.*, 2010, Porter *et al.*, 2004c) has been demonstrated by many independent studies (Table 8.3). These encouraging results demonstrate the feasibility of the use of these elements in order to enhance the biological properties of biomaterials. For example, Patel *et al.* (2002) compared the *in vivo* bone formation ability of pure HAp and Si-substituted HAp in a rabbit model. They demonstrated that Si-substituted HAp had significantly more bone ingrowth (37.5% ± 5.9) than that for pure HAp (22.0% ± 6.5). The percentage of bone/implant coverage was also significantly higher for Si-substituted HAp (59.8% ± 7.3) compared to HAp (47.1% ± 3.6) (Patel *et al.*, 2002). The mechanism by which the addition of certain amounts of Si ions improves the osteoconductivity of Ca-P ceramics is still not yet fully understood. It was reported that the *in vivo* solubility of Si-substituted HAp ceramics is higher than that of phase pure HAp ceramics due to the changes in the grain boundary structure observed by transmission electron microscope (TEM) (Porter *et al.*, 2003., 2004a, b.), which was thought to ascribe the enhanced osteoconductivity of the Si-substituted HAp (Porter *et al.*, 2003., 2004a, b.). However, other researchers believe that it is arbitrary to attribute the enhanced biological performance to the released Si ions as there is no experimental evidence that the released Si ions are at therapeutic concentrations and directly link to the enhanced biological performance (Bohner, 2009). Therefore, further studies are required to clarify the mechanism by which the addition of bioactive trace elements benefits the biological performance of the Ca-P ceramics. The beneficial effects of trace elements are dose-dependent (Ni *et al.*, 2010; Wernike *et al.*, 2010) and determining the effects of the trace elements on the degradation behaviours of Ca-P ceramics would be of great importance for designing a Ca-P based material with controlled release of bioactive trace elements.

Besides the chemical modification, another possible way to enhance the bioactivity of the Ca-P ceramics is to alter the ceramic surface with micro- and nanoscale topography known to facilitate bone ingrowth and enhance osseointegration (Martin *et al.*, 1995). Osteoblasts cultured on a micron-scale rough surface was reported to generate an osteogenic microenvironment to regulate bone remodelling by releasing local factors to promote osteoblast differentiation and inhibit osteoclast activity (Boyan *et al.*, 2003). Nanostructured surfaces are known to significantly affect cell adhesion and proliferation (Popat *et al.*, 2007). Enhanced cellular proliferation of marrow stromal cells (MSCs) was demonstrated on the nanotubular surfaces of anodized Ti compared to the flat Ti surfaces (Popat *et al.*, 2007). Webster *et al.* (2000) observed increased osteoblast proliferation on nanophase ceramics which was ascribed to the enhanced initial protein interaction with the surfaces via nano-sized grains. Therefore, Ca-P ceramics with hierarchically defined macro-, micro- and nanoscale structures are promising for bone regeneration applications.

Although nanophase ceramics have enhanced biocompatibility compared to conventional ceramics, their surface structure is less hierarchical, compared to the surface structure of the substratum (ECM) in which cells live *in vivo* (Lutolf and Hubbell, 2005). In our previous work, we used a simple solution based method to convert the surface grains of BCP composed of HAp and TCP into bipyramid hexagonal HAp single crystals (Plate II (a) and (b) in the colour section between pages 186 and 187) (Wang *et al.*, 2012b). The water contact angle on the ceramic with single crystal surface is much lower than that of the as-sintered ceramic (insets in Plate II (a) and (b)). We demonstrated that adhesion of primary human osteoblasts (HOBs) was better on the ceramic with HAp single crystal surface compared to that on the as-sintered ceramic, as shown in Plate II (c) and (d) (Wang *et al.*, 2012b). Therefore, Ca-P ceramics with hierarchically defined micro- and nanoscale structures are promising for bone regeneration applications.

The incorporation of biological molecules into Ca-P ceramics or onto the surface layer of the Ca-P ceramics can also enhance the osteoconductivity of Ca-P ceramics. Growth factors such as bone morphogenetic proteins (BMPs), transforming growth factor-β, insulin-like growth factor and fibroblast growth factor have been shown to act as local regulators of cellular activity and are potent mitogenes (Maus *et al.*, 2008). Therefore, immobilization of bioactive growth factors may lead to rapid cell functions, including cellular proliferation and differentiation activity thus accelerating the tissue regeneration (Alam *et al.*, 2001; Maus *et al.*, 2008; Wernike *et al.*, 2010; Schnettler *et al.*, 2003).

8.4.2 Enhancement of mechanical properties

Various methods have been developed to enhance the mechanical strength of Ca-P ceramics. One effective way is to incorporate a second reinforcing phase with high mechanical strength into the ceramic. Zirconia (Choi *et al.*, 1998; Delgado *et al.*, 1999; Li *et al.*, 1996.), alumina (Jun *et al.*, 2003; Knepper *et al.*, 1998.) and Ni_3Al (Knepper *et al.*, 1998) have been shown to greatly improve the mechanical strength of Ca-P ceramics. Glass (Lopes *et al.*, 1999; Knowles and Bonfield, 1993; Santos *et al.*, 1994), silver (Chaki and Wang, 1994; Zhang *et al.*, 1997), carbon fibres (Knepper *et al.*, 1998; Park and Vasilos, 1997), carbon nanotubes (CNTs) (Low *et al.*, 2011; White *et al.*, 2007) and HAp whisker (Suchanek *et al.*, 1997) have all been used as reinforcing phases. While the addition of these second phases enhances the mechanical properties of the Ca-P ceramics, it also raises certain concerns. For example, reinforcing bioinert phases, such as zirconia and alumina, will decrease the bioactivity of the Ca-P ceramic. Additionally, these phases are chemically stable compared to the degradable Ca-P ceramics, therefore, the ceramic grains of the reinforced phase may be pulverized into particulates after the Ca-P phase is bioresorbed, possibly leading to particle-induced inflammatory reactions. Compared to zirconia and alumina, carbon nanofibres (CNFs) and CNTs do not impact on the overall bioactivity of the Ca-P due to their small dimensions and potential bioactivity reported by some researchers (Price *et al.*, 2003). However, the potential toxicity of the released CNFs and CNTs warrants special attention.

Nanophase ceramics have attracted a great deal of attention in the field of orthopaedic and tissue engineering due to their enhanced biocompatibility and bioactivity compared to conventional bioceramics (Kay *et al.*, 2002; Webster *et al.*, 2001.). Mechanically, nanophase ceramics also surpass conventional ceramics with coarse grains (Webster, 2001). Ahn *et al.* (2001) reported that nanostructured HAp-dense ceramics have 70% improvement over the best conventional HAp in terms of the critical 3-point bending strength. Therefore, controlling grain size at the nanoscale is another promising way to produce Ca-P ceramics with improved mechanical strength and enhanced biological properties. In order to obtain a nanophase Ca-P ceramic, the sintering temperature for densification of the ceramic should be as low as possible to avoid the overgrowth of the grains. Wang and co-workers (Wang and Shaw, 2007) reported that the sintering temperature for nanophase HAp bioceramics can be dramatically reduced by using nano-sized, rod-shaped HAp single crystal powders as starting materials because the side facets (prism faces, {010}) of single crystals offer a fast diffusion path with a low activation energy. Therefore, dense HAp bodies with an average grain diameter of 67.1 nm can be obtained at temperatures as low as 850°C.

The other alternative ways to overcome the brittleness of Ca-P ceramics are to (1) combine them with biocompatible polymers to produce inorganic/organic composite scaffolds and (2) coat the Ca-P porous scaffolds with biocompatible or bioactive polymers (Bretcanu *et al.*, 2012; Fang *et al.*, 2012; Zhou *et al.*, 2012). Besides the enhancement of mechanical properties, scaffolds prepared or modified with this method can be endowed with drug delivery capability using polymers which can be loaded with drugs or bioactive molecules. However, this topic is beyond the scope of this chapter and readers with an interest can refer to the related references. Some review articles (Boissard *et al.*, 2009; Mohamad Yunos *et al.*, 2008; Rezwan *et al.*, 2006; Wu, 2009) are provided here for convenience of the readers.

8.5 Calcium silicate (Ca-Si) based bioceramics and their applications in biomedical fields

Ca-Si based materials are another important class of bioactive materials for potential use in bone regeneration. These are degradable and their dissolution products (Ca and Si) are beneficial for the growth and functions of bone cells. Ca-Si based bioactive materials can be divided into four groups: bioactive glass, bioactive ceramics, glass-ceramics and cements. The beneficial effects of Ca-Si based materials on inducing bone formation were first found on glass in the system of SiO_2-CaO-Na_2O-P_2O_5 by Hench *et al.* in the early 1970s (Hench *et al.*, 1971). Glasses with this composition were able to bond to soft and hard tissues forming a carbonated HAp layer when exposed to biological fluid (Hench 1991); hence, it was named bioglass. A number of different types of bioactive glasses have been developed and some have been used in clinic such as 45S5 Bioglass®. Crystallographically, glass is composed of amorphous phase while ceramic consists of crystalline phase. Glass can be partially recrystallized by heat treatment and form glass-ceramics.

Glass-ceramics consisting of both amorphous and crystalline phases have attracted great interests in the biomaterials field due to their improved mechanical properties compared to bioactive glass (Thompson and Hench, 1998). In addition, their resistance to abrasion and the coefficient of thermal expansion can be easily manipulated (Tulyaganov *et al.*, 2006) and their degradation rate or ion release behaviour can be tailored using post-heat treatment, by controlling the extent of crystallization (Tulyaganov *et al.*, 2006). Fabrication, mechanical properties, their degradation and osteoconductivity of Ca-Si based glass and glass-ceramics on cellular responses have been well documented (Cerruti and Sahai, 2006; Gorustovich *et al.*, 2010; Höland, 1997; Hoppe *et al.*, 2011; Thompson and Hench, 1998, Sola *et al.*, 2011). Ca-Si based materials have also been studied as potential bone cements for the fixation of prostheses to living bone in total joint arthroplasty (Chen *et al.*, 2010a; Ding *et al.*, 2009; Guo *et al.*, 2007; Zhao *et al.*, 2005).

Stoichiometric calcium silicate ($CaSiO_3$) has two crystalline modifications: wollastonite (sometimes called wollastonite I or β-$CaSiO_3$) and pseudo-wollastonite (also called α-$CaSiO_3$). Wollastonite is low-temperature modification exhibiting triclinic structure with a space group of $\alpha P30$ while pseudo-wollastonite exhibits monoclinic structure with a space group of *C2/c* (Schmetterer and Masset, 2012). Dicalcium silicate (Ca_2SiO_4) has four different modifications: γ-phase, stable at room temperature, α'- and α-phases, stable at high temperature and β-phase which is not thermodynamically stable (Schmetterer and Masset). Tricalcium silicate (Ca_3SiO_5) is a phase stable between 1250 and 1900°C. At or above 1900°C the phase will dissociate into the CaO and Ca_2SiO_4 (Welch and Gutt, 1959).

$CaSiO_3$, Ca_2SiO_4 and Ca_3SiO_5 have the ability to induce apatite formation in SBF not only in the form of ceramics (Gou *et al.*, 2005; Lin *et al.*, 2009; Sainz *et al.*, 2010) but also in the form of powders (Gou and Chang, 2004; Hazar, 2007; Zhao and Chang, 2004) and ceramic coatings (Liu and Ding, 2001; Liu *et al.*, 2001, 2002, 2004b., 2005, 2008; Xue *et al.*, 2004a). Even in the same physical form, their ability to induce apatite formation varies drastically due to the difference in their degradation properties. For example, plasma sprayed CaO-SiO_2 based coatings (e.g., $CaSiO_3$, Ca_2SiO_4 and Ca_3SiO_5) are revealed to illustrate different bioactivity (ability to form apatite *in vitro*). Hydroxyapatite was formed on the surface of plasma sprayed $CaSiO_3$ coatings after immersion in SBF solution for one day (Liu and Ding, 2001; Liu *et al.*, 2004b). For plasma sprayed Ca_2SiO_4 coatings, some apatite particles were observed on their surfaces after only one hour while a dense apatite layer was formed after one day, indicating the superior bioactivity of plasma sprayed Ca_2SiO_4 coatings (Liu *et al.*, 2005). Compared to $CaSiO_3$ and Ca_2SiO_4, Ca_3SiO_5 is less studied as ceramic or ceramic coatings. The main property that is considered interesting to biomedical material scientists refers to its self-setting ability (Zhao *et al.*, 2005).

The main mechanisms for the bioactivity of CaO-SiO_2 based ceramic (including CaO-SiO_2 based ceramic coatings) are shown in Fig. 8.1 and depicted as follows: in the first step, Ca^{2+} exchanges with H^+ in the SBF solution causing the release of Ca^{2+}. The dissolution of Ca and Si is incongruent, with Ca released preferentially relative to Si ions (Green and Luttge, 2006; Peters *et al.*, 2004) thus leading to a leached layer rich in silanol (≡Si–O–H). In the second step, Ca^{2+} ions in the SBF solution are electrostatically attracted to the newly formed silanol-rich layer which is negatively charged (Liu *et al.*, 2004a). The pH value of the leached solution increases due to preferential dissolution of Ca^{2+} ions, which increases the supersaturation with respect to apatite, thus resulting in the apatite precipitation (Liu *et al.*, 2004a). In the third step, the Ca^{2+} ions begin to attach by electrostatic attraction to the negatively charged surface decorated by ≡Si-O^-, followed by adsorption of PO_4^{3-}/HPO_4^{2-} on Ca^{2+} ions. Finally, a layer of bone-like apatite is formed on

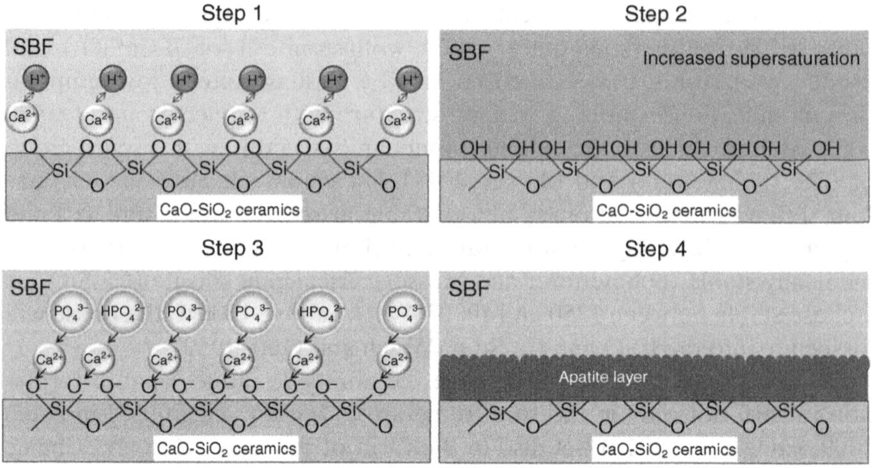

8.1 Schematic illustration of the processes of apatite formation on the $CaO-SiO_2$ ceramics in SBF.

the surface of $CaO-SiO_2$ based ceramics. During these processes, the weight change of ceramic or coating, variation in the pH values, ion composition of the leach solution and apatite formation ability are tightly interrelated and depend largely on the reactivity of the materials. Compared to $CaSiO_3$ coatings, the dissolution rate of Ca_2SiO_4 coatings is higher and may account for the much easier formation of apatite on Ca_2SiO_4 than on $CaSiO_3$ coatings.

Several *in vitro* cell studies have been carried out to study the biocompatibility of $CaO-SiO_2$ based ceramics. Ni and co-workers (Ni *et al.*, 2007) investigated the effects of the $CaSiO_3$ ceramics on the cell attachment, proliferation and differentiation of rat osteoblast-like cells. Enhanced attachment and proliferation were found on the cells cultured on the $CaSiO_3$ ceramics, compared to β-TCP ceramics (Ni *et al.*, 2007). Effects of the released Ca and Si ions on the osteoblast activity caused by the degradation of the ceramics were emphasized in this work, which is consistent with the studies reported by others (Carlisle, 1970, 1986; Hoppe *et al.*, 2011; MacDonald, 2000; Sun *et al.*, 2009a). Sun *et al.* (2009b) reported that the dissolution products from plasma sprayed Ca_2SiO_4 coatings enhance the expression of osteoblast-related genes and promote differentiation of MG63 cells at the initial period in agreement with the cell responses to bioactive glass. However, when used in the form of 3D scaffolds, $CaO-SiO_2$ ceramics degrade at a much higher rate as the specific surface area of a porous scaffold is significantly higher than that of ceramic coatings or discs. This high degradation rate is likely to cause the collapse of the scaffold structure prior to the formation of sufficient bone ECM. Additionally, the large deviation in pH values from the physiological level caused by excessive dissolution products of $CaO-SiO_2$ ceramics scaffolds may negatively affect the surrounding cells (Kohn *et al.*,

2002; Ramp *et al.*, 1994). Therefore, the dissolution rate should be properly controlled or adjusted for improved biological outcomes. Moreover, CaO-SiO_2 ceramics are not mechanically strong enough for bone regeneration for load-bearing applications. In order to improve the mechanical properties and modulate the degradation rate of CaO-SiO_2 based ceramics, new types of ceramics based on ternary system CaO-M_xO_y-SiO_2 (where M represents bioactive elements for bone regeneration) were developed (Luo *et al.*, 2012; Wu *et al.*, 2005a, 2006a,b).

8.6 Approaches to improve the performance of Ca-Si based bioceramics

Similar to the Ca-P ceramics, drawbacks, such as the insufficient bioactivity and the brittleness, also exist in the Ca-Si based ceramics. To overcome these drawbacks, the most popular approach is to add the third bioactive element, e.g. magnesium (Mg), Zn, and zirconium (Zr) to form a ternary compound. This approach is not only capable of endowing the Ca-P based ceramics with an ability to release bioactive ions essential for bone growth, thus enhance their bioactivity, but also of significantly improving their mechanical strength.

8.6.1 Effects of Mg on the properties of Ca-Si based bioceramics

Mg is one of the most important trace elements in mammalian biology. It is most abundant, amounting to around 6 mol%, in cartilage and bone tissue during the initial phases of osteogenesis while it tends to disappear in the mature bone (Bigi *et al.*, 1992). It was reported that Mg deficiency adversely affects all stages of skeletal metabolism, causing cessation of bone growth and decrease in osteoblastic and osteoclastic activity (Landi *et al.*, 2008). Its effects on enhancing osteoconductivity and osteoinductivity on Ca-P ceramics have been confirmed (Landi *et al.*, 2008). The addition of Mg can also enhance the biological properties of CaO-SiO_2 ceramics whilst at the same time improving their mechanical properties (Kobayashi *et al.*, 1997; Ou *et al.*, 2008).

Incorporation of Mg in the CaO-SiO_2 system will result in the formation of pure phase products or produce composites of two or more phases depending on the proportions of CaO:SiO_2:MgO and on the sintering temperatures. Pure phase ceramics in CaO-SiO_2-MgO including diopside ($CaMgSi_2O_6$) (Iwata *et al.*, 2004; Kamitakahara *et al.*, 2006; Kobayashi *et al.*, 1997; Nonami, 1992; Nonami and Tsutsumi, 1999; Wu and Chang, 2007), bredigite ($Ca_7MgSi_4O_{16}$) (Landi *et al.*, 2008; Luo *et al.*, 2012; Wu and Chang, 2007; Wu *et al.*, 2007.), akermanite ($Ca_2MgSi_2O_7$) (Chen *et al.*, 2010b), merwinite ($Ca_3MgSi_2O_8$) (Hafezi-Ardakani *et al.*, 2011; Ou *et al.*, 2008.) and monticellite ($CaMgSiO_4$)

(Chen *et al.*, 2008., 2010b) have been investigated for their potential application in orthopaedics. Diopside shows good mechanical properties with a bending strength of 300 MPa and a fracture toughness of 3.5 MPa $m^{1/2}$, higher than those of $CaSiO_3$ ceramics with similar density and 2–3 times higher than those of dense HAp ceramics (Nonami and Tsutsumi, 1999). Nakajima *et al.* (1989) reported that diopside ceramics can induce apatite formation in SBF solution and closely bond to the bone tissue when implanted in rabbits (Miake *et al.*, 1995; Nakajima, 1990; Nonami and Tsutsumi, 1999) but their degradation rate is extremely low compared to those of HAp and $CaSiO_3$ ceramics, thus making it suboptimal for their use in bone regeneration (Nonami and Tsutsumi, 1999). Nevertheless, diopside ceramics may be good candidates for coating in orthopaedic applications where high chemical stability is required. Other CaO-SiO_2-MgO based ceramics have been widely investigated including akermanite and bredigite (Wu and Chang, 2007; Wu *et al.*, 2007). Wu and co-workers (Wu and Chang, 2007; Wu *et al.*, 2007) compared the degradation and bioactivity of akermanite, bredigite and diopside ceramics. Their results showed that the activation energy for the release of Si ions increased, and the degradation rate and the ability for apatite formation declined from bredigite to diopside ceramics as the content of Mg increased. The dissolution products from all these three ceramics were found to stimulate osteoblast proliferation that is dependent on the dosage used. Yin's group investigated the mechanical properties and *in vitro* bioactivity of merwinite and monticellite ceramics (Chen *et al.*, 2008; Ou *et al.*, 2008). The authors found that both exhibit mechanical properties superior to $CaSiO_3$ and HAp ceramics and can induce apatite formation in SBF.

8.6.2 Effects of Zn on the properties of Ca-Si based bioceramics

Due to the critical role of Zn ions in bone metabolism, as discussed in detail earlier (Section 8.4.1), addition of Zn into the CaO-SiO_2 system can lead to the formation of new mineral phase. Hardystonite ($Ca_2ZnSi_2O_7$) is a representative ceramic in the CaO-SiO_2-ZnO system (Wu *et al.*, 2005b). Zreiqat's group carried out several studies to explore the potential biomedical applications for hardystonite (Wu, 2009; Wu *et al.*, 2005b; Zreiqat *et al.*, 2010). The main advantages of hardystonite are that it contains Zn in its crystal lattice and the release of Zn, as well as the slowing down of Ca and Si ion release compared to the $CaSiO_3$ ceramic. The dissolution of hardystonite will not result in an alkaline microenvironment thus avoiding the potential dangers on cells, compared to that encountered for the $CaSiO_3$ ceramic. In addition, its improved mechanical properties compared to the $CaSiO_3$ ceramics (Table 8.4) and enhanced proliferation and differentiation of primary HOBs were also demonstrated, suggesting its good biocompatibility (Wu *et al.*,

Table 8.4 Mechanical properties of ceramics in CaO-M_xO_y-SiO_2 systems (M: Mg, Zn and Zr) in comparison with $CaSiO_3$ and HAp ceramics

Name	Molecular formula	Mechanical properties			References
		Bending strength (MPa)	Fracture toughness (MPa m$^{1/2}$)	Young's modulus (GPa)	
Bredigite	$Ca_7MgSi_4O_{16}$	156 ± 6	1.57 ± 0.12	43.0 ± 4.53	(Luo *et al.*, 2012, Wu and Chang, 2007; Wu *et al.*, 2007; Landi *et al.*, 2008)
Akermanite	$Ca_2MgSi_2O_7$	141.8 ± 2.3	1.53 ± 0.10	56.2 ± 5.4	(Chen *et al.*, 2010b)
Diopside	$CaMgSi_2O_6$	300	3.5	170	(Iwata *et al.*, 2004; Kobayashi *et al.*, 1997; Kamitakahara *et al.*, 2006; Nonami, 1992; Nonami and Tsutsumi, 1999b; Wu & Chang, 2007)
Merwinite	$Ca_3MgSi_2O_8$	151.2 ± 5.7	1.72 ± 0.11	31 ± 2	(Ou *et al.*, 2008; Hafezi-Ardakani *et al.*, 2011)
Monticellite	$CaMgSiO_4$	163.9 ± 3.6	1.65 ± 0.12	45.5 ± 4.1	(Chen *et al.*, 2008, 2010b)
Hardystonite	$Ca_2ZnSi_2O_7$	136.4 ± 3.9	1.24 ± 0.03	N/A	(Wu *et al.*, 2005b)
Wollastonite	$CaSiO_3$	95	<1.0	N/A	(Wu, 2009)
Hydroxyapatite	$Ca_{10}(PO_4)_6(OH)_2$	115–200	1.0	80–110	(Kokubo *et al.*, 2003)

2008). However, the main disadvantage of hardystonite is its low degradation rate, whereby the biological effects of the trace element ions (in the hardystonite) cannot be fully utilized. In contrast, as we mentioned earlier, $CaSiO_3$ is highly degradable compared to hardystonite and fast degradation is likely to result in mechanical deterioration of the scaffolds and high alkalinity of the cellular microenvironment. Both $CaSiO_3$ and hardystonite have benefits and weakness and cannot, on their own, meet the requirements of optimal materials for bone regeneration. Therefore, in an attempt to address their drawbacks, $CaSiO_3$ has been incorporated into hardystonite to produce a composite material whereby degradation and mechanical properties can be tailored by adjusting the proportion of $CaSiO_3$ and hardystonite (Wang *et al.*, 2012a). Figure 8.2 shows scanning electron micrographs (SEM) of the $CaSiO_3$ scaffold, hardystonite and $CaSiO_3$/hardystonite composite scaffold. Under high magnification, it can be seen that some amount of glass phase was formed in the grain boundary, resulting in enhanced mechanical properties. *In vitro* study showed enhanced cell attachment and differentiation of primary HOBs when cultured on the composite scaffold, compared to those on the $CaSiO_3$ and hardystonite scaffolds. In order to improve the bioactivity of hardystonite, we recently developed Sr-incorporated hardystonite (Sr-HT) (Zreiqat *et al.*, 2010) and showed that this material is degradable and able to release four bioactive ions (Zn, Ca, Si and Sr) resulting in better *in vivo* bone formation (Zreiqat *et al.*, 2010).

8.6.3 Effects of Zr on the properties of Ca-Si based bioceramics

Baghdadite ($Ca_3ZrSi_2O_9$) is one of the representative bioceramics in the CaO-ZrO_2-SiO_2 system for potential application in bone regeneration. Ramaswamy *et al.* (2008) investigated the responses of osteoblasts, osteoclasts and endothelial cells on baghdadite ceramic discs. Baghdadite ceramics not only promote proliferation and induce differentiation of HOBs but also enhance the osteoclast activity, indicating that baghdadite may promote bone remodelling processes while accelerating new bone formation (Ramaswamy *et al.*, 2008). Additionally, it also supported the attachment of endothelial cells with expressed endothelial cell markers of ZO-1 and VE-Cadherin. In the *in vivo* studies using Wistar rats carried out by Luo *et al.* (2012), it was reported that the baghdadite spheres induced a higher rate of new bone formation in the femur defects than β-TCP and diopside spheres. The expression levels of collagen-I and osteopontin were higher for baghdadite compared to β-TCP and diopside spheres. Moreover, we have demonstrated that baghdadite scaffolds with highly porous structure have the ability to repair critical-sized bone defects in a rabbit radius under normal load after 12 weeks of implantation (Roohani-

8.2 SEM micrographs of the hardystonite-$CaSiO_3$ (a, b), hardystonite (c, d) and $CaSiO_3$ (e, f) scaffolds, and micrographs on the right show the sintering behaviors of the scaffolds (b, d and f). (*Source*: Reprinted from Wang *et al.*, 2012a, with permission from Elsevier.)

Esfahani *et al.*, 2012), suggesting its potential use in the repair of critical-sized bony defects.

8.7 Summary and future trends

Although Ca-P and Ca-Si based ceramics are bioactive and osteoconductive compared to bioinert ceramics, such as ZrO_2 and Al_2O_3, their osteoconductivity and mechanical properties remain suboptimal for bone regeneration.

Improving the mechanical properties and enhancing the osteoconductivity of these ceramic materials are of the utmost priority before their possible translation into the clinic. Mechanical properties can be improved by incorporating other stronger ceramic materials, either through compositing or coating with biocompatible polymers. Osteoconductivity can be enhanced by incorporation of bioactive trace elements (Si, Zn, Mg, Si and Sr) or biomolecules and growth factors or by topographical modifications at both micron and nanoscale.

Those methods being capable of improving more than one property may lead to better biological and mechanical outcomes of biomaterials. For example, when using biocompatible polymers to improve the mechanical properties of the ceramics, the first choice would involve the use of those polymers that are suitable as drug carriers. Therefore, the ceramics can be endowed with the ability of drug delivery while their mechanical properties are improved. Generation of micron and nanoscale topographical features on ceramic surfaces may be one option to enhance the biological performance of ceramics (coatings or scaffolds). Critically, methods which can integrate both the chemical and topographical modification in one single process should be developed.

Substitution by and incorporation of trace elements are promising ways to enhance the properties of ceramics for bone regeneration application. The beneficial effects of trace elements vary and are dose-dependent. Therefore, the degradation of ceramics doped with trace elements should be well controlled to maintain the released trace element ions at therapeutic concentrations. Moreover, dynamic cell culture is more promising in determining the effects of released trace element ions on the cell responses, as ions are easily accumulated in the static culture condition being different from the *in vivo* environment.

Material selection and design are the first step and the most important for producing ceramic scaffolds for bone regeneration. However, an ideal scaffold for bone regeneration should not only have optimal chemical composition but should also have an optimal structure which can promote bone ingrowth. Pore size, porosity, thickness and microstructure of struts (pore wall) and interconnectivity have significant influence on the mechanical responses and functional performance (Burg *et al.*, 2000; Gauthier *et al.*, 1998; Gerhardt and Boccaccini, 2010; Otsuki *et al.*, 2006;. Mallick *et al.*, 2012). We encourage the reader to refer to the excellent recent review papers on the structural design of scaffolds for bone regeneration (Hollister, 2005, 2009, Hutmacher, 2001; Salerno *et al.*, 2007; Yoo, 2011., 2012) for more information.

8.8 References

Ahn, E. S., Gleason, N. J., Nakahira, A. and Ying, J. Y. (2001). Nanostructure processing of hydroxyapatite-based bioceramics. *Nano Letters*, **1**, 149–153. (Doi: 10.1021/Nl0055299).

Alam, M. I., Asahina, I., Ohmamiuda, K., Takahashi, K., Yokota, S. and Enomoto, S. (2001). Evaluation of ceramics composed of different hydroxyapatite to tricalcium phosphate ratios as carriers for rhBMP-2. *Biomaterials*, **22**, 1643–1651. (Doi: 10.1016/S0142-9612(00)00322-7).

Albrektsson, T. and Johansson, C. (2001). Osteoinduction, osteoconduction and osseointegration. *European Spine Journal*, **10**, S96–S101. (Doi: 10.1007/s005860100282).

Arinzeh, T. L., Tran, T., Mcalary, J. and Daculsi, G. (2005). A comparative study of biphasic calcium phosphate ceramics for human mesenchymal stem-cell-induced bone formation. *Biomaterials*, **26**, 3631–3638. (Doi: 10.1016/j.biomaterials.2004.09.035).

Arumugam, M. Q., Ireland, D. C. and Brooks, R. A. (2004). Orthosilicic acid increases collagen type I mRNA expression in human bone-derived osteoblasts *in vitro*. *Key Engineering Materials*, **254–2**, 869–872. (Doi: 10.4028/www.scientific.net/KEM.254-256.869).

Arvidson, K., Abdallah, B. M., Applegate, L. A., Baldini, N., Cenni, E., Gomez-Barrena, E., Granchi, D., Kassem, M., Konttinen, Y. T., Mustafa, K., Pioletti, D. P., Sillat, T. and Finne-Wistrand, A. (2011). Bone regeneration and stem cells. *Journal of Cellular and Molecular Medicine*, **15**, 718–746. (Doi: 10.1111/j.1582-4934.2010.01224.x).

Barinov, S. M. (2010). Calcium phosphate-based ceramic and composite materials for medicine. *Russian Chemical Reviews*, **79**, 13–29. (Doi 10.1070/Rc2010v079n01abeh004098).

Bigi, A., Foresti, E., Gregorini, R., Ripamonti, A., Roveri, N. and Shah, J. S. (1992). The role of magnesium on the structure of biological apatites. *Calcified Tissue International*, **50**, 439–444. (Doi: 10.1007/BF00296775).

Boccaccini, A. R., Erol, M., Stark, W. J., Mohn, D., Hong, Z. K. and Mano, J. F. (2010). Polymer/bioactive glass nanocomposites for biomedical applications: A review. *Composites Science and Technology*, **70**, 1764–1776. (Doi: 10.1016/j.compscitech.2010.06.002).

Bodde, E. W., Wolke, J. G., Kowalski, R. S. and Jansen, J. A. (2007). Bone regeneration of porous beta-tricalcium phosphate (Conduit TCP) and of biphasic calcium phosphate ceramic (Biosel) in trabecular defects in sheep. *Journal of Biomedical Materials Research A*, **82**, 711–722. (Doi: 10.1002/jbm.a.30990).

Bohner, M. (2009). Silicon-substituted calcium phosphates – A critical view. *Biomaterials*, **30**, 6403–6406. (Doi: 10.1016/j.biomaterials.2009.08.007).

Boissard, C. I. R., Bourban, P. E., Tami, A. E., Alini, M. and Eglin, D. (2009). Nanohydroxyapatite/poly(ester urethane) scaffold for bone tissue engineering. *Acta Biomaterialia*, **5**, 3316–3327. (Doi: 10.1016/j.actbio.2009.05.001).

Botelho, C. M., Brooks, R. A., Best, S. M., Lopes, M. A., Santos, J. D., Rushton, N. and Bonfield, W. (2006). Human osteoblast response to silicon-substituted hydroxyapatite. *Journal of Biomedical Materials Research Part A*, **79A**, 723–730. (Doi: 10.1002/Jbm.A.30806).

Bouler, J. M., Trecant, M., Delecrin, J., Royer, J., Passuti, N. and Daculsi, G. (1996). Macroporous biphasic calcium phosphate ceramics: Influence of five synthesis parameters on compressive strength. *Journal of Biomedical Materials Research*, **32**, 603–609. (Doi: 10.1002/(SICI)1097-4636(199612)32:4<603::AID-JBM13>3.0.CO;2-E).

Boyan, B. D., Lossdorfer, S., Wang, L., Zhao, G., Lohmann, C. H., Cochran, D. L. and Schwarts, Z. (2003). Osteoblasts generate an osteogenic microenvironment when grown on surfaces with rough microtopographies. *Europen Cells and Materials*, **5**, 11–12. (ww.w.ecmjournal.org/journal/supplements/vol005supp02/pdf/vol005supp02a10.pdf).

Bretcanu, O., Boccaccini, A. R. and Salih, V. (2012). Poly-DL-lactic acid coated Bioglass((R)) scaffolds: Toughening effects and osteosarcoma cell proliferation. *Journal of Materials Science*, **47**, 5661–5672. (Doi: 10.1007/s10853-012-6315-4).

Burg, K. J. L., Porter, S. and Kellam, J. F. (2000). Biomaterial developments for bone tissue engineering. *Biomaterials*, **21**, 2347–2359. (Doi: 10.1016/S0142-9612(00)00102-2).

Calasans-Maia, M., Rossi, A. M., Dias, E. P., Santos, S. R. A., Ascoli, F. and Granjeiro, J. M. (2008). Stimulatory effect on osseous repair of zinc-substituted hydroxyapatite: Histological study in rabbit's tibia. *Bioceramics 20*, **361–363**, 1269–1272. (Doi:10.4028/www.scientific.net/KEM.361-363.1269).

Canalis, E., Hott, M., Deloffre, P., Tsouderos, Y. and Marie, P. J. (1996). The divalent strontium salt S12911 enhances bone cell replication and bone formation *in vitro*. *Bone*, **18**, 517–523. (Doi: 10.1016/8756-3282(96)00080-4).

Capuccini, C., Torricelli, P., Boanini, E., Gazzano, M., Giardino, R. and Bigi, A. (2009). Interaction of Sr-doped hydroxyapatite nanocrystals with osteoclast and osteoblast-like cells. *Journal of Biomedical Materials Research Part A*, **89A**, 594–600. (Doi: 10.1002/Jbm.A.31975).

Carlisle, E. M. (1970). Silicon: A possible factor in bone calcification. *Science*, **167**, 279–280. (Doi:10.1126/science.167.3916.279).

Carlisle, E. M. (1986). Silicon as an essential trace-element in animal nutrition. *Ciba Foundation Symposia*, **121**, 123–139. (Doi: 10.1002/9780470513323.ch8).

Carrodeguas, R. G. and De Aza, S. (2011). alpha-Tricalcium phosphate: Synthesis, properties and biomedical applications. *Acta Biomaterialia*, **7**, 3536–3546. (Doi: 10.1016/j.actbio.2011.06.019).

Cerruti, M. and Sahai, N. (2006). Silicate biomaterials for orthopaedic and dental implants. *Medical Mineralogy and Geochemistry*, **64**, 283–313. (Doi: 10.2138/rmg.2006.64.9).

Chaki, T. K. and Wang, P. E. (1994). Densification and strengthening of silver-reinforced hydroxyapatite-matrix composite prepared by sintering. *Journal of Materials Science-Materials in Medicine*, **5**, 533–542. (Doi: 10.1007/BF00124886).

Chen, C. C., Lai, M. H., Wang, W. C. and Ding, S. J. 2010a. Properties of anti-washout-type calcium silicate bone cements containing gelatin. *Journal of Materials Science-Materials in Medicine*, **21**, 1057–1068. (Doi: 10.1007/s10856-009-3948-1).

Chen, X., Ou, J., Wei, Y., Huang, Z., Kang, Y. and Yin, G. (2010b). Effect of MgO contents on the mechanical properties and biological performances of bioceramics in the MgO-CaO-SiO_2 system. *Journal of Materials Science: Materials in Medicine*, **21**, 1463–1471. (Doi: 10.1007/s10856-010-4025-5).

Chen, X. C., Ou, J., Kang, Y. Q., Huang, Z. B., Zhu, H. Y., Yin, G. F. and Wen, H. M. (2008). Synthesis and characteristics of monticellite bioactive ceramic. *Journal of Materials Science-Materials in Medicine*, **19**, 1257–1263. (Doi: 10.1007/s10856-007-3233-0).

Choi, J. W., Kong, Y. M., Kim, H. E. and Lee, I. S. (1998). Reinforcement of hydroxyapatite bioceramic by addition of Ni_3Al and Al_2O_3. *Journal of the American Ceramic Society*, **81**, 1743–1748. (Doi: 10.1111/j.1151-2916.1998.tb02543.x).

Chow, L. C. (2000). Calcium phosphate cements: Chemistry, properties, and applications. *Mineralization in Natural and Synthetic Biomaterials*, **599**, 27–37. (Doi: 10.1557/PROC-599-27).

Chow, L. C. and Takagi, S. (2001). A natural bone cement – A laboratory novelty led to the development of revolutionary new biomaterials. *Journal of Research of the National Institute of Standards and Technology*, **106**, 1029–1033. (Doi: 10.6028/jres.106.053).

Cihlar, J., Buchal, A. and Trunec, M. (1999). Kinetics of thermal decomposition of hydroxyapatite bioceramics. *Journal of Materials Science*, **34**, 6121–6131. (Doi: 10.1016/S8756-3282(02)00950-X).

Combes, C. and Rey, C. (2010). Amorphous calcium phosphates: Synthesis, properties and uses in biomaterials. *Acta Biomaterialia*, **6**, 3362–3378. (Doi: 10.1016/j.actbio.2010.02.017).

Costantino, P. D., Chaplin, J. M., Wolpoe, M. E., Catalano, P. J., Sen, C., Bederson, J. B. and Govindaraj, S. (2000). Applications of fast-setting hydroxyapatite cement: Cranioplasty. *Otolaryngology Head and Neck Surgery*, **123**, 409–412. (Doi: 10.1067/mhn.2000.107679).

Daculsi, G., Laboux, O., Malard, O. and Weiss, P. (2003). Current state of the art of biphasic calcium phosphate bioceramics. *Journal of Materials Science-Materials in Medicine*, **14**, 195–200. (www.sumisan.com/emailing/5.pdf).

Daculsi, G., LeGeros, R. Z., Heughebaert, M. and Barbieux, I. (1990). Formation of carbonate-apatite crystals after implantation of calcium phosphate ceramics. *Calcified Tissue International*, **46**, 20–27. (Doi: 10.1007/BF02555820).

Delgado, J. A., Morejon, L., Martinez, S., Ginebra, M. P., Carlsson, N., Fernandez, E., Planell, J. A., Clavaguera-Mora, M. T. and Rodriguez-Viejo, J. (1999). Zirconia-toughened hydroxyapatite ceramic obtained by wet sintering. *Journal of Materials Science-Materials in Medicine*, **10**, 715–719. (Doi: 10.1023/A:1008923121172).

Ding, S. J., Shie, M. Y. and Wang, C. Y. (2009). Novel fast-setting calcium silicate bone cements with high bioactivity and enhanced osteogenesis *in vitro*. *Journal of Materials Chemistry*, **19**, 1183–1190. (Doi: 10.1039/B819033j).

Dorozhkin, S. V. (2010). Amorphous calcium (ortho)phosphates. *Acta Biomaterialia*, **6**, 4457–4475. (Doi: 10.1016/j.actbio.2010.06.031).

Dorozhkin, S. V. and Epple, M. (2002). Biological and medical significance of calcium phosphates. *Angewandte Chemie International Edition*, **41**, 3130–3146. (Doi: 10.1002/1521-3773(20020902)41:17<3130::aid-anie3130>3.0.co;2-1).

Driessens, F. C. M., Planell, J. A., Boltong, M. G., Khairoun, I. and Ginebra, M. P. (1998). Osteotransductive bone cements. *Proceedings of the Institution of Mechanical Engineers Part H-Journal of Engineering in Medicine*, **212**, 427–435. (Doi: 10.1243/0954411981534196).

Driessens, F. C. M., Ramselaar, M. M. A., Schaeken, H. G., Stols, A. L. H., Vanmullem, P. J. and Dewijn, J. R. (1992). Chemical-reactions of calcium-phosphate implants after implantation *in vivo*. *Journal of Materials Science-Materials in Medicine*, **3**, 413–417. (Doi: 10.1007/BF00701237).

Ducheyne, P., Radin, S. and King, L. (1993). The effect of calcium phosphate ceramic composition and structure on *in vitro* behavior. I. Dissolution. *Journal of Biomedical Materials Research*, **27**, 25–34. (Doi: 10.1002/jbm.820270105).

Enderle, R., Gotz-Neunhoeffer, F., Gobbels, M., Muller, F. A. and Greil, P. (2005). Influence of magnesium doping on the phase transformation temperature of beta-TCP ceramics examined by Rietveld refinement. *Biomaterials*, **26**, 3379–3384. (Doi: 10.1016/j.biomaterials.2004.09.017).

Fang, T., Wen, J., Zhou, J., Shao, Z. and Dong, J. (2012). Poly (ε-caprolactone) coating delays vancomycin delivery from porous chitosan/β-tricalcium phosphate composites. *Journal of Biomedical Materials Research Part B: Applied Biomaterials*, **100B**, 1803–1811. (Doi: 10.1002/jbm.b.32747).

Frayssinet, P., Rouquet, N., Tourenne, F., Fages, J., Hardy, D. and Bonel, G. (1993). Cell-degradation of calcium-phosphate ceramics. *Cells and Materials*, **3**, 383–394.

Garcia, F., Arias, J. L., Mayor, B., Pou, J., Rehman, I., Knowles, J., Best, S., Leon, B., Perez-Amor, M. and Bonfield, W. (1998). Effect of heat treatment on pulsed laser deposited amorphous calcium phosphate coatings. *Journal of Biomedical Materials Research*, **43**, 69–76. (Doi: 10.1002/(SICI)1097-4636(199821)43:1<69::AID-JBM8>3.0.CO;2-K).

Gauthier, O., Bouler, J. M., Aguado, E., Pilet, P. and Daculsi, G. (1998). Macroporous biphasic calcium phosphate ceramics: Influence of macropore diameter and macroporosity percentage on bone ingrowth. *Biomaterials*, **19**, 133–139. (Doi: 10.1016/S0142-9612(97)00180-4).

Gerhardt, L. C. and Boccaccini, A. R. (2010). Bioactive glass and glass-ceramic scaffolds for bone tissue engineering. *Materials*, **3**, 3867–3910. (Doi: 10.3390/Ma3073867).

Ghosh, S. K., Nandi, S. K., Kundu, B., Datta, S., De, D. K., Roy, S. K. and Basu, D. (2008). *In vivo* response of porous hydroxyapatite and beta-tricalcium phosphate prepared by aqueous solution combustion method and comparison with bioglass scaffolds. *Journal of Biomedical Materials Research B Applied Biomaterials*, **86**, 217–227. (Doi: 10.1002/jbm.b.31009).

Ginebra, M.-P., Canal, C., Espanol, M., Pastorino, D. and Montufar, E. B. (2012). Calcium phosphate cements as drug delivery materials. *Advanced Drug Delivery Reviews*, **64**, 1090–1110. (Doi: 10.1016/j.addr.2012.01.008).

Gorustovich, A. A., Roether, J. A. and Boccaccini, A. R. (2010). Effect of bioactive glasses on angiogenesis: A review of *in vitro* and *in vivo* evidences. *Tissue Engineering Part B-Reviews*, **16**, 199–207. (Doi: 10.1089/ten.teb.2009.0416).

Gou, Z., Chang, J., Zhai, W. and Wang, J. (2005). Study on the self-setting property and the *in vitro* bioactivity of beta-Ca_2SiO_4. *Journal of Biomedical Materials Research B Applied Biomaterials*, **73**, 244–251. (Doi: 10.1002/jbm.*b.30203)*.

Gou, Z. G. and Chang, J. (2004). Synthesis and *in vitro* bioactivity of dicalcium silicate powders. *Journal of the European Ceramic Society*, **24**, 93–99. (Doi: 10.1016/S0955-2219(03)00320-0).

Green, E. and Luttge, A. (2006). Incongruent dissolution of wollastonite measured with vertical scanning interferometry. *American Mineralogist*, **91**, 430–434. (Doi: 10.2138/am.2006.1968).

Gross, K. A. and Rodriguez-Lorenzo, L. M. (2004). Sintered hydroxyfluorapatites. Part II: Mechanical properties of solid solutions determined by microindentation. *Biomaterials*, **25**, 1385–1394. (Doi: 10.1016/S0142-9612(03)00636-7).

Guo, H., Wei, J., Yuan, Y. and Liu, C. S. (2007). Development of calcium silicate/calcium phosphate cement for bone regeneration. *Biomedical Materials*, **2**, S153–S159. (Doi: 10.1088/1748-6041/2/3/S13).

Hafezi-Ardakani, M., Moztarzadeh, F., Rabiee, M. and Talebi, A. R. (2011). Synthesis and characterization of nanocrystalline merwinite (Ca3Mg(SiO4)

(2)) via sol-gel method. *Ceramics International*, **37**, 175–180. (Doi: 10.1016/j.ceramint.2010.08.034).

Hazar, A. B. Y. (2007). Preparation and *in vitro* bioactivity of CaSiO3 powders. *Ceramics International*, **33**, 687–692. (Doi: 10.1016/j.ceramint.2006.12.013).

He, W. H., Tang, R. K. and Xu, X. R. (2011). Synthesis of hollow amorphous calcium phosphate for drug delivery. *Chinese Journal of Inorganic Chemistry*, **27**, 1233–1238.

Hench, L. L. (1991). Bioceramics: From concept to clinic. *Journal of the American Ceramic Society*, **74**, 1487–1510. (Doi: 10.1111/j.1151-2916.1991.tb07132.x).

Hench, L. L., Splinter, R. J., Allen, W. C. and Greenlee, T. K. (1971). Bonding mechanisms at the interface of ceramic prosthetic materials. *Journal of Biomedical Materials Research*, **5**, 117–141. (Doi: 10.1002/jbm.820050611).

Heymann, D., Guicheux, J. and Rousselle, A. V. (2001). Ultrastructural evidence *in vitro* of osteoclast-induced degradation of calcium phosphate ceramic by simultaneous resorption and phagocytosis mechanisms. *Histology and Histopathology*, **16**, 37–44. (http://hdl.handle.net/10201/20645).

Heymann, D., Pradal, G. and Benahmed, M. (1999). Cellular mechanisms of calcium phosphate ceramic degradation. *Histology and Histopathology*, **14**, 871–877. (http://revistas.um.es/hh/article/view/128751/119851).

Höland, W. (1997). Biocompatible and bioactive glass-ceramics – state of the art and new directions. *Journal of Non-Crystalline Solids*, **219**, 192–197. (Doi: 10.1016/S0022-3093(97)00329-3).

Hollister, S. J. (2005). Porous scaffold design for tissue engineering. *Nature Materials*, **4**, 518–524. (Doi: 10.1038/Nmat1421).

Hollister, S. J. (2009). Scaffold design and manufacturing: From concept to clinic. *Advanced Materials*, **21**, 3330–3342. (Doi: 10.1002/adma.200802977).

Hoppe, A., Guldal, N. S. and Boccaccini, A. R. (2011). A review of the biological response to ionic dissolution products from bioactive glasses and glass-ceramics. *Biomaterials*, **32**, 2757–2774. (Doi: 10.1016/j.biomaterials.2011.01.004).

Hsu, H. C., Chiu, C. Y., Tuan, W. H. and Lee, H. Y. (2008). Structural stability of calcium phosphate cement during aging in water. *Materials Science and Engineering C-Biomimetic and Supramolecular Systems*, **28**, 429–433. (Doi: 10.1016/j.msec.2007.04.004).

Huan, Z. and Chang, J. (2009). Novel bioactive composite bone cements based on the beta-tricalcium phosphate-monocalcium phosphate monohydrate composite cement system. *Acta Biomaterialia*, **5**, 1253–1264. (Doi: 10.1016/j.actbio.2008.10.006).

Hutmacher, D. W. (2001). Scaffold design and fabrication technologies for engineering tissues – state of the art and future perspectives. *Journal of Biomaterials Science-Polymer Edition*, **12**, 107–124. (Doi: 10.1163/156856201744489).

Hyakuna, K., Yamamuro, T., Kotoura, Y., Oka, M., Nakamura, T., Kitsugi, T., Kokubo, T. and Kushitani, H. (1990). Surface-reactions of calcium-phosphate ceramics to various solutions. *Journal of Biomedical Materials Research*, **24**, 471–488. (Doi: 10.1002/jbm.820240406).

Ito, T., Saito, M., Uchino, T., Senna, M., Iafisco, M., Prat, M., Rimondini, L. and Otsuka, M. (2012). Preparation of injectable auto-forming alginate gel containing simvastatin with amorphous calcium phosphate as a controlled release medium and their therapeutic effect in osteoporosis model rat. *Journal of*

Materials Science-Materials in Medicine, **23**, 1291–1297. (Doi: 10.1007/s10856-012-4597-3).

Iwata, N. Y., Lee, G. H., Tokuoka, Y. and Kawashima, N. (2004). Sintering behavior and apatite formation of diopside prepared by coprecipitation process. *Colloids and Surfaces B-Biointerfaces*, **34**, 239–245. (Doi: 10.1016/j.colsurfb.2004.01.007).

Jones, J. R. and Hench, L. L. (2003). Regeneration of trabecular bone using porous ceramics. *Current Opinion in Solid State and Materials Science*, **7**, 301–307. (Doi: 10.1016/j.cossms.2003.09.012).

Jun, Y. K., Kim, W. H., Kweon, O. K. and Hong, S. H. (2003). The fabrication and biochemical evaluation of alumina reinforced calcium phosphate porous implants. *Biomaterials*, **24**, 3731–3739. (Doi: 10.1016/S0142-9612(03)00248-5).

Kamitakahara, M., Ohtsuki, C., Kozaka, Y., Ogata, S., Tanihara, M. and Miyazaki, T. (2006). Preparation of porous glass-ceramics containing whitlockite and diopside for bone repair. *Journal of the Ceramic Society of Japan*, **114**, 82–86. (Doi: 10.2109/jcersj.114.82).

Kay, S., Thapa, A., Haberstroh, K. M. and Webster, T. J. (2002). Nanostructured polymer/nanophase ceramic composites enhance osteoblast and chondrocyte adhesion. *Tissue Engineering*, **8**, 753–761. (Doi: 10.1089/10763270260424114).

Kimakhe, S., Heymann, D., Guicheux, J., Pilet, P., Giumelli, B. and Daculsi, G. (1998). Polymyxin B inhibits biphasic calcium phosphate degradation induced by lipopolysaccharide-activated human monocytes/macrophages. *Journal of Biomedical Materials Research,* **40,** 336–340. (Doi: 10.1002/(SICI)1097-4636(199805)40:2<336::AID-JBM19>3.0.CO;2-J).

Kishi, S. and Yamaguchi, M. (1994). Inhibitory effect of zinc compounds on osteoclast-like cell formation in mouse marrow cultures. *Biochemical Pharmacology*, **48**, 1225–1230. (Doi: doi:10.1016/0006-2952(94)90160-0).

Klein, C. P. A. T., Driessen, A. A., Degroot, K. and Vandenhooff, A. (1983). Biodegradation behavior of various calcium-phosphate materials in bone tissue. *Journal of Biomedical Materials Research*, **17**, 769–784. (Doi: 10.1002/jbm.820170505).

Knepper, M., Milthorpe, B. K. and Moricca, S. (1998). Interdiffusion in short-fibre reinforced hydroxyapatite ceramics. *Journal of Materials Science-Materials in Medicine*, **9**, 589–596. (Doi: 10.1023/A:1008965626695).

Knowles, J. C. and Bonfield, W. (1993). Development of a glass-reinforced hydroxyapatite with enhanced mechanical-properties – The effect of glass composition on mechanical-properties and its relationship to phase-changes. *Journal of Biomedical Materials Research*, **27**, 1591–1598. (Doi: 10.1002/jbm.820271217).

Kobayashi, T., Okada, K., Kuroda, T. and Sato, K. (1997). Osteogenic cell cytotoxicity and biomechanical strength of the new ceramic Diopside. *Journal of Biomedical Materials Research*, **37**, 100–107. (Doi: 10.1002/(sici)1097-4636(199710)37:1<100::aid-jbm12>3.0.co;2-k).

Koerten, H. K. and van der Meulen, J. (1999). Degradation of calcium phosphate ceramics. *Journal of Biomedical Materials Research*, **44**, 78–86. (Doi: 10.1002/(SICI)1097-4636(199901)44:1<78::AID-JBM9>3.0.CO;2-6).

Kohn, D. H., Sarmadi, M., Helman, J. I. and Krebsbach, P. H. (2002). Effects of pH on human bone marrow stromal cells *in vitro*: Implications for tissue engineering of bone. *Journal of Biomedical Materials Research*, **60**, 292–299. (Doi: 10.1002/jbm.10050).

Kohri, M., Miki, K., Waite, D. E., Nakajima, H. and Okabe, T. (1993). *In vitro* stability of biphasic calcium phosphate ceramics. *Biomaterials*, **14**, 299–304. (Doi: 10.1016/0142-9612(93)90122-I).

Kokubo, T., Kim, H. M. and Kawashita, M. (2003). Novel bioactive materials with different mechanical properties. *Biomaterials*, **24**, 2161–2175. (Doi: 10.1016/S0142-9612(03)00044-9).

Kong, Y. M., Kim, H. E. and Kim, H. W. (2008). Phase conversion of tricalcium phosphate into Ca-deficient apatite during sintering of hydroxyapatite-tricalcium phosphate biphasic ceramics. *Journal of Biomedical Materials Research Part B-Applied Biomaterials*, **84B**, 334–339. (Doi: 10.1002/Jbm.B.30876).

Kothapalli, C., Wei, M., Vasiliev, A. and Shaw, M. T. (2004). Influence of temperature and concentration on the sintering behavior and mechanical properties of hydroxyapatite. *Acta Materialia*, **52**, 5655–5663. (Doi: 10.1016/j.actamat.2004.08.027).

Kveton, J. F. and Coelho, D. H. (2004). Hydroxyapatite cement in temporal bone surgery: A 10 year experience. *Laryngoscope*, **114**, 33–37. (Doi: 10.1097/00005537-200401000-00005).

Kwon, S.-H., Jun, Y.-K., Hong, S.-H. and Kim, H.-E. (2003). Synthesis and dissolution behavior of β-TCP and HA/β-TCP composite powders. *Journal of the European Ceramic Society*, **23**, 1039–1045. (Doi: 10.1016/s0955-2219(02)00263-7).

Kwon, S. Y., Kim, Y. S., Woo, Y. K., Kim, S. S. and Park, J. B. (1997). Hydroxyapatite impregnated bone cement: *In vitro* and *in vivo* studies. *Bio-medical Materials and Engeering*, **7**, 129–140. (Doi: 10.3233/BME-1997-7205).

Landi, E., Logroscino, G., Proietti, L., Tampieri, A., Sandri, M. and Sprio, S. (2008). Biomimetic Mg-substituted hydroxyapatite: From synthesis to *in vivo* behaviour. *Journal of Materials Science-Materials in Medicine*, **19**, 239–247. (Doi: 10.1007/s10856-006-0032-y).

Langer, R. and Vacanti, J. P. (1993). Tissue engineering. *Science*, **260**, 920–926. (Doi: 10.1126/science.8493529).

Lavernia, C. and Schoenung, J. M. (1991). Calcium-phosphate ceramics as bone substitutes. *American Ceramic Society Bulletin*, **70**, 95–100.

Le Huec, J. C., Schaeverbeke, T., Clement, D., Faber, J. and Le Rebeller, A. (1995). Influence of porosity on the mechanical resistance of hydroxyapatite ceramics under compressive stress. *Biomaterials*, **16**, 113–118. (Doi: 10.1016/0142-9612(95)98272-g).

Lee, T. M., Wang, B. C., Yang, Y. C., Chang, E. and Yang, C. Y. (2001). Comparison of plasma-sprayed hydroxyapatite coatings and hydroxyapatite/tricalcium phosphate composite coatings: *In vivo* study. *Journal of Biomedical Materials Research*, **55**, 360–367. (Doi: 10.1002/1097-4636(20010605)55:3<360::AID-JBM1040>3.0.CO;2-Q [pii]).

LeGeros, R. Z. and LeGeros, J. P. (1993). An introduction to bioceramics. *In:* Hench, L. L. and Wilson, J. (eds.) *An Introduction to Bioceramics*. World Scientific, Singapore.

LeGeros, R. Z. and LeGeros, J. P. (2006). Calcium phosphate biomaterials: An update. *International Journal of Oral-Medical Sciences*, **4**, 117–123. (Doi: 10.5466/ijoms.4.117).

LeGeros, R. Z., Lin, S., Rohanizadeh, R., Mijares, D. and LeGeros, J. P. (2003). Biphasic calcium phosphate bioceramics: Preparation, properties and applications. *Journal of Materials Science: Materials in Medicine*, **14**, 201–209. (Doi: 10.1023/A:1022872421333).

Li, J., Liao, H. and Hermansson, L. (1996). Sintering of partially-stabilized zirconia and partially-stabilized zirconia-hydroxyapatite composites by hot isostatic pressing and pressureless sintering. *Biomaterials*, **17**, 1787–1790. (Doi: 10.1016/0142-9612(95)00356-8).

Li, Y., Zhang, X. and de Groot, K. (1997). Hydrolysis and phase transition of alpha-tricalcium phosphate. *Biomaterials*, **18**, 737–741. (Doi: 10.1016/S0142961296002037).

Liao, C. J., Lin, F. H., Chen, K. S. and Sun, J. S. (1999). Thermal decomposition and reconstitution of hydroxyapatite in air atmosphere. *Biomaterials*, **20**, 1807–1813. (Doi: 10.1016/S0142-9612(99)00076-9).

Lim, D. J., Park, B. K., Jong, W. G., Lee, K. N., Franceschi, R., Kim, S. H., Lee, S. E. and Koh, J. T. (2010). Regulation of bone formation using rapamycin-induced BMP2 expression system: Influence of implanted cell number. *Molecular & Cellular Toxicology*, **6**, 187–194. (Doi: 10.1007/s13273-002-0010-5).

Lin, Q., Li, Y. B., Lan, X. H., Lu, C. H., Chen, Y. X. and Xu, Z. Z. (2009). The apatite formation ability of CaF_2 doping tricalcium silicates in simulated body fluid. *Biomedical Materials*, **4**. (Doi: 10.1088/1748-6041/4/4/045005).

Liu, X., Ding, C. and Chu, P. K. (2004a). Mechanism of apatite formation on wollastonite coatings in simulated body fluids. *Biomaterials*, **25**, 1755–1761. (Doi: 10.1016/S0142961203006835).

Liu, X. Y. and Ding, C. X. (2001). Phase compositions and microstructure of plasma sprayed wollastonite coating. *Surface & Coatings Technology*, **141**, 269–274.

Liu, X. Y., Ding, C. X. and Chu, P. K. (2004b). Mechanism of apatite formation on wollastonite coatings in simulated body fluids. *Biomaterials*, **25**, 1755–1761. (Doi: 10.1016/j.biomaterials.2003.08.024).

Liu, X. Y., Ding, C. X. and Wang, Z. Y. (2001). Apatite formed on the surface of plasma-sprayed wollastonite coating immersed in simulated body fluid. *Biomaterials*, **22**, 2007–2012.

Liu, X. Y., Morra, M., Carpi, A. and Li, B. (2008). Bioactive calcium silicate ceramics and coatings. *Biomedicine & Pharmacotherapy*, **62**, 526–529. (Doi: 10.1016/j.biopha.2008.07.051).

Liu, X. Y., Tao, S. Y. and Ding, C. X. (2002). Bioactivity of plasma sprayed dicalcium silicate coatings. *Biomaterials*, **23**, 963–968. (Doi: 10.1016/S0142-9612(01)00210-1).

Liu, X. Y., Xie, Y. T., Ding, C. X. and Chu, P. K. (2005). Early apatite deposition and osteoblast growth on plasma-sprayed dicalcium silicate coating. *Journal of Biomedical Materials Research Part A*, **74A**, 356–365. (Doi: 10.1002/Jbm.A.30273).

Lobo, S. E. and Arinzeh, T. L. (2010). Biphasic calcium phosphate ceramics for bone regeneration and tissue engineering applications. *Materials*, **3**, 815–826. (Doi: 10.3390/Ma3020815).

Locardi, B., Pazzaglia, U. E., Gabbi, C. and Profilo, B. (1993). Thermal-behavior of hydroxyapatite intended for medical applications. *Biomaterials*, **14**, 437–441. (Doi: 10.1016/0142-9612(93)90146-S).

Lopes, M. A., Monteiro, F. J. and Santos, J. D. (1999). Glass-reinforced hydroxyapatite composites: Fracture toughness and hardness dependence on microstructural characteristics. *Biomaterials*, **20**, 2085–2090. (Doi: 10.1016/S0142-9612(99)00112-X).

Low, K. L., Tan, S. H., Zein, S. H. S., McPhail, D. S. and Boccaccini, A. R. (2011). Optimization of the mechanical properties of calcium phosphate/multi-walled carbon nanotubes/bovine serum albumin composites using response surface methodology. *Materials & Design*, **32**, 3312–3319. (Doi: 10.1016/j.matdes.2011.02.022).

Lu, J. X., Descamps, M., Dejou, J., Koubi, G., Hardouin, P., Lemaitre, J. and Proust, J. P. (2002). The biodegradation mechanism of calcium phosphate biomaterials in bone. *Journal of Biomedical Materials Research*, **63**, 408–412. (Doi: 10.1002/jbm.10259).

Luo, T., Wu, C. and Zhang, Y. (2012). The *in vivo* osteogenesis of Mg or Zr-modified silicate-based bioceramic spheres. *Journal of Biomedical Materials Research Part A*, **100**, 2269–2277. (Doi: 10.1002/jbm.a.34161).

Lutolf, M. P. and Hubbell, J. A. (2005). Synthetic biomaterials as instructive extracellular microenvironments for morphogenesis in tissue engineering. *Nature Biotechnology*, **23**, 47–55. (Doi: 10.1038/Nbt1055).

MacDonald, R. S. (2000). The role of zinc in growth and cell proliferation. *The Journal of Nutrition*, **130**, 1500S-1508S.

Mallick, K. K., Winnett, J., van Grunsven, W., Lapworth, J. and Reilly, G. C. (2012). Three-dimensional porous bioscaffolds for bone tissue regeneration: Fabrication via adaptive foam reticulation and freeze casting techniques, characterization, and cell study. *Journal of Biomedical Materials Research A*, **100**, 2948–2959. (Doi: 10.1002/jbm.a.34238).

Martin, J. Y., Schwartz, Z., Hummert, T. W., Schraub, D. M., Simpson, J., Lankford, J., Dean, D. D., Cochran, D. L. and Boyan, B. D. (1995). Effect of titanium surface-roughness on proliferation, differentiation, and protein-synthesis of human osteoblast-like cells (Mg63). *Journal of Biomedical Materials Research*, **29**, 389–401. (Doi: 10.1002/jbm.820290314).

Maus, U., Andereya, S., Gravius, S., Ohnsorge, J. A., Niedhart, C. and Siebert, C. H. (2008). BMP-2 incorporated in a tricalcium phosphate bone substitute enhances bone remodeling in sheep. *Journal of Biomaterials Applications*, **22**, 559–576. (Doi: 10.1177/0885328207083311).

Maxian, S. H., Zawadsky, J. P. and Dunn, M. G. (1993). Mechanical and histological-evaluation of amorphous calcium-phosphate and poorly crystallized hydroxyapatite coatings on titanium implants. *Journal of Biomedical Materials Research*, **27**, 717–728. (Doi: 10.1002/jbm.820270604).

Metsger, D. S., Rieger, M. R. and Foreman, D. W. (1999). Mechanical properties of sintered hydroxyapatite and tricalcium phosphate ceramic. *Journal of Materials Science: Materials in Medicine*, **10**, 9–17. (Doi: doi:10.1016/j.actbio.2008.02.031).

Miake, Y., Yanagisawa, T., Yajima, Y., Noma, H., Yasui, N. and Nonami, T. (1995). High resolution and analytical electron microscopic studies of new crystals induced by a bioactive ceramic (Diopside). *Journal of Dental Research*, **74**, 1756–1763. (Doi: 10.1177/00220345950740110701).

Mohamad Yunos, D., Bretcanu, O. and Boccaccini, A. (2008). Polymer-bioceramic composites for tissue engineering scaffolds. *Journal of Materials Science*, **43**, 4433–4442. (Doi: 10.1007/s10853-008-2552-y).

Muralithran, G. and Ramesh, S. (2000). The effects of sintering temperature on the properties of hydroxyapatite. *Ceramics International*, **26**, 221–230. (Doi: 10.1016/S0272-8842(99)00046-2).

Nagano, M., Nakamura, T., Kokubo, T., Tanahashi, M. and Ogawa, M. (1996). Differences of bone bonding ability and degradation behaviour *in vivo* between amorphous calcium phosphate and highly crystalline hydroxyapatite coating. *Biomaterials*, **17**, 1771–1777. (Doi: 10.1016/0142-9612(95)00357-6).

Nakajima, S. (1990). Experimental studies of healing process on reinforcement ceramic implantation in rabbit mandible. *Shikwa Gakuho*, **90**, 525–553.

Nakajima, S., Harada, Y., Kurihara, Y., Wakatsuki, T. and Norma, H. (1989). Physicochemical characteristics of new reinforcement ceramic implants. *Shikwa Gakuho*, **89**, 1709–1717.

Narayanan, R., Seshadri, S. K., Kwon, T. Y. and Kim, K. H. (2008). Calcium phosphate-based coatings on titanium and its alloys. *Journal of Biomedical Materials Research Part B-Applied Biomaterials*, **85B**, 279–299. (Doi: 10.1002/Jbm.B.30932).

Neumann, M. and Epple, M. (2006). Composites of calcium phosphate and polymers as bone substitution materials. *European Journal of Trauma*, **32**, 125–131. (Doi: 10.1007/s00068-006-6044-y).

Ni, G. X., Lin, J. H., Chiu, P. K., Li, Z. Y. and Lu, W. W. (2010). Effect of strontium-containing hydroxyapatite bone cement on bone remodeling following hip replacement. *Journal of Materials Science Materials in Medicine*, **21**, 377–384. (Doi: 10.1007/s10856-009-3866-2).

Ni, S. Y., Chang, J., Chou, L. and Zhai, W. Y. (2007). Comparison of osteoblast-like cell responses to calcium silicate and tricalcium phosphate ceramics *in vitro*. *Journal of Biomedical Materials Research Part B-Applied Biomaterials*, **80B**, 174–183. (Doi: 10.1002/Jbm.B.30582).

Nonami, T. (1992). Developmental-study of diopside for use as implant material. *Tissue-Inducing Biomaterials*, **252**, 87–92. (Doi: 10.1557/PROC-252-87).

Nonami, T. and Tsutsumi, S. (1999). Study of diopside ceramics for biomaterials. *Journal of Materials Science-Materials in Medicine*, **10**, 475–479. (Doi: 10.1023/A:1008996908797).

Nyan, M., Sato, D., Kihara, H., Machida, T., Ohya, K. and Kasugai, S. (2009). Effects of the combination with alpha-tricalcium phosphate and simvastatin on bone regeneration. *Clinical Oral Implants Research*, **20**, 280–287. (Doi: 10.1111/j.1600-0501.2008.01639.x).

Otsuki, B., Takemoto, M., Fujibayashi, S., Neo, M., Kokubo, T. and Nakamura, T. (2006). Pore throat size and connectivity determine bone and tissue ingrowth into porous implants: Three-dimensional micro-CT based structural analyses of porous bioactive titanium implants. *Biomaterials*, **27**, 5892–5900. (Doi: 10.1016/j.biomaterials.2006.08.013).

Ou, J., Kang, Y. Q., Huang, Z. B., Chen, X. C., Wu, J., Xiao, R. C. and Yin, G. F. (2008). Preparation and *in vitro* bioactivity of novel merwinite ceramic. *Biomedical Materials*, **3**. (Doi: 10.1088/1748-6041/3/1/015015).

Panzavolta, S., Bracci, B., Rubini, K. and Bigi, A. (2011). Optimization of a biomimetic bone cement: Role of DCPD. *Journal of Inorganic Biochemistry*, **105**, 1060–1065. (Doi: 10.1016/j.jinorgbio.2011.05.006).

Park, J. B. and Bronzino, J. D. (eds.) (2003). *Biomaterials Principles and Applications*, Boca Raton: CRC Press.

Park, K. and Vasilos, T. (1997). Characteristics of carbon fibre-reinforced calcium phosphate composites fabricated by hot pressing. *Journal of Materials Science Letters*, **16**, 985–987.

Patel, N., Best, S. M., Bonfield, W., Gibson, I. R., Hing, K. A., Damien, E. and Revell, P. A. (2002). A comparative study on the *in vivo* behavior of hydroxyapatite and silicon substituted hydroxyapatite granules. *Journal of Materials Science: Materials in Medicine*, **13**, 1199–1206. (Doi: 10.1023/A:1021114710076).

Peters, S. C., Blum, J. D., Driscoll, C. T. and Likens, G. E. (2004). Dissolution of wollastonite during the experimental manipulation of Hubbard Brook Watershed 1. *Biogeochemistry*, **67**, 309–329. (Doi: 10.1023/B:BIOG.0000015787.44175.3f).

Pina, S., Vieira, S. I., Rego, P., Torres, P. M., da Cruz e Silva, O. A., da Cruz e Silva, E. F. and Ferreira, J. M. (2010). Biological responses of brushite-forming Zn- and ZnSr- substituted beta-tricalcium phosphate bone cements. *European Cells and Materials*, **20**, 162–177. (http://scholar.google.com.au/scholar?hl=enandas_sdt=0,5andq=Biological+responses+of+brushite-forming+Zn-+and+Zn-+substituted+beta-tricalcium+phosphate+bone+cements).

Popat, K. C., Leoni, L., Grimes, C. A. and Desai, T. A. (2007). Influence of engineered titania nanotubular surfaces on bone cells. *Biomaterials*, **28**, 3188–3197. (Doi: 10.1016/j.biomaterials.2007.03.020).

Porter, A. E., Best, S. M. and Bonfield, W. (2004a). Ultrastructural comparison of hydroxyapatite and silicon-substituted hydroxyapatite for biomedical applications. *Journal of Biomedical Materials Research Part A*, **68A**, 133–141. (Doi: 10.1002/jbm.a.20064).

Porter, A. E., Botelho, C. M., Lopes, M. A., Santos, J. D., Best, S. M. and Bonfield, W. (2004b). Ultrastructural comparison of dissolution and apatite precipitation on hydroxyapatite and silicon-substituted hydroxyapatite *in vitro* and *in vivo*. *Journal of Biomedical Materials Research Part A*, **69A**, 670–679. (Doi: 10.1002/Jbm.A.30035).

Porter, A. E., Patel, N., Skepper, J. N., Best, S. M. and Bonfield, W. (2003). Comparison of *in vivo* dissolution processes in hydroxyapatite and silicon-substituted hydroxyapatite bioceramics. *Biomaterials*, **24**, 4609–4620. (Doi: S0142961203003557).

Porter, A. E., Patel, N., Skepper, J. N., Best, S. M. and Bonfield, W. (2004c). Effect of sintered silicate-substituted hydroxyapatite on remodelling processes at the bone-implant interface. *Biomaterials*, **25**, 3303–3314. (Doi: 10.1016/j.biomaterials.2003.10.006).

Price, R. L., Waid, M. C., Haberstroh, K. M. and Webster, T. J. (2003). Selective bone cell adhesion on formulations containing carbon nanofibers. *Biomaterials*, **24**, 1877–1887. (Doi: 10.1016/S0142-9612(02)00609-9).

Radin, S. R. and Ducheyne, P. (1993). The effect of calcium-phosphate ceramic composition and structure on *in-vitro* behavior. II. Precipitation. *Journal of Biomedical Materials Research*, **27**, 25–34.

Ramaswamy, Y., Wu, C., Van Hummel, A., Combes, V., Grau, G. and Zreiqat, H. (2008). The responses of osteoblasts, osteoclasts and endothelial cells to zirconium modified calcium-silicate-based ceramic. *Biomaterials*, **29**, 4392–4402. (Doi: 10.1016/j.biomaterials.2008.08.006).

Ramp, W. K., Lenz, L. G. and Kaysinger, K. K. (1994). Medium pH modulates matrix, mineral, and energy metabolism in cultured chick bones and osteoblast-like cells. *Bone and Mineral*, **24**, 59–73. (Doi: 10.1016/s0169-6009(08)80131-6).

Ravaglioli, A. and Krajewski, A. (eds.) (1992). *Bioceramics, Materials, Properties and Applications* London: Chapman and Hall.

Raynaud, S., Champion, E., Lafon, J. P. and Bernache-Assollant, D. (2002). Calcium phosphate apatites with variable Ca/P atomic ratio III. Mechanical properties and degradation in solution of hot pressed ceramics. *Biomaterials*, **23**, 1081–1089. (Doi: 10.1016/S0142-9612(01)00220-4).

Reffitt, D. M., Ogston, N., Jugdaohsingh, R., Cheung, H. F., Evans, B. A., Thompson, R. P., Powell, J. J. and Hampson, G. N. (2003). Orthosilicic acid stimulates collagen type 1 synthesis and osteoblastic differentiation in human osteoblast-like cells *in vitro*. *Bone*, **32**, 127–135. (Doi: 10.1016/S8756-3282(02)00950-X).

Rey, C., Combes, C., drouet, C. and Grossin, D. (2011). Bioactive ceramics: Physical chemistry, *Comprehensive Biomaterials*, **1**, 187–221. (Doi: 10.1016/B978-0-08-055294-1.00178-1).

Rezwan, K., Chen, Q. Z., Blaker, J. J. and Boccaccini, A. R. (2006). Biodegradable and bioactive porous polymer/inorganic composite scaffolds for bone tissue engineering. *Biomaterials*, **27**, 3413–3431. (Doi: 10.1016/j.biomaterials.2006.01.039).

Roohani-Esfahani, S. I., Dunstan, C. R., Davies, B., Pearce, S., Williams, R. and Zreiqat, H. (2012). Repairing a critical-sized bone defect with highly porous modified and unmodified baghdadite scaffolds. *Acta Biomaterialia*, **8**, 4162–4172. (Doi: 10.1016/j.actbio.2012.07.036).

Ryu, H. S., Youn, H. J., Hong, K. S., Chang, B. S., Lee, C. K. and Chung, S. S. (2002). An improvement in sintering property of beta-tricalcium phosphate by addition of calcium pyrophosphate. *Biomaterials*, **23**, 909–914. (Doi: 10.1016/S0142-9612(01)00201-0).

Safronova, T. V., Putlyaev, V. I., Avramenko, O. A., Shekhirev, M. A. and Veresov, A. G. (2011). Ca-deficient hydroxyapatite powder for producing tricalcium phosphate based ceramics. *Glass and Ceramics*, **68**, 28–32. (Doi: 10.1007/s10717-011-9315-y).

Sainz, M. A., Pena, P., Serena, S. and Caballero, A. (2010). Influence of design on bioactivity of novel $CaSiO_3$-$CaMg(SiO_3)_2$ bioceramics: *In vitro* simulated body fluid test and thermodynamic simulation. *Acta Biomaterialia*, **6**, 2797–2807. (Doi: 10.1016/j.actbio.2010.01.003).

Saito, T., Kin, Y. and Koshino, T. (2002). Osteogenic response of hydroxyapatite cement implanted into the femur of rats with experimentally induced osteoporosis. *Biomaterials*, **23**, 2711–2716.(Doi: 10.1016/S0142-9612(02)00003-0).

Salerno, A., Oliviero, M., Di Maio, E., Iannace, S. and Netti, P. A. (2007). Design and preparation of μ-bimodal porous scaffold for tissue engineering. *Journal of Applied Polymer Science*, **106**, 3335–3342. (Doi: 10.1002/App.26881).

Sandstead, H. H. (2000). Zinc: Growth, development, and function. *The Journal of Trace Elements in Experimental Medicine*, **13**, 41–49. (Doi: 10.1002/(SICI)1520-670X(2000)13:1<41::AID-JTRA6>3.0.CO;2-G).

Santos, J. D., Knowles, J. C., Reis, R. L., Monteiro, F. J. and Hastings, G. W. (1994). Microstructural characterization of glass-reinforced hydroxyapatite composites. *Biomaterials*, **15**, 5–10. (Doi: 10.1016/0142-9612(94)90188-0).

Schmetterer, C. and Masset, P. (2012). Heat capacity of compounds in the CaO-SiO2 system – A Review. *Journal of Phase Equilibria and Diffusion*, **33**, 261–275. (Doi: 10.1007/s11669-012-0046-8).

Schnettler, R., Alt, V., Dingeldein, E., Pfefferle, H. J., Kilian, O., Meyer, C., Heiss, C. and Wenisch, S. (2003). Bone ingrowth in bFGF-coated hydroxyapatite ceramic implants. *Biomaterials*, **24**, 4603–4608. (Doi: 10.1016/S0142961203003545).

Sola, A., Bellucci, D., Cannillo, V. and Cattini, A. (2011). Bioactive glass coatings: A review. *Surface Engineering*, **27**, 560–572. (Doi: 10.1179/1743294410y.0000000008).

Sous, M., Bareille, R., Rouais, F., Clement, D., Amedee, J., Dupuy, B. and Baquey, C. (1998). Cellular biocompatibility and resistance to compression of macroporous beta-tricalcium phosphate ceramics. *Biomaterials*, **19**, 2147–2153. (Doi: 10.1016/S0142-9612(98)00118-5).

Suchanek, W., Yashima, M., Kakihana, M. and Yoshimura, M. (1997). Hydroxyapatite/hydroxyapatite-whisker composites without sintering additives: Mechanical properties and microstructural evolution. *Journal of the American Ceramic Society*, **80**, 2805–2813. (Doi: 10.1111/j.1151-2916.1997.tb03197.x).

Sun, J., Wei, L., Liu, X., Li, J., Li, B., Wang, G. and Meng, F. (2009a). Influences of ionic dissolution products of dicalcium silicate coating on osteoblastic proliferation, differentiation and gene expression. *Acta Biomaterialia*, **5**, 1284–1293. (Doi: 10.1016/j.actbio.2008.10.011).

Sun, J. Y., Li, J. Y., Liu, X. Y., Wei, L., Wang, G. C. and Meng, F. H. (2009b). Proliferation and gene expression of osteoblasts cultured in DMEM containing the ionic products of dicalcium silicate coating. *Biomedicine & Pharmacotherapy*, **63**, 650–657. (Doi: 10.1016/j.biopha.2009.01.007).

Thompson, I. D. and Hench, L. L. (1998). Mechanical properties of bioactive glasses, glass-ceramics and composites. *Proceedings of the Institution of Mechanical Engineers Part H-Journal of Engineering in Medicine*, **212**, 127–136. (Doi: 10.1243/0954411981533908).

Toquet, J., Rohanizadeh, R., Guicheux, J., Couillard, S., Passuti, N., Daculsi, G. and Heymann, D. (1999). Osteogenic potential *in vitro* of human bone marrow cells cultured on macroporous biphasic calcium phosphate ceramic. *Journal of Biomedical Materials Research*, **44**, 98–108. (Doi: 10.1002/(SICI)1097-4636(199901)44:1<98::AID-JBM11>3.3.CO;2-G).

Tulyaganov, D. U., Agathopoulos, S., Ventura, J. M., Karakassides, M. A., Fabrichnaya, O. and Ferreira, J. M. F. (2006). Synthesis of glass-ceramics in the CaO-MgO-SiO_2 system with B_2O_3, P_2O_5, Na_2O and CaF_2 additives. *Journal of the European Ceramic Society*, **26**, 1463–1471. (Doi: 10.1016/j.jeurceramsoc.2005.02.009).

Viswanath, B., Raghavan, R., Gurao, N. P., Ramamurty, U. and Ravishankar, N. (2008). Mechanical properties of tricalcium phosphate single crystals grown by molten salt synthesis. *Acta Biomaterialia*, **4**, 1448–1454. (Doi: 10.1016/j.actbio.2008.02.031).

Wang, G., Lu, Z., Dwarte, D. and Zreiqat, H. (2012a). Porous scaffolds with tailored reactivity modulate *in-vitro* osteoblast responses. *Materials Science and Engineering: C*, **32**, 1818–1826. (Doi: 10.1016/j.msec.2012.04.068).

Wang, G., Lu, Z., Xie, K. Y., Lu, W. Y., Roohani-Esfahani, S. I., Kondyurin, A. and Zreiqat, H. (2012b). A facile method to *in situ* formation of hydroxyapatite single crystal architecture for enhanced osteoblast adhesion. *Journal of Materials Chemistry*, **22**, 19081–19087. (Doi: 10.1039/c2jm34367c).

Wang, G. and Zreiqat, H. (2010). Functional coatings or films for hard-tissue applications. *Materials*, **3**, 3994–4050. (Doi: 10.3390/ma3073994).

Wang, J. W. and Shaw, L. L. (2007). Morphology-enhanced low-temperature sintering of nanocrystalline hydroxyapatite. *Advanced Materials*, **19**, 2364–2369. (Doi: 10.1002/adma.200602333).

Webster, T. J. (2001). Nanophase ceramics: The future orthopedic and dental implant material. *Advances in Chemical Engineering*, **27**, 125–166. (Doi: 10.1016/S0065-2377(01)27005-7).

Webster, T. J., Ergun, C., Doremus, R. H., Siegel, R. W. and Bizios, R. (2000). Enhanced functions of osteoblasts on nanophase ceramics. *Biomaterials*, **21**, 1803–1810. (Doi: doi:10.1016/S0142-9612(00)00075-2).

Webster, T. J., Schadler, L. S., Siegel, R. W. and Bizios, R. (2001). Mechanisms of enhanced osteoblast adhesion on nanophase alumina involve vitronectin. *Tissue Engineering*, **7**, 291–301. (Doi: 10.1089/10763270152044152).

Welch, J. H. and Gutt, W. (1959). Tricalcium silicate and its stability within the system CaO-SiO_2. *Journal of the American Ceramic Society*, **42**, 11–15. (Doi: 10.1111/j.1151-2916.1959.tb09135.x).

Wernike, E., Montjovent, M. O., Liu, Y., Wismeijer, D., Hunziker, E. B., Siebenrock, K. A., Hofstetter, W. and Klenke, F. M. (2010). VEGF incorporated into calcium phosphate ceramics promotes vascularisation and bone formation *in vivo*. *European Cells & Materials*, **19**, 30–40. (www.ecmjournal.org/journal/papers/vol019/pdf/v019a04.pdf).

White, A. A., Best, S. M. and Kinloch, I. A. (2007). Hydroxyapatite-carbon nanotube composites for biomedical applications: A review. *International Journal of Applied Ceramic Technology*, **4**, 1–13. (Doi: 10.1111/j.1744-7402.2007.02113.x).

Woodard, J. R., Hilldore, A. J., Lan, S. K., Park, C. J., Morgan, A. W., Eurell, J. A. C., Clark, S. G., Wheeler, M. B., Jamison, R. D. and Wagoner Johnson, A. J. (2007). The mechanical properties and osteoconductivity of hydroxyapatite bone scaffolds with multi-scale porosity. *Biomaterials*, **28**, 45–54. (Doi: 10.1016/j.biomaterials.2006.08.021).

Wu, C. and Chang, J. (2007). Degradation, bioactivity, and cytocompatibility of diopside, akermanite, and bredigite ceramics. *Journal of Biomedical Materials Research Part B-Applied Biomaterials*, **83B**, 153–160. (Doi: 10.1002/Jbm.B.30779).

Wu, C., Chang, J., Ni, S. and Wang, J. (2006a). *In vitro* bioactivity of akermanite ceramics. *Journal of Biomedical Materials Research A*, **76**, 73–80. (Doi: 10.1002/jbm.a.30496).

Wu, C., Chang, J., Wang, J., Ni, S. and Zhai, W. (2005a). Preparation and characteristics of a calcium magnesium silicate (bredigite) bioactive ceramic. *Biomaterials*, **26**, 2925–2931. (Doi: 10.1016/j.biomaterials.2004.09.019).

Wu, C., Chang, J., Zhai, W., Ni, S. and Wang, J. (2006b). Porous akermanite scaffolds for bone tissue engineering: Preparation, characterization, and *in vitro* studies. *Journal of Biomedical Materials Research Part B: Applied Biomaterials*, **78B**, 47–55. (Doi: 10.1002/jbm.b.30456).

Wu, C., Ramaswamy, Y., Chang, J., Woods, J., Chen, Y. and Zreiqat, H. (2008). The effect of Zn contents on phase composition, chemical stability and cellular bioactivity in Zn-Ca-Si system ceramics. *Journal of Biomedical Materials Research B: Applied Biomaterials*, **87**, 346–353. (Doi: 10.1002/jbm.b.31109).

Wu, C. T. (2009). Methods of improving mechanical and biomedical properties of Ca-Si-based ceramics and scaffolds. *Expert Review of Medical Devices*, **6**, 237–241. (Doi: 10.1586/Erd.09.3).

Wu, C. T., Chang, J. and Zhai, W. Y. (2005b). A novel hardystonite bioceramic: Preparation and characteristics. *Ceramics International*, **31**, 27–31. (Doi: 10.1016/j.ceramint.2004.02.008).

Wu, C. T., Chang, J., Zhai, W. Y. and Ni, S. Y. (2007). A novel bioactive porous bredigite ($Ca_7MgSi_4O_{16}$) scaffold with biomimetic apatite layer for bone tissue engineering. *Journal of Materials Science-Materials in Medicine*, **18**, 857–864. (Doi: 10.1007/s10856-006-0083-0).

Xue, W. C., Liu, X. Y., Zheng, X. B. and Ding, C. X. (2004a). Plasma-sprayed diopside coatings for biomedical applications. *Surface & Coatings Technology*, **185**, 340–345. (Doi: 10.1016/j.surfcoat.2003.12.018).

Xue, W. C., Tao, S. Y., Liu, X. Y., Zheng, X. B. and Ding, C. X. (2004b). *In vivo* evaluation of plasma sprayed hydroxyapatite coatings having different crystallinity. *Biomaterials*, **25**, 415–421. (Doi: 10.1016/S0142-9612(03)00545-3).

Yamaguchi, M. and Miwa, H. (1991). Stimulatory effect of beta-alanyl-L-histidinato zinc on bone formation in tissue culture. *Pharmacology*, **42**, 230–240. (Doi: 10.1159/000138802).

Yamaguchi, M., Oishi, H. and Suketa, Y. (1987). Stimulatory effect of zinc on bone formation in tissue culture. *Biochemical Pharmacology*, 36, 4007–4012. (Doi: 10.1016/0006-2952(87)90471-0).

Yang, S. F., Leong, K. F., Du, Z. H. and Chua, C. K. (2001). The design of scaffolds for use in tissue engineering. Part 1. Traditional factors. *Tissue Engineering*, **7**, 679–689. (Doi: doi:10.1089/107632701753337645).

Yang, S. F., Leong, K. F., Du, Z. H. and Chua, C. K. (2002). The design of scaffolds for use in tissue engineering. Part II. Rapid prototyping techniques. *Tissue Engineering*, **8**, 1–11. (Doi: 10.1089/107632702753503009).

Yoo, D. J. (2011). Computer-aided porous scaffold design for tissue engineering using triply periodic minimal surfaces. *International Journal of Precision Engineering and Manufacturing*, **12**, 61–71. (Doi: 10.1007/s12541-011-0008-9).

Yoo, D. J. (2012). Heterogeneous porous scaffold design for tissue engineering using triply periodic minimal surfaces. *International Journal of Precision Engineering and Manufacturing*, **13**, 527–537. (Doi: 10.1007/s12541-012-0068-5).

Yuan, H., Van Den Doel, M., Li, S., Van Blitterswijk, C. A., De Groot, K. and De Bruijn, J. D. (2002). A comparison of the osteoinductive potential of two calcium phosphate ceramics implanted intramuscularly in goats. *Journal of Materials Science: Materials in Medicine*, **13**, 1271–1275. (Doi: 10.1023/A:1021191432366).

Yuan, H., Yang, Z., De Bruij, J. D., De Groot, K. and Zhang, X. (2001). Material-dependent bone induction by calcium phosphate ceramics: A 2.5-year study in dog. *Biomaterials*, **22**, 2617–2623. (Doi: 10.1016/S0142-9612(00)00450-6).

Zhang, X., Gubbels, G. H. M., Terpstra, R. A. and Metselaar, R. (1997). Toughening of calcium hydroxyapatite with silver particles. *Journal of Materials Science*, **32**, 235–243. (Doi: 10.1023/A:1018568308926).

Zhang, Y., Liu, Y., Ji, X. B., Banks, C. E. and Zhang, W. (2011). Flower-like hydroxyapatite modified carbon paste electrodes applicable for highly sensitive detection of heavy metal ions. *Journal of Materials Chemistry*, **21**, 7552–7554. (Doi: 10.1039/C1jm10949a).

Zhao, J., Liu, Y., Sun, W. B. and Zhang, H. (2011). Amorphous calcium phosphate and its application in dentistry. *Chemistry Central Journal*, **5**, 40. (Doi: 10.1186/1752-153X-5-40).

Zhao, W. Y. and Chang, J. (2004). Sol-gel synthesis and *in vitro* bioactivity of tricalcium silicate powders. *Materials Letters*, **58**, 2350–2353. (Doi: 10.1016/j.matlet.2004.02.045).

Zhao, W. Y., Wang, J. Y., Zhai, W. Y., Wang, Z. and Chang, J. (2005). The self-setting properties and *in vitro* bioactivity of tricalcium silicate. *Biomaterials*, **26**, 6113–6121. (Doi: 10.1016/j.biomaterials.2005.04.025).

Zhou, H., Lawrence, J. G. and Bhaduri, S. B. (2012). Fabrication aspects of PLA-CaP/PLGA-CaP composites for orthopedic applications: A review. *Acta Biomaterialia*, **8**, 1999–2016. (Doi: 10.1016/j.actbio.2012.01.031).

Zreiqat, H., Ramaswamy, Y., Wu, C., Paschalidis, A., Lu, Z., James, B., Birke, O., McDonald, M., Little, D. and Dunstan, C. R. (2010). The incorporation of strontium and zinc into a calcium–silicon ceramic for bone tissue engineering. *Biomaterials*, **31**, 3175–3184. (Doi: 10.1016/j.biomaterials.2010.01.024).

Part III

Biomaterials for bone repair and regeneration

9 Cartilage grafts for bone repair and regeneration

C. S. BAHNEY and R. S. MARCUCIO, University of California, San Francisco, USA

DOI: 10.1533/9780857099037.3.219

Abstract: During embryonic development and fracture repair the majority of bone is formed indirectly from a cartilage intermediate in a process called endochondral ossification. In contrast, the majority of therapeutic strategies to promote bone regeneration aim to stimulate direct osteogenesis through intramembranous ossification. Clinically this has led to failures associated with osteonecrosis and poor osseointegration of the bone graft. In this chapter we discuss how using cartilage to stimulate bone repair may be a better therapeutic strategy than current bone grafting techniques.

Key words: endochondral ossification, tissue engineering, bone regeneration, cartilage graft.

9.1 Introduction

Although bone has a great capacity for repair, there are a number of clinical indications that require the healing process to be augmented through the use of osteoconductive or osteoinductive grafts to stimulate bone regeneration. Current bone grafting technologies are associated with a number of clinical drawbacks including osteonecrosis and poor integration of the graft with the host tissue. One reason for this could be that bone grafts promote repair through intramembranous ossification, or direct bone formation. However, the majority of bones in the body form and are repaired through the process of endochondral ossification, or indirect bone formation that proceeds through a cartilage intermediate. Developing new technologies that promote bone regeneration through an endochondral repair mechanism may help to overcome observed clinical failures (Fig. 9.1a).

A cartilage graft developed in cell-laden, three-dimensional biomaterial scaffolds is one potential technique by which endochondral repair can be promoted (Fig. 9.1b). Cartilage is an inherently bioactive substrate that promotes vascularization and mineralization of the extracellular matrix (ECM).

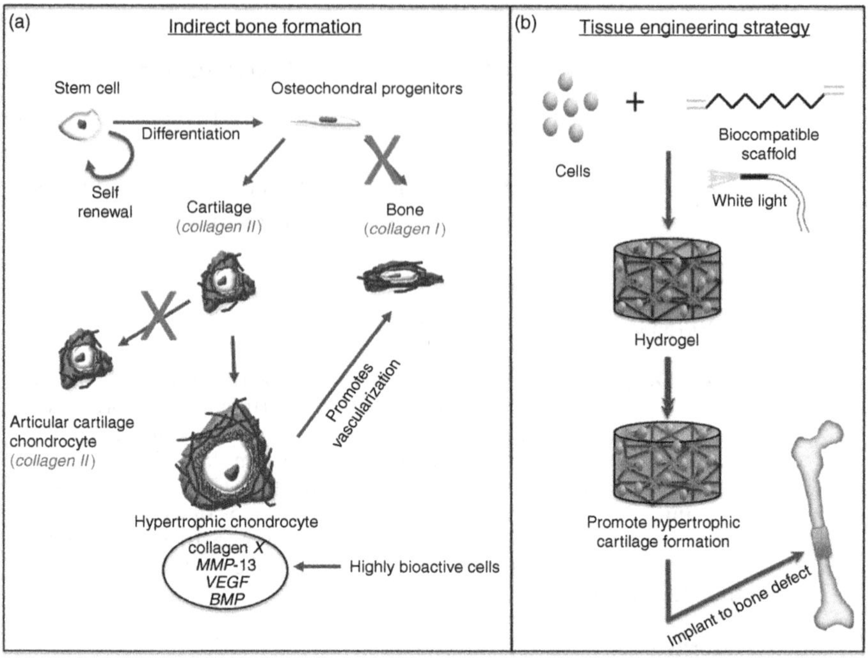

9.1 Schematic diagram of endochondral bone formation: (a) Indirect bone formation from stem cells can occur by first promoting chondrogenesis of osteochondral progenitors, then allowing hypertrophic maturation of the chondrocytes in a manner analogous to endochondral bone formation; then hypertrophic chondrocytes will promote vascularized bone formation by the secretion of bioactive factors. (b) A tissue engineering strategy for endochondral bone formation can be designed to promote hypertrophic cartilage development by stem cells photoencapsulated into a three-dimensional biocompatible scaffold using white light (see also, Bahney *et al.*, 2011b). These scaffolds could then be transplanted into bone defects to promote vascularize bone repair.

Capitalizing on these features of cartilage in a graft could help to address some of the current problems with bone grafting. The concept of using endochondral repair has been largely overlooked as an effective clinical strategy. This chapter will present relevant information that serves as foundational data enabling translation of this approach into a viable patient treatment.

9.2 Current problems associated with bone grafting

Autografted bone remains the gold standard technique for augmenting bone regeneration. However, this technique is significantly limited by availability of tissue, need for a second surgical site, and possible donor site morbidity. A number of allograft strategies have been developed as a result of the limitations associated with autografted bone. De-vitalized bone and de-mineralized bone substitute (DBS) are the most common allograft materials and have now been commercially processed into a number of different forms for application either as a structural graft or a morselized void filler. These products promote intramembranous bone regeneration by the osteoconductive and osteoinductive nature of the biomaterial.

Clinical failures associated with allografts have been extensively documented in both animal and human studies. The main problem is that the tissue remains inert: the grafted bone fails to revitalize, causing osteonecrosis, poor osseointegration, and an eventual loss in structural integrity of the bone graft (Benevenia *et al.*, 2000; Lee *et al.*, 2000; Enneking and Campanacci, 2001; Brigman *et al.*, 2004; Wheeler and Enneking, 2005). Contributing to these failures is that injuries requiring grafting to augment bone regeneration often occur in areas where there is compromised vasculature. This low blood supply creates a hypoxic microenvironment without good supply of nutrients and metabolites, which together produce a hostile transplant setting for enabling revitalization of the bone. Osteoblast activity is essential to create a viable and integrated bone regenerate when relying on intramembranous bone repair. A fully vascularized bone autograft, such as tissue explanted from the fibula, has now been used with some success to address these failures. Alternatively, development of synthetic scaffolds that can deliver bioactive factors or cells to help promote regeneration of a vascularized bone are emerging techniques that are being actively researched to address some of these concerns (Kaigler *et al.*, 2006; Leach *et al.*, 2006; Meijer *et al.*, 2007).

9.3 Cartilage grafts: an alternative to bone grafting

Using cartilage to stimulate bone regeneration via endochondral repair is an alternative grafting approach with the potential to overcome some of the limitations of bone grafting technologies that are currently available. Endochondral ossification is the normal mechanism by which bones form and the process by which the majority of fractures heal. The goal of using cartilage grafts to repair bone is to capitalize on these normal developmental and repair sequences to efficiently promote formation of a vascularized bone. This section will detail the perceived benefits of using cartilage instead of bone to promote regeneration and review some of the pre-clinical studies that have investigated the conversion of cartilage to bone.

9.3.1 Advantages of cartilage grafts

Innate qualities of chondrocytes, such as their adaptation for function at low oxygen tension along with their ability to promote vascularization and tissue mineralization, position cartilage to offer clinical advantages as a tissue graft for bone repair. Cartilage is a highly unique tissue in that it normally exists under avascular and hypoxic conditions. This could be advantageous to stimulating bone repair in clinical situations where there is restricted blood supply at the damage site. Presently, non-unions are 36% more likely when there is vascular injury concomitant with the fracture (Dickson *et al.*, 1995). The ability for cartilage to survive under these hypoxic conditions could improve initial graft viability and thus promote improved healing.

Hypertrophic chondrocytes are a specialized cell within growth plate and fracture callus cartilage that is highly bioactive. Of relevance to promoting bone formation, hypertrophic chondrocytes secrete factors that stimulate vascular invasion, osteogenesis, and mineralization. Expression of pro-angiogenic factors, such as vascular endothelial growth factor (VEGF), promotes neovascularization during embryonic long bone development (Gerber *et al.*, 1999; Carlevaro *et al.*, 2000; Colnot and Helms, 2001) and will help reduce the likelihood of osteonecrosis in the graft tissue. Hypertrophic chondrocytes also synthesize matrix metalloproteinase-13 (MMP-13) (Johansson *et al.*, 1997; Behonick *et al.*, 2007), bone morphogenetic proteins (BMPs) (Yu *et al.*, 2010), bone sialoprotein (BSP) (Bianco *et al.*, 1993; Gerstenfeld and Shapiro, 1996), and exhibits alkaline phosphatase activity (Gerstenfeld *et al.*, 2002; Scotti *et al.*, 2010). These are all factors that are beneficial in promoting bone formation and mineralization of the tissue. Bioactivity of the cartilage graft may also stimulate a host response to help promote osteointegration of the graft and the host bone.

9.3.2 Pre-clinical studies using cartilage for bone regeneration

The clinical application of cartilage grafts for bone regeneration is not a recognized treatment protocol. However, interest in this alternative approach is generating research and pre-clinical studies aimed at understanding the potential for endochondral bone regeneration. Using hypertrophic cartilage explants obtained from the cephalic region of the chicken sternum, Oliveira *et al.* (2009) showed that cartilage forms bone when placed ectopically on the dorsal side of nude mice. Articular chondrocytes have also been tested for their bone forming capacity with mixed results. Case *et al.* (2003) seeded rabbit articular chondrocytes onto bioresorbable, non-woven poly(glycolic acid) (PGA) felt disks and found that these cartilage implants would promote bone formation in the rabbit, especially when placed under dynamically loaded conditions. However, no bone repair was observed in a separate study where bovine articular chondrocytes were grown on PGA mesh scaffolds and implanted into cranial defects in nude mice (Vacanti *et al.*, 1995).

Since articular chondrocytes are considered to be a stable phenotype, developmentally protected from the process of hypertrophy and endochondral ossification, it is not clear what contributed to the different results in these two studies nor whether articular chondrocytes would ultimately be a good choice of cells (see discussion in Section 9.5.2 Chondrocytes). One factor that may have allowed bone formation in the first study (Case *et al.*, 2003) is that the chondrocytes were expanded in monolayer for three passages prior to use; this is widely accepted to cause phenotypic modulation of articular chondrocytes and result in the production of collagen I (von der Mark *et al.*, 1977; Dell'Accio *et al.*, 2001; Darling and Athanasiou, 2005). Differences in the developmental origin or vascularity of the bone defect could also have contributed to these inconsistent results.

Using mesenchymal stem cells (MSCs) to generate cartilage for bone repair is a more practical strategy for clinical therapies (as discussed in Section 9.5.2 Mesenchymal stem cells). MSCs can be stimulated to differentiate into cartilage using a three-dimensional culture and induction with transforming growth factor-β (TGF-β) and dexamethasone (Johnstone *et al.*, 1998). It has long been appreciated that the cartilage formed under these conditions is prone to hypertrophic maturation rather than forming a hyaline cartilage phenotype (Johnstone *et al.*, 1998; Pelttari *et al.*, 2006, 2008). Setting the framework for the potential success of using the hypertrophic cartilage produced by chondrogenic differentiation of human MSCs to repair bone are two studies that demonstrate bone can be formed from cartilage pellets implanted subcutaneously on the dorsal side of immunocompromised rodents (Scotti *et al.*, 2010; Farrell *et al.*, 2011) or following *in vitro* osteogenic stimulation (Muraglia *et al.*, 2000). To date there are only two studies that evaluate the use of cartilage as a precursor for bone regeneration in clinically meaningful models. In one study, Huang *et al.* (2006) demonstrated that MSCs differentiated into cartilage *in vitro* on esterified hyaluronan and gelatin scaffolds could reconstitute a complete bone, along with its articular cartilage, using a rabbit lunate arthroplasty model. More recently, cartilage grafts were tested in a murine model of a segmental bone defect and the data show that cartilage stimulates formation of a vascularized bone graft that integrates as well as a living bone autograft and much better than allograft (Bahney *et al.*, 2013).

9.4 Conversion of cartilage to bone

The concept of using a cartilage graft to repair bone capitalizes on the endochondral repair mechanisms through which the majority of fractures heal (Fig. 9.1). During fracture repair cartilage within the fracture callus mineralizes, becomes vascularized, and then remodels into a new bone that resembles the native bone in form and function. The molecular mechanisms underlying endochondral bone repair are not entirely clear. Previously this process has been presumed to parallel endochondral ossification during embryonic development of long bones. However, a close study of the fracture callus during endochondral repair provides evidence that a continued maturation of hypertrophic chondrocytes into osteocytes may also occur during healing (Descalzi Cancedda *et al.*, 1992; Roach, 1992; Scammell and Roach, 1996; Bianco *et al.*, 1998; Xing *et al.*, 2010; Bahney *et al.*, 2013). This section includes a discussion of endochondral ossification during both development and fracture repair.

9.4.1 Endochondral ossification during development

The majority of the molecular details describing endochondral ossification are understood in the context of embryonic long bone formation (Kronenberg, 2003). In this process, mesenchymal cells condense and undergo chondrogenic differentiation to form a cartilaginous outline of the future bone. Chondrocytes within this anlagen organize into morphologically and functionally distinct domains corresponding to their maturation state. Proliferating chondrocytes in the growth plate are help elongate the bone and are distinguished by their flattened morphology, columnar organization, and genetic expression of collagen II and aggrecan. In a complex regulatory process controlled in part by the feedback loop between Indian hedgehog (IHH) and parathyroid hormone related protein (PTHrP), chondrocytes exit the proliferating pool and transit toward a hypertrophic state. PTHrP signaling from the perichondrium is believed to be directly responsible for inhibiting maturation of chondrocytes in the proliferating pool (Iwamoto *et al.*, 1995; O'Keefe *et al.*, 1997), while IIH expression in the prehypertrophic chondrocytes acts via molecular mediators Patched and the Gli family of transcription factors in a poorly understood manner to activate PTHrP in the perichondrial cells.

Ossification begins with the formation of a bony collar by the perichondrial cells adjacent to the hypertrophic chondrocytes in the mid-diaphysis of the elongating bone. Hypertrophic chondrocytes are characterized by their enlarged round shape and expression of type X collagen. The hypertrophic chondrocytes also play a large role in lengthening the bone (Cooper *et al.*, 2013), and promote mineralization and vascularization of the cartilage through the expression of bioactive molecules such as VEGF, MMP-13, BSP, and alkaline phosphatase (Bianco *et al.*, 1993; Johansson *et al.*, 1997; Gerber *et al.*, 1999).

Subsequent ossification of the cortical region of the bone is coordinated by the vasculature. Invasion of the vascular network into the cartilage anlagen brings osteogenic stem cells, osteoblasts, and osteoclasts that will produce the bone matrix and remodel the tissue. During embryonic ossification, hypertrophic chondrocytes undergo apoptosis. Vascularization is a key step in triggering apoptosis of the hypertrophic chondrocyte, perhaps by providing pro-apoptotic signals (Gerber *et al.*, 1999). This process of apoptosis is required in part to create marrow space in the diaphysis of the long bone, but may also generate empty lacunae for the new osteocytes to reside.

The transient chondrocytes located in the epiphyseal plate of long bones seem to be phenotypically distinct from the permanent chondrocytes located in articular cartilage. Articular chondrocytes are the single cellular component of the cartilage that lines the ends of the long bones, and they represent a stable phenotype characterized by the production of collagen II, but not collagens I or X. The origin of the articular chondrocyte and molecular mechanisms governing articular cartilage development are not well understood (Ito and Kida, 2000; Pacifici *et al.*, 2005; Khan *et al.*, 2007; Koyama *et al.*, 2008).

9.4.2 Fracture repair

Fractures can heal through either endochondral or intramembranous ossification, with fracture stability and the resulting mechanical microenvironment considered major factors determining the mode of healing (Le *et al.*, 2001; Perren, 2002; Thompson *et al.*, 2002). Instability at the fracture site promotes endochondral bone repair through a cartilage intermediate (Plate III in the colour section between pages 186 and 187), whereas rigidly stabilized fractures heal by direct bone formation. Bone repair through either pathway results in a bone regenerate that nearly perfectly resembles the original tissue in form and function.

Relevant to this chapter, only endochondral fracture repair will be discussed in detail. This healing process has four distinct phases: hematoma formation and inflammation, soft (cartilage) callus formation (Plate IIIa and d), hard callus formation (Plate IIIb and e), and remodelling (Plate IIIc and f). As with most injuries, the initial phase of hematoma formation and inflammation is initiated immediately as the fracture disrupts the local vasculature. A fracture hematoma is formed and contains bone debris created by the fracture and also traps erythrocytes and immune cells from the spilled vasculature and disrupted marrow space. This hematoma forms a bioactive structure that triggers the immune response. The immune response plays a critical role in modulating the rate and extent of bone repair (Mizuno *et al.*, 1990; Grundnes and Reikeras, 1993; Hauser *et al.*, 1997; Xing *et al.*, 2010). The subsequent formation of a cartilage callus is achieved by chondrogenic differentiation of progenitor cells mobilized from the periosteum (Colnot, 2009). Chondrocytes undergo hypertrophic maturation inside the fracture callus (Plate IIIb, d) and this change initiates vascular invasion through the production of VEGF. Conversion of the cartilage callus into a hard (bone) callus is coordinated by the vasculature and the matrix is gradually ossified (Plate IIIc-f) and eventually remodeled into the standard canalicular architecture of the native bone.

The process through which cartilage is converted into bone in the fracture callus has not been well studied. This is largely because endochondral repair is presumed to be analogous to endochondral ossification during long bone formation. However, there are also technical challenges that accompany the careful fate-mapping of the chondrocytes during endochondral repair. To support the similarity between endochondral development and repair, Maes *et al.* (2007, 2010) used a genetic system to track the role of immature osteoblasts during fracture repair and found invasion of osteocytes plays a role in bone repair. There may also be significant differences that are not fully elucidated at the present time, in part because of the difficulty in following and uniquely identifying hypertrophic chondrocytes during endochondral repair (see Section 9.4.3). Apoptosis of chondrocytes occurs during endochondral repair, but it is not clear to what extent this occurs, and whether this clears the matrix for invasion of osteocytes or is just part of the remodeling process. Fastidious cell death assays are very difficult, but some studies suggest that apoptosis is not the only fate of the hypertrophic chondrocyte in the fracture callus (Roach, 1997). Direct transformation of chondrocytes into bone has also been proposed to contribute to bone regeneration during repair (Plate IIIe). This concept has known many different names – transdifferentiation, chondrocyte metaplasia, transchondroid differentiation, chondroid bone, neoplastic bone formation – with both *in vitro* and *in vivo* studies providing evidence that chondrocytes are capable of this phenotypic transformation (Roach, 1992; Aigner *et al.*, 2000; Hall, 2005; Wlodarski *et al.*, 2006). For convenience, we will discuss this concept in terms of 'chondrocyte transdifferentiation' in the following section.

9.4.3 Chondrocyte transdifferentiation

It is possible that the chondrocyte has much more plasticity during fracture repair than previously appreciated, and that chondrocytes may directly contribute to the bone regenerate by transforming into osteocytes. *In vitro* transdifferentiation of both mesenchymal and embryonic stem cells (ESCs) has shown that fully differentiated cells can switch fate within the skeletal lineage (Hegert *et al.*, 2002; Song and Tuan, 2004; Muraglia *et al.*, 2000). Detailed mechanisms for transdifferentiation in these systems were not provided but Song and Tuan (2004) suggest that the differentiated MSCs switch fate by returning or de-differentiating to a more primitive cell fate prior to switching lineages.

Transdifferentiation has also been suggested as a mechanism for *in vivo* bone formation during repair and disease pathologies. In a study of mesenchymal chondrosarcomas, Aigner *et al.* (2000) used *in situ* detection techniques to trace a transformation of chondrocytes into bone and suggest that transdifferentiation, or in their words 'neoplastic osteogenesis', occurs in these malignant tumors. Transformation of cartilage into bone has also been suggested to occur during fracture repair (Roach, 1992; Scammell and Roach, 1996; Bianco *et al.*, 1998; Bahney *et al.*, 2013). The mechanism that underlies cartilage to bone transformation during endochondral repair is not clear. Hypertrophic chondrocytes and early osteoblasts share many of the same gene expression profiles; for example, hypertrophic chondrocytes express most of the classic bone markers including collagen type I, osteocalcin, osteopontin, BSP, and alkaline phosphatase (Mark *et al.*, 1988; Bianco *et al.*, 1993; Lian *et al.*, 1993; Gerstenfeld and Shapiro, 1996). These data suggest that phenotypic transformation may just be a normal part of maturation. Alternatively, hypertrophic chondrocytes in the fracture callus could transdifferentiate by de-differentiating to a progenitor state *in vivo* as proposed by *in vitro* studies by Song and Tuan: in support of this we have recently found that chondrocytes at the chondro–osseous transition in the fracture callus show immunohistochemical evidence of the pluripotent stem cell marker OCT4A (Bahney *et al.*, 2013). However, the role of activating these pluripotent genes is not clear.

9.5 Generating cartilage grafts

Since preliminary studies suggest that chondrocytes directly contribute to the regeneration of bone during endochondral repair (Scotti *et al.*, 2010; Bahney *et al.*, 2013), a cell-based grafting therapy may produce the best clinical outcomes. This is analogous to the improved clinical results observed with bone autografts as compared to allografts, which rely on osteoconductive or osteoinductive scaffolds to promote bone regeneration in the absence of living cells. For this therapy a tissue engineering strategy that combines chondrogenic cells with a three-dimensional scaffold is likely to be the most effective technology for the clinical translation of cartilage grafts to repair large bone defects. Cartilage grafts are currently used clinically only in the repair of articular cartilage defects. In the following section, potential strategies for using cartilage grafts to promote endochondral bone repair are discussed.

9.5.1 Auto- and allograft cartilage transplantation

Presently, the gold standard bone grafting technique is bone autograft, most commonly harvested from the iliac crest and then morsilized. An analogous technique could be used to prepare cartilage grafts for bone regeneration. However, there are only a few post-natal sources of cartilage that could be considered as candidates for transplantation. Articular cartilage 'plugs' from low weight-bearing locations in the joint have been transplanted to weight-bearing regions to create a mosaicplasty that can resurface the joint and treat degenerative osteoarthritis. This procedure has shown some clinical success (Kish and Hangody, 2004). However, autologous articular cartilage transplantations are significantly limited by availability of graft tissues, donor site morbidity, and a disruption of the normal joint biomechanics that could result in a further propagation of osteoarthritis (Lee *et al.*, 2000; Bedi *et al.*, 2010).

This approach poses additional limitations to repair a bone defect. Critical sized bone defects are likely to be much larger than those in the articular cartilage surface, and therefore the amount of cartilage required for therapeutic application will be proportionally larger. Consequently, finding an appropriate source for harvesting cartilage becomes a challenge, firstly in an effort to minimize the effect on healthy tissues, and secondly in obtaining the correct phenotype from the cartilage. Costal cartilage may be another post-natal source for cartilage autografts: this tissue would have more appropriate size dimensions available for the treatment of segmental bone defects. Both of these adult cartilage sources are hyaline, or permanent, cartilage that will not go through the hypertrophic maturation observed in the growth plate except under disease conditions. Whether this hyaline phenotype can be re-engineered to undergo hypertrophy and endochondral ossification for bone repair is presently unknown.

Allograft cartilage would significantly expand tissue availability and there is precedence for its use to treat articular cartilage defects. Osteochondral allografts have been used to treat focal ostochondral defects in the knee and have demonstrated good to excellent results according to the Merle d'Aubigné and Postel scale (Bedi *et al.*, 2010). Commercially, Zimmer also sells a product (DeNovo® NT) that is a juvenile cartilage allograft consisting of cartilage particulates and living cells (Adkisson *et al.*, 2010b). Whether the phenotype and volume/form of these allograft cartilage tissues would be appropriate for repairing a bone defect has never been tested. There is also an unresolved concern over the immunogenicity of cartilage allografts. Some immune response following transplantation of fresh cartilage allografts is indicated due to production of antigens to cartilage specific proteins (Phipatanakul *et al.*, 2004), but whether this response will cause graft rejection or remains subclinical is unclear (Adkisson *et al.*, 2010a).

9.5.2 Cell sources for cartilage tissue engineering

Combining a scaffold with a cell source is a tissue engineering technique that could be used to create cartilage grafts as an alternative approach to a living tissue transplant discussed above (Fig. 9.1b). However, the correct choice of cells for generating cartilage with the appropriate phenotype is not entirely clear. There are at least three potential cell sources/types that could be used in this application, each associated with benefits and drawbacks. Potential cell sources for cartilage engineering include chondrocytes, MSCs, and ESCs. A detailed review follows in the next subsection.

Chondrocytes

Similar to the use of cartilage tissue transplantation discussed previously, autologous chondrocyte transplant (ACT) techniques have already been developed for the treatment of articular cartilage defects. The technical basis for ACT procedures was first developed in a rabbit model (Grande *et al.*, 1989) and subsequently translated into patients (Brittberg *et al.*, 1994), where it has now gone through multiple levels of refinement. The general technique is to harvest articular cartilage in a first operation from a 'non-weight-bearing' area of the knee joint. However, since chondrocytes represent a very small fraction of the tissue (only 2–5% of the volume), the cells must be expanded *in vitro* to obtain the large numbers of cells requisite to facilitate tissue repair. In a second operation, the chondrocytes are transplanted back into the defect either as a bolus of cells confined to the defect by a periosteal flap (Brittberg *et al.*, 1994), or in combination with various scaffolds.

Although some ACT studies report significantly reduced pain post-operatively (Peterson *et al.*, 2000; Bentley *et al.*, 2003), there are a number of drawbacks with both the procedure and the repair tissue (Luyten *et al.*, 2001; Niemeyer *et al.*, 2008). Aside from the obvious disadvantage of two surgeries, in the majority of cases, the repair tissue that forms at the transplantation site resembles a fibrous scar tissue, fibrocartilage, rather than native hyaline cartilage (Roberts *et al.*, 2003; Tins *et al.*, 2005; Roberts *et al.*, 2009). The cause of the phenotypic change in the repair tissue is unclear, but one contributing factor could be that the chondrocytes de-differentiate when they are cultured in monolayer during cell expansion (von der Mark *et al.*, 1977; Dell'Accio *et al.*, 2001; Darling and Athanasiou, 2005; Barlic *et al.*, 2008). The fibrocartilagenous phenotype would be equally disadvantageous for bone regeneration applications. To consider hyaline chondrocytes as a viable candidate for endochondral bone regeneration, protocols that promote a transient or hypertrophic phenotype would need to be developed.

Mesenchymal stem cells

MSCs have chondrogenic potential and they offer a potential alternative to chondrocytes in cartilage engineering applications (Johnstone and Yoo, 2001; Caplan, 2005; Johnstone *et al.*, 2006; Pelttari *et al.*, 2008; Hwang and

Elisseeff, 2009). These cells represent an adult-derived multipotent cell population that can be isolated from multiple tissue sources including bone marrow (Johnstone *et al.*, 1998; Yoo *et al.*, 1998), adipose tissue (Zuk *et al.*, 2002), periosteum (Miura *et al.*, 1994; De Bari *et al.*, 2006), and synovium (De Bari *et al.*, 2001; Sakaguchi *et al.*, 2005). To be considered an MSC, a cell must, at a minimum, possess the ability to differentiate along adipose, cartilage, and bone lineages *in vitro* (Pittenger *et al.*, 1999; Dominici *et al.*, 2006). Differentiation towards other cell types including tendon/ligament (Young *et al.*, 1998), muscle (Dezawa *et al.*, 2005), cardiac (Orlic *et al.*, 2001), neuronal (Woodbury *et al.*, 2000), and stromal (Majumdar *et al.*, 1998) tissues have also been reported, but whether the MSCs engraft and differentiate directly into these cell types when transplanted into patients or animals or just stimulate repair within these tissues is unclear (Laflamme and Murry, 2005; Caplan and Dennis, 2006; Grassel and Ahmed, 2007).

The relative accessibility of MSCs from multiple tissue sources, and the ability to readily expand this cell population *in vitro,* offers a distinct advantage over autologous chondrocytes (Johnstone *et al.*, 2006; Pelttari *et al.*, 2008). Furthermore, MSCs appear to be immunoprivileged due to the absence of major histocompatibility complex (MHC) class II cell surface markers, making allograft cell therapies a viable clinical option (Le Blanc *et al.*, 2003). The MSCs themselves may be beneficial as transplant cells since they demonstrate both immunoregulatory and trophic effects on the host tissue. Immunosuppressive and anti-inflammatory cytokines such as interleukin-10 (Bartholomew *et al.*, 2002), nitric oxide (Sato *et al.*, 2007), and prostaglandins (Aggarwal and Pittenger, 2005) are produced by MSCs and can play a role in preventing host vs graft rejection through modulation of T-cells (Krampera *et al.*, 2003; Tse *et al.*, 2003). MSCs also have the capacity to stimulate host tissue repair through release of bioactive factors such as TGF-β, stem cell factor (SCF), insulin-like growth factor (IGF), epidermal growth factor (EGF), and granulocyte and macrophage colony stimulating factors (G/M-CSF) (Haynesworth *et al.*, 1996; Di Nicola *et al.*, 2002).

Chondrogenesis of MSCs was first demonstrated *in vitro* using a three-dimensional pellet culture system with a chemically defined medium containing TGF-β1 and dexamethasone to control differentiation (Johnstone *et al.*, 1998; Yoo *et al.*, 1998). More recently, MSCs have been used in combination with three-dimensional scaffolds to facilitate cartilage engineering applications suitable for large scale tissue defects. One of the major drawbacks of cartilage engineering from MSCs is that the standard differentiation protocols do not generate a stable (articular) chondrocyte phenotype in the resultant neocartilagenous tissue. Rather this engineered cartilage produces collagen I and X, at both an RNA and protein level, in addition to expression of the characteristic collagen II and aggrecan. This phenotype is more characteristic of the endochondral, or transitional, cartilage that is seen in the growth plate and is a distinct disadvantage for articular cartilage engineering applications where the potential for ossification of the tissue negatively impacts its therapeutic potential (Yoo *et al.*, 1998; Pelttari *et al.*, 2008). However, for cartilage engineering applications geared toward bone regeneration this phenotype is actually preferable.

Embryonic stem cells and induced pluripotent stem cells

The use of ESCs for cartilage tissue engineering is a new concept and preliminary evidence suggests that these cells may be yield a viable method for generating cartilaginous tissues (Kramer *et al.*, 2000; Nakayama *et al.*, 2003; Hwang *et al.*, 2006; Koay *et al.*, 2007). ESCs can be isolated from pluripotent cells of an early preimplantation embryo and propagated indefinitely in an undifferentiated state *in vitro* (Thomson *et al.*, 1998). Pluripotent cells resembling ESCs can also be derived experimentally by reprogramming of somatic cells by somatic cell nuclear transfer (SCNT) (Wilmut *et al.*, 1997; Byrne *et al.*, 2007) or direct reprogramming (induced pluripotent stem (iPS) cells) (Takahashi and Yamanaka, 2006; Yamanaka, 2009). Such pluripotent cells have the potential to play an important role in the future of cell replacement therapies since the patient's own somatic cells can be used for reprogramming.

Successful use of ESCs in tissue engineering applications requires tight control over the differentiation process. The dangerous consequence of this has been demonstrated by teratoma formation when undifferentiated ESCs were injected into the knee joint (Wakitani *et al.*, 2003, 2004; Nakajima *et al.*, 2008). To help consolidate tissue phenotype, chondrogenic inductive signals TGF-β and BMPs have been applied to ESCs to promote cartilage differentiation, but current protocols still produced heterotypic tissues following differentiation (Kramer *et al.*, 2000; Nakayama *et al.*, 2003; Koay *et al.*, 2007). Pioneering work in this field by the Elisseeff laboratory has suggested a step-wise ESC differentiation protocol, that first takes the cells to mesenchymal progenitors, may offer improved control (Lian *et al.*, 2007; Hwang *et al.*, 2008b; Yamashita *et al.*, 2010).

9.5.3 Biomaterial scaffolds

Generating cartilage grafts at a scale appropriate for translation to clinical defects requiring treatment will almost certainly require a biomaterial scaffold to support bone regeneration. These three-dimensionally networked, porous structures are critical in containing the cells and providing a framework for tissue development (Ratner and Bryant, 2004; Elisseeff *et al.*, 2006). Cells injected without a scaffold demonstrate poor viability and typically they diffuse, or are quickly cleared, from the injection site (Erlebacher *et al.*, 2002; Laflamme and Murry, 2005).

The breadth of scaffolds being investigated for cell-based cartilage and bone tissue engineering applications is outside the scope of this book chapter but a variety of good reviews are available (Ratner and Bryant, 2004; Khademhosseini and Langer, 2007; Huebsch and Mooney, 2009; Lutolf, 2009; Vinatier *et al.*, 2009a). What seems appropriate to highlight here is how the scaffold can support the biology in a variety of ways. The mechanical properties of the scaffold are important both in providing structural support to the defect, but also can be important in promoting differentiation of cells down a specific lineage (Discher *et al.*, 2005; Elisseeff *et al.*, 2006; Lutolf *et al.*, 2009). Other biophysical aspects of the scaffold, such as pore size, will impact the synthesis and localization of the ECM molecules critical to supporting bone and cartilage function (Bryant *et al.*, 2003; Buxton *et al.*, 2007; Kasten *et al.*, 2008; Erickson *et al.*, 2009). Biochemistry of the scaffold can also be modulated to create biodegradable and biomimetic scaffolds. By changing the type or amount of biodegradable linkages the degradation kinetics of the scaffold can be grossly controlled (Ifkovits and Burdick, 2007; Nicodemus and Bryant, 2008). More recently, bioresponsive scaffolds have been developed in which the cells mediate degradation of the scaffold to offer more tailored systems (West and Hubbell, 1999; Lutolf *et al.*, 2003; Wang *et al.*, 2005; Patterson and Hubbell, 2010; Bahney *et al.*, 2011a). Biomimetic scaffolds, or those in which key features of the ECM are retained or are engineered in, can also enhance cell function and ECM production (Kim *et al.*, 2005; Lutolf and Hubbell, 2005; Benoit *et al.*, 2008; Salinas and Anseth, 2008, 2009).

Engineering the appropriate scaffold can be achieved by using either natural (biological) or synthetic materials. Natural biomaterials, often made from ECM components such as collagens or hyaluronan, or from decellularized tissues, are advantageous because they can retain growth factor and cell binding sites that help promote tissue specific responses. However, these materials are difficult to modify and carry the risk of immunogenicity. A number of biocompatible synthetic materials have been developed that are advantageous, because their physical properties can be easily controlled. However, without modifications these scaffolds typically lack bioactivity. The ideal scaffold is likely to be a synthetic scaffold that incorporates aspects of the ECM, can deliver or sequester growth factors, and can be tuned to evolve with the developing tissue (Mann *et al.*, 2001; Ulijn *et al.*, 2007; Fan *et al.*, 2007; Ferreira *et al.*, 2007; Lutolf, 2009; Vinatier *et al.*, 2009a; Bahney *et al.*, 2011a; Irwin *et al.*, 2011).

9.6 Future trends

The concept of using cartilage grafts to repair bone defects holds significant promise to improve patient outcomes in the future. Cartilage forms a natural template for bone formation both during embryonic limb formation and endochondral bone repair. This cartilage anlagen is assimilated to hypoxic conditions typical of traumatic injury, and secretes bioactive factors that will promote vascularization and mineralization of the graft, thus presenting advantages over current clinical therapies.

The challenge moving towards a cartilage graft appropriate for clinical application is developing a strategy that can be translated from research to the patient. Important considerations for engineering an endochondral bone regenerate are: (i) generating a graft with the correct phenotype (hypertrophic cartilage) and (ii) creating a system that can fill the large/complex defects typical of those treated with a graft. Consequently, auto- and allografted cartilages are not likely to prove appropriate techniques for translation, but may serve an important pre-clinical role in validating this approach.

Cell-based cartilage engineering strategies may hold the most promise for translating this concept of endochondral bone regeneration into a viable patient therapy (Scotti *et al.*, 2010; Bahney *et al.*, 2013). Furthermore, adult osteochondral progenitors are likely to hold the best and most immediate promise for this purpose (see review by Bahney and Miclau, 2012), and can be isolated from either bone marrow or adipose tissue. These progenitor cells (MSCs) can be induced to undergo chondrogenesis in scaffolds such as biodegradable hydrogels where the cells are uniformly distributed throughout the structure during synthesis of the structure (Bryant *et al.*, 2008; Bahney *et al.*, 2011a). Hypertrophic cartilage constructs can then be created by culturing *in vitro* (Erickson *et al.*, 2009; Buxton *et al.*, 2011) or by transplanting *in vivo* with appropriate bioactive factors to induce controlled differentiation (Mann *et al.*, 2001; Hwang *et al.*, 2008a; Vinatier *et al.*, 2009b).

The next step in advancing this technology is to validate the hypothesis that hypertrophic cartilage grafts can repair bone defects in appropriate preclinical models. Defining the molecular mechanism of bone regeneration from a cartilage graft will also be key in designing appropriate therapeutic techniques. From these preliminary studies scaffold–cell systems optimized to promote endochondral bone regeneration can be tested. Despite the limited information available presently regarding the application of cartilage for endochondral bone regeneration, the bioengineering toolbox for translating this concept is available.

9.7 References

Adkisson, H. D., Milliman, C., Zhang, X., Mauch, K., Maziarz, R. T. and Streeter, P. R. (2010a). Immune evasion by neocartilage-derived chondrocytes: Implications for biologic repair of joint articular cartilage. *Stem Cell Res*, **4**, 57–68.

Adkisson, H. D. T., Martin, J. A., Amendola, R. L., Milliman, C., Mauch, K. A., Katwal, A. B., Seyedin, M., Amendola, A., Streeter, P. R. and Buckwalter, J. A. (2010b).

The potential of human allogeneic juvenile chondrocytes for restoration of articular cartilage. *Am J Sports Med,* **38**, 1324–1333.

Aggarwal, S. and Pittenger, M. F. (2005). Human mesenchymal stem cells modulate allogeneic immune cell responses. *Blood,* **105**, 1815–1822.

Aigner, T., Loos, S., Muller, S., Sandell, L. J., Unni, K. K. and Kirchner, T. (2000). Cell differentiation and matrix gene expression in mesenchymal chondrosarcomas. *Am J Pathol,* **156**, 1327–1335.

Bahney, C. S., Hsu, C. W., Yoo, J. U., West, J. L. and Johnstone, B. (2011a). A bioresponsive hydrogel tuned to chondrogenesis of human mesenchymal stem cells. *FASEB J,* **25**, 1486–1496.

Bahney, C. S., Hu, D. P., Taylor, A. J., Ferro, F., Britz, H. M., Hallgrimsson, B., Johnstone, B., Miclau, T. and Marcucio, R. (2013). Stem cell derived endochondral cartilage stimulates bone healing by tissue transformation. *J Bone Miner Res,* Nov. 21. doi: 10.1002/jbmr.2148. [Epub ahead of print].

Bahney, C.S., Lujan, T.J., Hsu, C.W., Bottlang, M., West, J.L. and Johnstone, B. (2011b). Visible light photoinitiation of mesenchymal stem cell-laden bioresponsive hydrogels. *Eur Cell Mater*, Jul 15, **22**, 43–55; discussion 55.

Bahney, C. S. and Miclau, T. (2012). Therapeutic potential of stem cells in orthopedics. *Indian J Orthop,* **46**, 4–9.

Barlic, A., Drobnic, M., Malicev, E. and Kregar-Velikonja, N. (2008). Quantitative analysis of gene expression in human articular chondrocytes assigned for autologous implantation. *J Orthop Res,* **26**, 847–853.

Bartholomew, A., Sturgeon, C., Siatskas, M., Ferrer, K., Mcintosh, K., Patil, S., Hardy, W., Devine, S., Ucker, D., Deans, R., Moseley, A. and Hoffman, R. (2002). Mesenchymal stem cells suppress lymphocyte proliferation *in vitro* and prolong skin graft survival *in vivo*. *Exp Hematol,* **30**, 42–48.

Bedi, A., Feeley, B. T. and Williams, R. J., 3rd (2010). Management of articular cartilage defects of the knee. *J Bone Joint Surg Am,* **92**, 994–1009.

Behonick, D. J., Xing, Z., Lieu, S., Buckley, J. M., Lotz, J. C., Marcucio, R. S., Werb, Z., Miclau, T. and Colnot, C. (2007). Role of matrix metalloproteinase 13 in both endochondral and intramembranous ossification during skeletal regeneration. *PLoS One,* **2**, e1150.

Benoit, D. S., Schwartz, M. P., Durney, A. R. and Anseth, K. S. (2008). Small functional groups for controlled differentiation of hydrogel-encapsulated human mesenchymal stem cells. *Nature Materials,* **7**, 816–823.

Bentley, G., Biant, L. C., Carrington, R. W., Akmal, M., Goldberg, A., Williams, A. M., Skinner, J. A. and Pringle, J. (2003). A prospective, randomised comparison of autologous chondrocyte implantation versus mosaicplasty for osteochondral defects in the knee. *J Bone Joint Surg Br,* **85**, 223–230.

Benevenia, J., Zimmerman, M., Keating, J., Cyran, F., Blacksin, M. and Parsons, J. R. (2000). Mechanical environment affects allograft incorporation. *J Biomed Mater Res,* **53**, 67–72.

Bianco, P., Cancedda, F. D., Riminucci, M. and Cancedda, R. (1998). Bone formation via cartilage models: the 'borderline' chondrocyte. *Matrix Biol,* **17**, 185–192.

Bianco, P., Riminucci, M., Bonucci, E., Termine, J. D. and Robey, P. G. (1993). Bone sialoprotein (BSP) secretion and osteoblast differentiation: relationship to bromodeoxyuridine incorporation, alkaline phosphatase, and matrix deposition. *J Histochem Cytochem,* **41**, 183–191.

Brittberg, M., Lindahl, A., Nilsson, A., Ohlsson, C., Isaksson, O. and Peterson, L. (1994). Treatment of deep cartilage defects in the knee with autologous chondrocyte transplantation. *N Engl J Med,* **331**, 889–895.

Brigman, B. E., Hornicek, F. J., Gebhardt, M. C. and Mankin, H. J. (2004). Allografts about the knee in young patients with high-grade sarcoma. *Clin Orthop Relat Res*, Apr. (421), 232–239.

Bryant, S., Nicodemus, G. D. and Villanueva, I. (2008). Designing 3D photopolymer hydrogels to regulate biomechanical cues and tissue growth for cartilage tissue engineering. *Pharm Res,* **25**, 2379–2386.

Bryant, S. J., Durand, K. L. and Anseth, K. S. (2003). Manipulations in hydrogel chemistry control photoencapsulated chondrocyte behavior and their extracellular matrix production. *J Biomed Mater Res A,* **67**, 1430–1436.

Buxton, A. N., Bahney, C. S., Yoo, J. U. and Johnstone, B. (2011). Temporal exposure to chondrogenic factors modulates human mesenchymal stem cell chondrogenesis in hydrogels. *Tissue Eng Part A,* **17**, 371–380.

Buxton, A. N., Zhu, J., Marchant, R., West, J. L., Yoo, J. U. and Johnstone, B. (2007). Design and characterization of poly(ethylene glycol) photopolymerizable semi-interpenetrating networks for chondrogenesis of human mesenchymal stem cells. *Tissue Eng,* **13**, 2549–2560.

Byrne, J. A., Pedersen, D. A., Clepper, L. L., Nelson, M., Sanger, W. G., Gokhale, S., Wolf, D. P. and Mitalipov, S. M. (2007). Producing primate embryonic stem cells by somatic cell nuclear transfer. *Nature,* **450**, 497–502.

Caplan, A. I. (2005). Review: mesenchymal stem cells: cell-based reconstructive therapy in orthopedics. *Tissue Eng,* **11**, 1198–1211.

Caplan, A. I. and Dennis, J. E. (2006). Mesenchymal stem cells as trophic mediators. *J Cell Biochem,* **98**, 1076–1084.

Carlevaro, M. F., Cermelli, S., Cancedda, R. and Descalzi Cancedda, F. (2000). Vascular endothelial growth factor (VEGF) in cartilage neovascularization and chondrocyte differentiation: auto-paracrine role during endochondral bone formation. *J Cell Sci,* **113** (Pt 1), 59–69.

Case, N. D., Duty, A. O., Ratcliffe, A., Muller, R. and Guldberg, R. E. (2003). Bone formation on tissue-engineered cartilage constructs *in vivo*: effects of chondrocyte viability and mechanical loading. *Tissue Eng,* **9**, 587–596.

Colnot, C. (2009). Skeletal cell fate decisions within periosteum and bone marrow during bone regeneration. *J Bone Miner Res,* **24**, 274–282.

Colnot, C. I. and Helms, J. A. (2001). A molecular analysis of matrix remodeling and angiogenesis during long bone development. *Mech Dev,* **100**, 245–250.

Cooper, K. L., Oh, S., Sung, Y., Dasari, R. R., Kirschner, M. W. and Tabin, C. J. (2013). Multiple phases of chondrocyte enlargement underlie differences in skeletal proportions. *Nature,* **495**, 375–378.

Darling, E. M. and Athanasiou, K. A. (2005). Rapid phenotypic changes in passaged articular chondrocyte subpopulations. *J Orthop Res,* **23**, 425–432.

De Bari, C., Dell'Accio, F., Tylzanowski, P. and Luyten, F. P. (2001). Multipotent mesenchymal stem cells from adult human synovial membrane. *Arthritis Rheum,* **44**, 1928–1942.

De Bari, C., Dell'Accio, F., Vanlauwe, J., Eyckmans, J., Khan, I. M., Archer, C. W., Jones, E. A., Mcgonagle, D., Mitsiadis, T. A., Pitzalis, C. and Luyten, F. P. (2006). Mesenchymal multipotency of adult human periosteal cells demonstrated by single-cell lineage analysis. *Arthritis Rheum,* **54**, 1209–1221.

Dell'Accio, F., De Bari, C. and Luyten, F. P. (2001). Molecular markers predictive of the capacity of expanded human articular chondrocytes to form stable cartilage *in vivo*. *Arthritis Rheum,* **44**, 1608–1619.
Descalzi Cancedda, F., Gentili, C., Manduca, P. and Cancedda, R. (1992). Hypertrophic chondrocytes undergo further differentiation in culture. *J Cell Biol,* **117**, 427–435.
Dezawa, M., Ishikawa, H., Itokazu, Y., Yoshihara, T., Hoshino, M., Takeda, S., Ide, C. and Nabeshima, Y. (2005). Bone marrow stromal cells generate muscle cells and repair muscle degeneration. *Science,* **309**, 314–317.
Di Nicola, M., Carlo-Stella, C., Magni, M., Milanesi, M., Longoni, P. D., Matteucci, P., Grisanti, S. and Gianni, A. M. (2002). Human bone marrow stromal cells suppress T-lymphocyte proliferation induced by cellular or nonspecific mitogenic stimuli. *Blood,* **99**, 3838–3843.
Dickson, K. F., Katzman, S. and Paiement, G. (1995). The importance of the blood supply in the healing of tibial fractures. *Contemp Orthop,* **30**, 489–493.
Discher, D. E., Janmey, P. and Wang, Y. L. (2005). Tissue cells feel and respond to the stiffness of their substrate. *Science,* **310**, 1139–1143.
Dominici, M., Le Blanc, K., Mueller, I., Slaper-Cortenbach, I., Marini, F., Krause, D., Deans, R., Keating, A., Prockop, D. and Horwitz, E. (2006). Minimal criteria for defining multipotent mesenchymal stromal cells. The International Society for Cellular Therapy position statement. *Cytotherapy,* **8**, 315–317.
Elisseeff, J., Ferran, A., Hwang, S., Varghese, S. and Zhang, Z. (2006). The role of biomaterials in stem cell differentiation: applications in the musculoskeletal system. *Stem Cells Dev,* **15**, 295–303.
Enneking, W. F. and Campanacci, D. A. (2001). Retrieved human allografts: a clinicopathological study. *J Bone Joint Surg Am,* **83-A**, 971–986.
Erickson, I. E., Huang, A. H., Chung, C., Li, R. T., Burdick, J. A. and Mauck, R. L. (2009). Differential maturation and structure-function relationships in mesenchymal stem cell- and chondrocyte-seeded hydrogels. *Tissue Eng Part A,* **15**, 1041–1052.
Erlebacher, A., Lukens, A. K. and Glimcher, L. H. (2002). Intrinsic susceptibility of mouse trophoblasts to natural killer cell-mediated attack *in vivo*. *Proc Natl Acad Sci U S A,* **99**, 16940–16945.
Fan, V. H., Tamama, K., Au, A., Littrell, R., Richardson, L. B., Wright, J. W., Wells, A. and Griffith, L. G. (2007). Tethered epidermal growth factor provides a survival advantage to mesenchymal stem cells. *Stem Cells,* **25**, 1241–1251.
Farrell, E., Both, S. K., Odorfer, K. I., Koevoet, W., Kops, N., O'brien, F. J., Baatenburg De Jong, R. J., Verhaar, J. A., Cuijpers, V., Jansen, J., Erben, R. G. and Van Osch, G. J. (2011). *In-vivo* generation of bone via endochondral ossification by *in-vitro* chondrogenic priming of adult human and rat mesenchymal stem cells. *BMC Musculoskelet Disord,* **12**, 31.
Ferreira, L. S., Gerecht, S., Fuller, J., Shieh, H. F., Vunjak-Novakovic, G. and Langer, R. (2007). Bioactive hydrogel scaffolds for controllable vascular differentiation of human embryonic stem cells. *Biomaterials,* **28**, 2706–2717.
Gerber, H. P., Vu, T. H., Ryan, A. M., Kowalski, J., Werb, Z. and Ferrara, N. (1999). VEGF couples hypertrophic cartilage remodeling, ossification and angiogenesis during endochondral bone formation. *Nat Med,* **5**, 623–628.
Gerstenfeld, L. C., Cruceta, J., Shea, C. M., Sampath, K., Barnes, G. L. and Einhorn, T. A. (2002). Chondrocytes provide morphogenic signals that selectively induce

osteogenic differentiation of mesenchymal stem cells. *J Bone Miner Res,* **17**, 221–230.

Gerstenfeld, L. C. and Shapiro, F. D. (1996). Expression of bone-specific genes by hypertrophic chondrocytes: implication of the complex functions of the hypertrophic chondrocyte during endochondral bone development. *J Cell Biochem,* **62**, 1–9.

Grande, D. A., Pitman, M. I., Peterson, L., Menche, D. and Klein, M. (1989). The repair of experimentally produced defects in rabbit articular cartilage by autologous chondrocyte transplantation. *J Orthop Res,* **7**, 208–218.

Grassel, S. and Ahmed, N. (2007). Influence of cellular microenvironment and paracrine signals on chondrogenic differentiation. *Front Biosci,* **12**, 4946–4956.

Hall, B. K. (2005). *Bones and Cartilage: Developmental and Evolutionary Skeletal Biology,* Academic Press. ISBN 0123190606.

Hauser, C. J., Zhou, X., Joshi, P., Cuchens, M. A., Kregor, P., Devidas, M., Kennedy, R. J., Poole, G. V. and Hughes, J. L. (1997). The immune microenvironment of human fracture/soft-tissue hematomas and its relationship to systemic immunity. *J Trauma,* **42**, 895–903; discussion 903–904.

Haynesworth, S. E., Baber, M. A. and Caplan, A. I. (1996). Cytokine expression by human marrow-derived mesenchymal progenitor cells *in vitro*: effects of dexamethasone and IL-1 alpha. *J Cell Physiol,* **166**, 585–592.

Hegert, C., Kramer, J., Hargus, G., Muller, J., Guan, K., Wobus, A. M., Muller, P. K. and Rohwedel, J. (2002). Differentiation plasticity of chondrocytes derived from mouse embryonic stem cells. *J Cell Sci,* **115**, 4617–4628.

Huang, J. I., Durbhakula, M. M., Angele, P., Johnstone, B. and Yoo, J. U. (2006). Lunate arthroplasty with autologous mesenchymal stem cells in a rabbit model. *J Bone Joint Surg Am,* **88**, 744–752.

Huebsch, N. and Mooney, D. J. (2009). Inspiration and application in the evolution of biomaterials. *Nature,* **462**, 426–432.

Hwang, N. S. and Elisseeff, J. (2009). Application of stem cells for articular cartilage regeneration. *J Knee Surg,* **22**, 60–71.

Hwang, N. S., Varghese, S. and Elisseeff, J. (2008a). Controlled differentiation of stem cells. *Adv Drug Deliv Rev,* **60**, 199–214.

Hwang, N. S., Varghese, S. and Elisseeff, J. (2008b). Derivation of chondrogenically-committed cells from human embryonic cells for cartilage tissue regeneration. *PLoS ONE,* **3**, e2498.

Hwang, N. S., Varghese, S., Zhang, Z. and Elisseeff, J. (2006). Chondrogenic differentiation of human embryonic stem cell-derived cells in arginine-glycine-aspartate-modified hydrogels. *Tissue Eng,* **12**, 2695–2706.

Ifkovits, J. L. and Burdick, J. A. (2007). Review: photopolymerizable and degradable biomaterials for tissue engineering applications. *Tissue Eng,* **13**, 2369–2385.

Irwin, E. F., Pollock, J. F., Schaffer, D. V. and Healy, K. (2011). *Designing Tunable Artificial Matrices for Stem Cell Culture*, Elsevier, Inc.

Ito, M. M. and Kida, M. Y. (2000). Morphological and biochemical re-evaluation of the process of cavitation in the rat knee joint: cellular and cell strata alterations in the interzone. *J Anat,* **197**, Pt **4**, 659–679.

Iwamoto, M., Shimazu, A. and Pacifici, M. (1995). Regulation of chondrocyte maturation by fibroblast growth factor-2 and parathyroid hormone. *J Orthop Res,* **13**, 838–845.

Johansson, N., SAARIALHO-Kere, U., Airola, K., Herva, R., Nissinen, L., Westermarck, J., Vuorio, E., Heino, J. and Kahari, V. M. (1997). Collagenase-3

(MMP-13) is expressed by hypertrophic chondrocytes, periosteal cells, and osteoblasts during human fetal bone development. *Dev Dyn,* **208**, 387–397.

Johnstone, B., Hering, T. M., Caplan, A. I., Goldberg, V. M. and Yoo, J. U. (1998). *In vitro* chondrogenesis of bone marrow-derived mesenchymal progenitor cells. *Exp Cell Res,* **238**, 265–272.

Johnstone, B. and Yoo, J. (2001). Mesenchymal cell transfer for articular cartilage repair. *Expert Opin Biol Ther,* **1**, 915–921.

Johnstone, B., Yoo, J. and Stewart, M. (2006). *Cell Sources for Cartilage Tissue Engineering,* Hoboken, Wiley.

Kaigler, D., Wang, Z., Horger, K., Mooney, D. J. and Krebsbach, P. H. (2006). VEGF scaffolds enhance angiogenesis and bone regeneration in irradiated osseous defects. *J Bone Miner Res,* **21**, 735–744.

Kasten, P., Beyen, I., Niemeyer, P., Luginbuhl, R., Bohner, M. and Richter, W. (2008). Porosity and pore size of beta-tricalcium phosphate scaffold can influence protein production and osteogenic differentiation of human mesenchymal stem cells: an *in vitro* and *in vivo* study. *Acta Biomater,* **4**, 1904–1915.

Khademhosseini, A. and Langer, R. (2007). Microengineered hydrogels for tissue engineering. *Biomaterials,* **28**, 5087–5092.

Khan, I. M., Redman, S. N., Williams, R., Dowthwaite, G. P., Oldfield, S. F. and Archer, C. W. (2007). The development of synovial joints. *Curr Top Dev Biol,* **79**, 1–36.

Kim, S., Chung, E. H., Gilbert, M. and Healy, K. E. (2005). Synthetic MMP-13 degradable ECMs based on poly(N-isopropylacrylamide-co-acrylic acid) semi-interpenetrating polymer networks. I. Degradation and cell migration. *J Biomed Mater Res A,* **75**, 73–88.

Kish, G. and Hangody, L. (2004). A prospective, randomised comparison of autologous chondrocyte implantation versus mosaicplasty for osteochondral defects in the knee. *J Bone Joint Surg Br,* **86**, 619; author reply 619–620.

Koay, E. J., Hoben, G. M. and Athanasiou, K. A. (2007). Tissue engineering with chondrogenically differentiated human embryonic stem cells. *Stem Cells,* **25**, 2183–2190.

Koyama, E., Shibukawa, Y., Nagayama, M., Sugito, H., Young, B., Yuasa, T., Okabe, T., Ochiai, T., Kamiya, N., Rountree, R. B., Kingsley, D. M., Iwamoto, M., Enomoto-Iwamoto, M. and Pacifici, M. (2008). A distinct cohort of progenitor cells participates in synovial joint and articular cartilage formation during mouse limb skeletogenesis. *Dev Biol,* **316**, 62–73.

Kramer, J., Hegert, C., Guan, K., Wobus, A. M., Muller, P. K. and Rohwedel, J. (2000). Embryonic stem cell-derived chondrogenic differentiation *in vitro*: activation by BMP-2 and BMP-4. *Mech Dev,* **92**, 193–205.

Krampera, M., Glennie, S., Dyson, J., Scott, D., Laylor, R., Simpson, E. and Dazzi, F. (2003). Bone marrow mesenchymal stem cells inhibit the response of naive and memory antigen-specific T cells to their cognate peptide. *Blood,* **101**, 3722–3729.

Kronenberg, H. M. (2003). Developmental regulation of the growth plate. *Nature,* **423**, 332–336.

Laflamme, M. A. and Murry, C. E. (2005). Regenerating the heart. *Nat Biotechnol,* **23**, 845–856.

Le, A. X., Miclau, T., Hu, D. and Helms, J. A. (2001). Molecular aspects of healing in stabilized and non-stabilized fractures. *J Orthop Res,* **19**, 78–84.

Le Blanc, K., Tammik, C., Rosendahl, K., Zetterberg, E. and Ringden, O. (2003). HLA expression and immunologic properties of differentiated and undifferentiated mesenchymal stem cells. *Exp Hematol,* **31**, 890–896.

Leach, J. K., Kaigler, D., Wang, Z., Krebsbach, P. H. and Mooney, D. J. (2006). Coating of VEGF-releasing scaffolds with bioactive glass for angiogenesis and bone regeneration. *Biomaterials,* **27**, 3249–3255.

Lee, C. R., Grodzinsky, A. J., Hsu, H. P., Martin, S. D. and Spector, M. (2000a). Effects of harvest and selected cartilage repair procedures on the physical and biochemical properties of articular cartilage in the canine knee. *J Orthop Res,* **18**, 790–799.

Lee, F. Y., Hazan, E. J., Gebhardt, M. C. and Mankin, H. J. (2000b). Experimental model for allograft incorporation and allograft fracture repair. *J Orthop Res,* **18**, 303–306.

Lenas, P., Moos, M. and Luyten, F. P. (2009). Developmental engineering: a new paradigm for the design and manufacturing of cell-based products. Part II: from genes to networks: tissue engineering from the viewpoint of systems biology and network science. *Tissue Eng Part B Rev,* **15**, 395–422.

Lian, J. B., Mckee, M. D., Todd, A. M. and Gerstenfeld, L. C. (1993). Induction of bone-related proteins, osteocalcin and osteopontin, and their matrix ultrastructural localization with development of chondrocyte hypertrophy *in vitro*. *J Cell Biochem,* **52**, 206–219.

Lian, Q., Lye, E., Suan Yeo, K., Khia Way Tan, E., Salto-Tellez, M., Liu, T. M., Palanisamy, N., El Oakley, R. M., Lee, E. H., Lim, B. and Lim, S. K. (2007). Derivation of clinically compliant MSCs from CD105+, CD24- differentiated human ESCs. *Stem Cells,* **25**, 425–436.

Lutolf, M. P. (2009). Biomaterials: Spotlight on hydrogels. *Nat Mater,* **8**, 451–453.

Lutolf, M. P., Gilbert, P. M. and Blau, H. M. (2009). Designing materials to direct stem-cell fate. *Nature,* **462**, 433–441.

Lutolf, M. P. and Hubbell, J. A. (2005). Synthetic biomaterials as instructive extracellular microenvironments for morphogenesis in tissue engineering. *Nat Biotechnol,* **23**, 47–55.

Lutolf, M. P., Lauer-Fields, J. L., Schmoekel, H. G., Metters, A. T., Weber, F. E., Fields, G. B. and Hubbell, J. A. (2003). Synthetic matrix metalloproteinase-sensitive hydrogels for the conduction of tissue regeneration: engineering cell-invasion characteristics. *Proc Natl Acad Sci U S A,* **100**, 5413–5418.

Luyten, F. P., Dell'Accio, F. and De Bari, C. (2001). Skeletal tissue engineering: opportunities and challenges. *Best Pract Res Clin Rheumatol,* **15**, 759–769.

Maes, C., Kobayashi, T. and Kronenberg, H. M. (2007). A novel transgenic mouse model to study the osteoblast lineage *in vivo*. *Ann N Y Acad Sci,* **1116**, 149–164.

Maes, C., Kobayashi, T., Selig, M. K., Torrekens, S., Roth, S. I., Mackem, S., Carmeliet, G. and Kronenberg, H. M. (2010). Osteoblast precursors, but not mature osteoblasts, move into developing and fractured bones along with invading blood vessels. *Dev Cell,* **19**, 329–344.

Majumdar, M. K., Thiede, M. A., Mosca, J. D., Moorman, M. and Gerson, S. L. (1998). Phenotypic and functional comparison of cultures of marrow-derived mesenchymal stem cells (MSCs) and stromal cells. *J Cell Physiol,* **176**, 57–66.

Mann, B. K., Schmedlen, R. H. and West, J. L. (2001). Tethered-TGF-beta increases extracellular matrix production of vascular smooth muscle cells. *Biomaterials,* **22**, 439–444.

Mark, M. P., Butler, W. T., Prince, C. W., Finkelman, R. D. and Ruch, J. V. (1988). Developmental expression of 44-kDa bone phosphoprotein (osteopontin) and bone gamma-carboxyglutamic acid (Gla)-containing protein (osteocalcin) in calcifying tissues of rat. *Differentiation,* **37**, 123–136.

Meijer, G. J., De Bruijn, J. D., Koole, R. and Van Blitterswijk, C. A. (2007). Cell-based bone tissue engineering. *PLoS Med,* **4**, e9.

Miura, Y., Fitzsimmons, J. S., Commisso, C. N., Gallay, S. H. and O'driscoll, S. W. (1994). Enhancement of periosteal chondrogenesis *in vitro*. Dose-response for transforming growth factor-beta 1 (TGF-beta 1). *Clin Orthop Relat Res*, 271–280.

Mizuno, K., Mineo, K., Tachibana, T., Sumi, M., Matsubara, T. and Hirohata, K. (1990). The osteogenetic potential of fracture haematoma. Subperiosteal and intramuscular transplantation of the haematoma. *J Bone Joint Surg Br,* **72**, 822–829.

Muraglia, A., Cancedda, R. and Quarto, R. (2000). Clonal mesenchymal progenitors from human bone marrow differentiate *in vitro* according to a hierarchical model. *J Cell Sci,* **113** (Pt 7), 1161–1166.

Nakajima, M., Wakitani, S., Harada, Y., Tanigami, A. and Tomita, N. (2008). *In vivo* mechanical condition plays an important role for appearance of cartilage tissue in ES cell transplanted joint. *J Orthop Res,* **26**, 10–17.

Nakayama, N., Duryea, D., Manoukian, R., Chow, G. and Han, C. Y. (2003). Macroscopic cartilage formation with embryonic stem-cell-derived mesodermal progenitor cells. *J Cell Sci,* **116**, 2015–2028.

Nicodemus, G. D. and Bryant, S. (2008). Cell encapsulation in biodegradable hydrogels for tissue engineering applications. *Tissue Eng Part B, Rev,* **14**, 149–165.

Niemeyer, P., Pestka, J. M., Kreuz, P. C., Erggelet, C., Schmal, H., Suedkamp, N. P. and Steinwachs, M. (2008). Characteristic complications after autologous chondrocyte implantation for cartilage defects of the knee joint. *Am J Sports Med,* **36**, 2091–2099.

O'keefe, R. J., Loveys, L. S., Hicks, D. G., Reynolds, P. R., Crabb, I. D., Puzas, J. E. and Rosier, R. N. (1997). Differential regulation of type-II and type-X collagen synthesis by parathyroid hormone-related protein in chick growth-plate chondrocytes. *J Orthop Res,* **15**, 162–174.

Oliveira, S. M., Mijares, D. Q., Turner, G., Amaral, I. F., Barbosa, M. A. and Teixeira, C. C. (2009). Engineering endochondral bone: *in vivo* studies. *Tissue Eng Part A,* **15**, 635–643.

Orlic, D., Kajstura, J., Chimenti, S., Jakoniuk, I., Anderson, S. M., Li, B., Pickel, J., Mckay, R., Nadal-Ginard, B., Bodine, D. M., Leri, A. and Anversa, P. (2001). Bone marrow cells regenerate infarcted myocardium. *Nature,* **410**, 701–705.

Pacifici, M., Koyama, E. and Iwamoto, M. (2005). Mechanisms of synovial joint and articular cartilage formation: recent advances, but many lingering mysteries. *Birth Defects Res C Embryo Today,* **75**, 237–248.

Patterson, J. and Hubbell, J. A. (2010). Enhanced proteolytic degradation of molecularly engineered PEG hydrogels in response to MMP-1 and MMP-2. *Biomaterials,* **31**, 7836–7845.

Pelttari, K., Steck, E. and Richter, W. (2008). The use of mesenchymal stem cells for chondrogenesis. *Injury,* **39**, Suppl **1**, S58–S65.

Pelttari, K., Winter, A., Steck, E., Goetzke, K., Hennig, T., Ochs, B. G., Aigner, T. and Richter, W. (2006). Premature induction of hypertrophy during *in vitro* chondrogenesis of human mesenchymal stem cells correlates with calcification and vascular invasion after ectopic transplantation in SCID mice. *Arthritis Rheum,* **54**, 3254–3266.

Perren, S. M. (2002). Evolution of the internal fixation of long bone fractures. The scientific basis of biological internal fixation: choosing a new balance between stability and biology. *J Bone Joint Surg Br,* **84**, 1093–1110.

Peterson, L., Minas, T., Brittberg, M., Nilsson, A., Sjogren-Jansson, E. and Lindahl, A. (2000). Two- to 9-year outcome after autologous chondrocyte transplantation of the knee. *Clin Orthop Relat Res,* May (374), 212–234.

Phipatanakul, W. P., Vandevord, P. J., Teitge, R. A. and Wooley, P. H. (2004). Immune response in patients receiving fresh osteochondral allografts. *Am J Orthop (Belle Mead NJ),* **33**, 345–348.

Pittenger, M. F., Mackay, A. M., Beck, S. C., Jaiswal, R. K., Douglas, R., Mosca, J. D., Moorman, M. A., Simonetti, D. W., Craig, S. and Marshak, D. R. (1999). Multilineage potential of adult human mesenchymal stem cells. *Science,* **284**, 143–147.

Ratner, B. and Bryant, S. (2004). Biomaterials: Where we have been and where we are going. *Annu. Rev. Biomed. Eng.,* **6**, 41–75.

Roach, H. I. (1992). Trans-differentiation of hypertrophic chondrocytes into cells capable of producing a mineralized bone matrix. *Bone Miner,* **19**, 1–20.

Roach, H. I. (1997). New aspects of endochondral ossification in the chick: chondrocyte apoptosis, bone formation by former chondrocytes, and acid phosphatase activity in the endochondral bone matrix. *J Bone Miner Res,* **12**, 795–805.

Roberts, S., Mccall, I. W., Darby, A. J., Menage, J., Evans, H., Harrison, P. E. and Richardson, J. B. (2003). Autologous chondrocyte implantation for cartilage repair: monitoring its success by magnetic resonance imaging and histology. *Arthritis Res Ther,* **5**, R60–R73.

Roberts, S., Menage, J., Sandell, L. J., Evans, E. H. and Richardson, J. B. (2009). Immunohistochemical study of collagen types I and II and procollagen IIA in human cartilage repair tissue following autologous chondrocyte implantation. *Knee,* **16**, 398–404.

Sakaguchi, Y., Sekiya, I., Yagishita, K. and Muneta, T. (2005). Comparison of human stem cells derived from various mesenchymal tissues: superiority of synovium as a cell source. *Arthritis Rheum,* **52**, 2521–2529.

Salinas, C. N. and Anseth, K. S. (2008). The influence of the RGD peptide motif and its contextual presentation in PEG gels on human mesenchymal stem cell viability. *J Tissue Eng Regen Med,* **2**, 296–304.

Salinas, C. N. and Anseth, K. S. (2009). Decorin moieties tethered into PEG networks induce chondrogenesis of human mesenchymal stem cells. *J Biomed Mater Res A* **90**(2), 456–464. doi: 10.1002/jbm.a.32112.

Sato, K., Ozaki, K., Oh, I., Meguro, A., Hatanaka, K., Nagai, T., Muroi, K. and Ozawa, K. (2007). Nitric oxide plays a critical role in suppression of T-cell proliferation by mesenchymal stem cells. *Blood,* **109**, 228–234.

Scammell, B. E. and Roach, H. I. (1996). A new role for the chondrocyte in fracture repair: endochondral ossification includes direct bone formation by former chondrocytes. *J Bone Miner Res,* **11**, 737–745.

Scotti, C., Tonnarelli, B., Papadimitropoulos, A., Scherberich, A., Schaeren, S., Schauerte, A., Lopez-Rios, J., Zeller, R., Barbero, A. and Martin, I. (2010). Recapitulation of endochondral bone formation using human adult mesenchymal stem cells as a paradigm for developmental engineering. *Proc Natl Acad Sci U S A,* **107**, 7251–7256.

Song, L. and Tuan, R. S. (2004). Transdifferentiation potential of human mesenchymal stem cells derived from bone marrow. *FASEB J,* **18**, 980–982.

Takahashi, K. and Yamanaka, S. (2006). Induction of pluripotent stem cells from mouse embryonic and adult fibroblast cultures by defined factors. *Cell,* **126**, 663–676.

Thompson, Z., Miclau, T., Hu, D. and Helms, J. A. (2002). A model for intramembranous ossification during fracture healing. *J Orthop Res,* **20**, 1091–1098.

Thomson, J. A., Itskovitz-Eldor, J., Shapiro, S. S., Waknitz, M. A., Swiergiel, J. J., Marshall, V. S. and Jones, J. M. (1998). Embryonic stem cell lines derived from human blastocysts. *Science,* **282**, 1145–1147.

Tins, B. J., Mccall, I. W., Takahashi, T., Cassar-Pullicino, V., Roberts, S., Ashton, B. and Richardson, J. (2005). Autologous chondrocyte implantation in knee joint: MR imaging and histologic features at 1-year follow-up. *Radiology,* **234**, 501–508.

Tse, W. T., Pendleton, J. D., Beyer, W. M., Egalka, M. C. and Guinan, E. C. (2003). Suppression of allogeneic T-cell proliferation by human marrow stromal cells: implications in transplantation. *Transplantation,* **75**, 389–397.

Ulijn, R. V., Bibi, N., Jayawarna, V., Thornton, P. E., Todd, S. J., Mart, R. J., Smith, A. and Gough, J. E. (2007). Bioresponsive hydrogels. *Mater Today,* **10**, 40–48.

Vacanti, C. A., Kim, W., Upton, J., Mooney, D. and Vacanti, J. P. (1995). The efficacy of periosteal cells compared to chondrocytes in the tissue engineered repair of bone defects. *Tissue Eng,* **1**, 301–308.

Vinatier, C., Mrugala, D., Jorgensen, C., Guicheux, J. and Noel, D. (2009a). Cartilage engineering: a crucial combination of cells, biomaterials and biofactors. *Trends Biotechnol,* **27**, 307–314.

Vinatier, C., Mrugala, D., Jorgensen, C., Guicheux, J. and Noël, D. (2009b). Cartilage engineering: a crucial combination of cells, biomaterials and biofactors. *Trends Biotechnol,* **27**, 307–314.

Von Der Mark, K., Gauss, V., Von Der Mark, H. and Muller, P. (1977). Relationship between cell shape and type of collagen synthesised as chondrocytes lose their cartilage phenotype in culture. *Nature,* **267**, 531–532.

Wakitani, S., Aoki, H., Harada, Y., Sonobe, M., Morita, Y., Mu, Y., Tomita, N., Nakamura, Y., Takeda, S., Watanabe, T. K. and Tanigami, A. (2004). Embryonic stem cells form articular cartilage, not teratomas, in osteochondral defects of rat joints. *Cell Transplant,* **13**, 331–336.

Wakitani, S., Takaoka, K., Hattori, T., Miyazawa, N., Iwanaga, T., Takeda, S., Watanabe, T. K. and Tanigami, A. (2003). Embryonic stem cells injected into the mouse knee joint form teratomas and subsequently destroy the joint. *Rheumatology (Oxford),* **42**, 162–165.

Wang, D. A., Williams, C. G., Yang, F., Cher, N., Lee, H. and Elisseeff, J. H. (2005). Bioresponsive phosphoester hydrogels for bone tissue engineering. *Tissue Eng,* **11**, 201–213.

West, J. and Hubbell, J. (1999). Polymeric biomaterials with degradation sites for protease involved in cell migration. *Macromolecules,* **32**, 241–244.

Wilmut, I., Schnieke, A. E., Mcwhir, J., Kind, A. J. and Campbell, K. H. (1997). Viable offspring derived from fetal and adult mammalian cells. *Nature,* **385**, 810–813.

Woodbury, D., Schwarz, E. J., Prockop, D. J. and Black, I. B. (2000). Adult rat and human bone marrow stromal cells differentiate into neurons. *J Neurosci Res,* **61**, 364–370.

Wheeler, D. L. and Enneking, W. F. (2005). Allograft bone decreases in strength *in vivo* over time. *Clin Orthop Relat Res*, Jun (435), 36–42.

Xing, Z., Lu, C., Hu, D., Yu, Y. Y., Wang, X., Colnot, C., Nakamura, M., Wu, Y., Miclau, T. and Marcucio, R. S. (2010). Multiple roles for CCR2 during fracture healing. *Dis Model Mech,* **3**, 451–458.

Yamanaka, S. (2009). A fresh look at iPS cells. *Cell,* **137**, 13–17.

Yamashita, A., Nishikawa, S. and Rancourt, D. E. (2010). Identification of five developmental processes during chondrogenic differentiation of embryonic stem cells. *PLoS One,* **5**, e10998.

Yoo, J. U., Barthel, T. S., Nishimura, K., Solchaga, L., Caplan, A. I., Goldberg, V. M. and Johnstone, B. (1998). The chondrogenic potential of human bone-marrow-derived mesenchymal progenitor cells. *J Bone Joint Surg Am,* **80**, 1745–1757.

Young, R. G., Butler, D. L., Weber, W., Caplan, A. I., Gordon, S. L. and Fink, D. J. (1998). Use of mesenchymal stem cells in a collagen matrix for Achilles tendon repair. *J Orthop Res,* **16**, 406–413.

Yu, Y. Y., Lieu, S., Lu, C., Miclau, T., Marcucio, R. S. and Colnot, C. (2010). Immunolocalization of BMPs, BMP antagonists, receptors, and effectors during fracture repair. *Bone,* **46**, 841–851.

Zuk, P. A., Zhu, M., Ashjian, P., De Ugarte, D. A., Huang, J. I., Mizuno, H., Alfonso, Z. C., Fraser, J. K., Benhaim, P. and Hedrick, M. H. (2002). Human adipose tissue is a source of multipotent stem cells. *Mol Biol Cell,* **13**, 4279–4295.

10

Chitosan for bone repair and regeneration

J. VENKATESAN and S. K. KIM, Pukyong National University, Republic of Korea

DOI: 10.1533/9780857099037.3.244

Abstract: In recent years, there has been extensive research into using artificial bone to treat bone loss or failure. Several biomaterials have been used in the artificial preparative method, including polymers, ceramics and metals. The natural polymer chitosan is one of the most extensively studied biomaterials for bone tissue engineering, due to its biocompatibility and biodegradability. In this chapter, we discuss methods of applying chitosan derivatives, grafting with different polymers and surface immobilization. We explore the use of chitosan composites with alginate and nanostructured hydroxyapatite, as well as chitosan with stem cell interaction for orthopedic treatment. Discussion of these areas will highlight the many potential applications of chitosan as a biomaterial for bone tissue engineering.

Key words: chitosan derivatives, alginate, hydroxyapatite (HAp), stem cells, bone tissue engineering.

10.1 Introduction

Natural bone is a hierarchical structure made up of fibrillar collagen (organic) and nanohydroxyapatite (nHAp) (inorganic) ($Ca_{10}(PO_4)_6(OH)_2$) (Fig. 10.1). Bone provides the structural arrangement of vertebrates and also maintains calcium and phosphate levels for human metabolic functions (Glimcher, 2006). Bone defects can arise in a number of ways, including motor accidents, birth defects, low calcium levels, gangrene, etc. Surgical treatment frequently requires implantation of a temporary or permanent prosthesis. This implantation still presents a challenge to orthopedic surgeons, especially in the case of large bone defects. In recent years, there have been significant developments in the treatment of bone defects using cell based tissue engineering models. There are two main methods of bone replacement: autograft and allograft. However, both of these methods have disadvantages. From a clinical perspective, the use of artificial scaffolds is attractive as it minimizes patient discomfort, risk of infection, scar formation and the cost of treatment. Polymers, ceramics and metals are currently being used as scaffolds for the treatment of bone fracture or defects. However, composite scaffolds

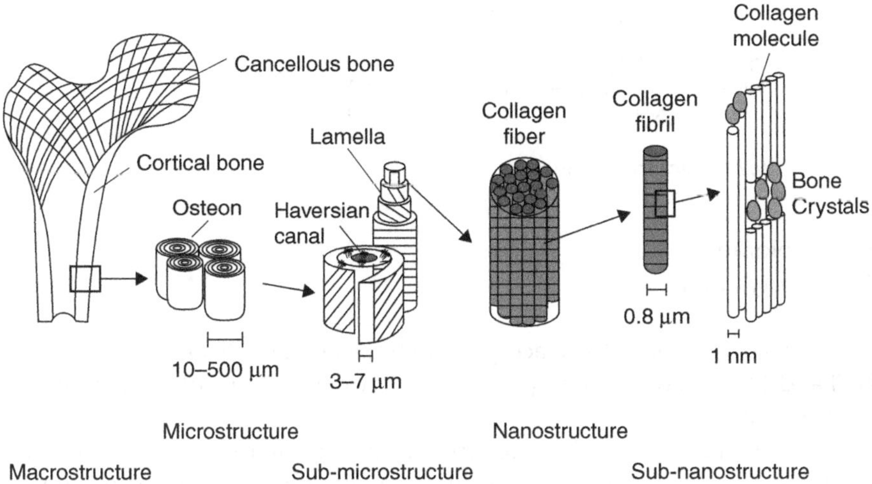

10.1 Hierarchical structure of bone. (*Source*: Reprinted from Nyman *et al.*, 2005, Copyright 2005, with permission from Elsevier.)

combining these materials are being investigated in order to better mimic the mineral component and the microstructure of natural bone.

10.2 Natural polymers – chitin and chitosan

Chitosan is the linear polysaccharide (Fig. 10.2). It is produced from chitin, a natural polysaccharide found in crabs, shrimp, lobsters, coral, jellyfish, butterflies, ladybugs, mushrooms and fungi. Marine crustacean shells are widely used as primary sources for the production of chitosan. For this reason, crabs and shrimp are marine species of great commercial importance. They can be found in the tropical and subtropical waters of the Pacific, Atlantic and Indian oceans.

10.2.1 Preparation of chitosan

Chitosan can be isolated directly from the cell walls of certain fungi, but is usually prepared from chitin. Chemical hydrolysis and enzymatic hydrolysis methods have been widely used for the isolation of chitosan from marine crustacean shells, as they are quite inexpensive.

In the chemical hydrolysis method, four main steps are involved in producing chitosan from marine crustacean shells. They are:

1. demineralization
2. deproteinization

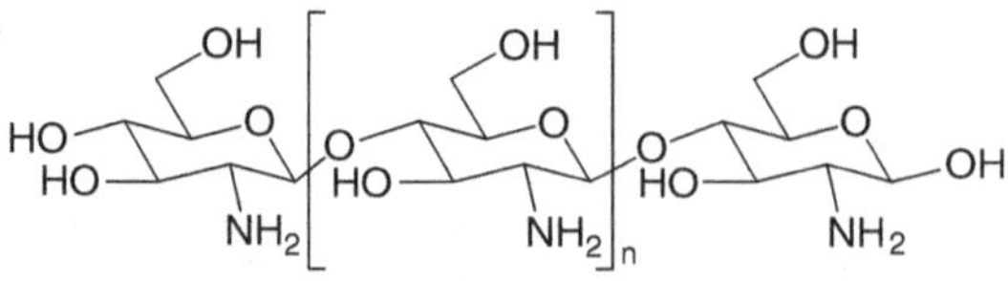

10.2 Structure of chitosan.

3. discoloration
4. deacetylation.

To produce 1 kg of 70% deacetylated chitosan from shrimp shells, 6.3 kg HCl and 1.8 kg NaOH are required (Madhavan and Nair, 1974; Shahidi and Abuzaytoun, 2005).

The enzymatic method could provide an alternative to this chemical production method. The degree of deacetylation and molecular weight of the chitosan depends on the source and preparation method. The molecular weight ranges from 300 to over 1000 kDa and the degree of deacetylation from 30% to 95%. Chitosan can be degraded by enzymes to form various small molecular weight fragments. Chitosanase has been used to prepare chitosan with a small molecular weight via methods using batch reactors, column reactors and ultrafiltration membrane reactors (Kim and Rajapakse, 2005). The chitosan has been extracted by various enzymes such as lysozyme, snailase, neutral protease and novel chitin deacetylase. The average molecular weight of chitosan obtained using the enzymatic method (267.97 kDa) has been reported to be far greater than that obtained by the chemical method (84.04 kDa) (No and Meyers, 1995; Cai *et al.*, 2006). The molecular weight of the chitosan used is of great significance in various bone related diseases.

10.2.2 Properties of chitosan

Chitosan is a copolymer consisting of β-(1→4)-2-acetamido-d-glucose and β-(1→4)-2-amino-D-glucose unit linkages. Over the past two decades, chitosan has been used in biomedical applications due to its high biocompatibility, biodegradability, porous structure, suitability for cell ingrowth, osteoconduction and intrinsic antibacterial nature. Low molecular weight chitosan has been known to carry out many biological activities including:

- acting as an antibacterial agent (Jeon and Kim, 2000; Jeon *et al.*, 2001; Eaton *et al.*, 2008; Fernandes *et al.*, 2008, 2009)
- acting as an antioxidant (Je *et al.*, 2004; Fernandes *et al.*, 2010)
- acting as a matrix metalloproteinase inhibitor (Kim and Kim, 2006; Rajapakse *et al.*, 2006; Van *et al.*, 2006)

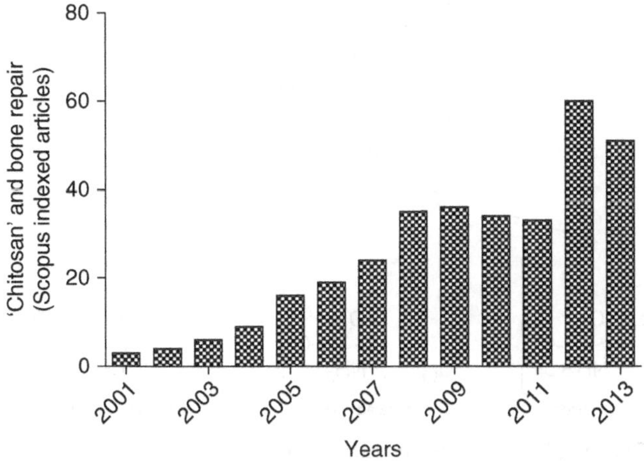

10.3 Publication in Scopus cited articles.

- having anti-diabetic and anti-HIV properties (Liu *et al.*, 2007; Artan *et al.*, 2010)
- acting as an anti-inflammatory agent (Yang *et al.*, 2010)
- enabling drug delivery (Liu *et al.*, 2007).

Because of this, it is frequently used for cartilage tissue engineering, wound healing and orthopedic applications. There is a growing interest in composite preparation for bone tissue engineering and the creation of artificial organs.

Chitosan can be easily modified into films, fibers, beads, sponges and more complex shapes for orthopedic treatment. Because of its cationic nature, chitosan attracts various negatively charged proteoglycans. Porous materials have a highly significant role in the bone implantation process, and porous chitosan structures can be formed by freezing and lyophilizing chitosan acetic acid solutions in suitable molds. Chitosan has been combined with a variety of materials such as HAp, alginate, hyaluronic acid, calcium phosphate, poly (methyl methacrylate), poly-L-lactic acid and growth factors for potential application in orthopedics.

The frequency of publications about chitosan and bone repair in the Scopus articles has increased year by year, indicating that there has been a significant development in the amount of research about chitosan for bone repair (Fig. 10.3).

10.3 Chitosan derivatives for bone tissue engineering

Chitosan is a natural and biodegradable polymer. It has the repeat unit of a $-NH_2$ group in a C-2 position and –OH in C-3 and C-6 position which

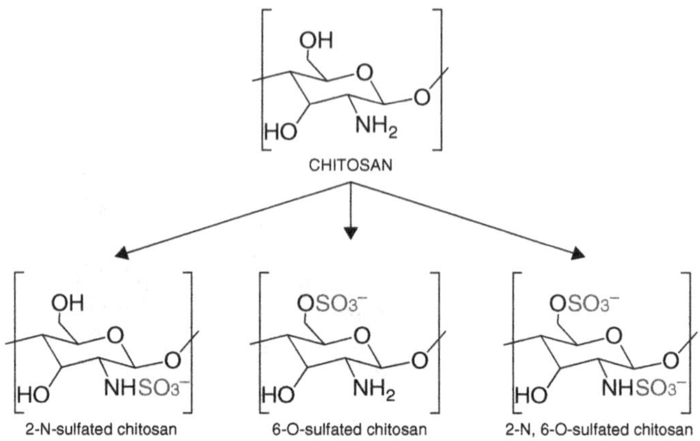

10.4 Chemical structures of chitosan modified with different sulfate groups. 2-N-sulfated chitosan (2SCS); 6-O-sulfated chitosan (6SCS); 2-N, 6-O-sulfated chitosan (26SCS). (*Source*: Adapted from Zhou *et al.*, 2009, Copyright 2009, with permission from Elsevier.)

can be specifically modified by chemical grafting or enzymatic modification (Sashiwa and Aiba, 2004). The bioactivity of chitosan is limited in pharmaceutical and biomedical applications, due to its poor solubility in water and low absorbability in *in vivo* systems. Chemical modification of chitosan can rectify this, making it soluble in water and therefore suitable for use in various biological and biomedical applications (Zhang *et al.*, 2010).

Research has shown that chitosan can support the growth and phenotypic expression of osteoblast and chondrocytes, making it suitable for the repair of bone and cartilage (Lahiji *et al.*, 2000; Fakhry *et al.*, 2004). Chitosan unregulated genes such as Type I collagen, integrin-binding sialoprotein, osteopontin, osteonectin and osteocalcin are associated with calcium binding and mineralization. When tested, they showed higher mineral deposition and calcium content (Mathews *et al.*, 2011). The degree of deacetylation of chitosan is a major factor affecting its bioactivity (Amaral *et al.*, 2006).

Chitosan can be modified via sulfation (Fig. 10.4) and phosphorylation (Fig. 10.5). The modification of chitosan can increase the solubility and bioavailability, making it more suitable for various biological and biomedical applications.

Chitosan film has been modified with dexamethasone (DEX) via the amino-alcohol forming reaction between the amino group of chitosan and the carbonyl group of DEX (Fig. 10.6). Assays of alkaline phosphatase activity indicate that DEX-chitosan shows an increase of about 3.5-fold compared to chitosan (Chiang *et al.*, 2012).

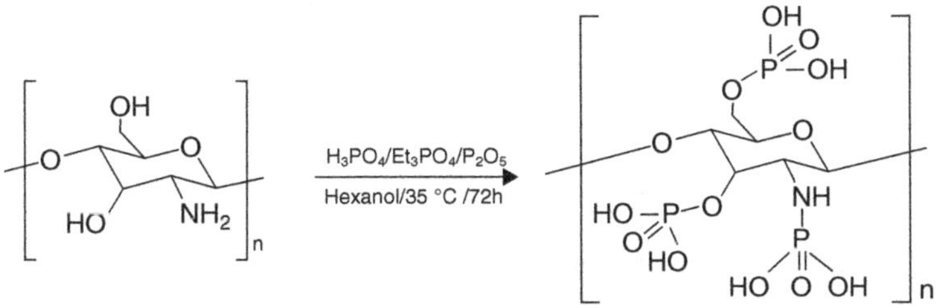

10.5 Synthesis of P-chitosan (Wang *et al.*, 2001, 2002; Nishi, 1986; Jayakumar and Selvamurugan, 2008; Jayakumar and Nagahama, 2008; Prabaharan, 2008; Jayakumar *et al.*, 2006).

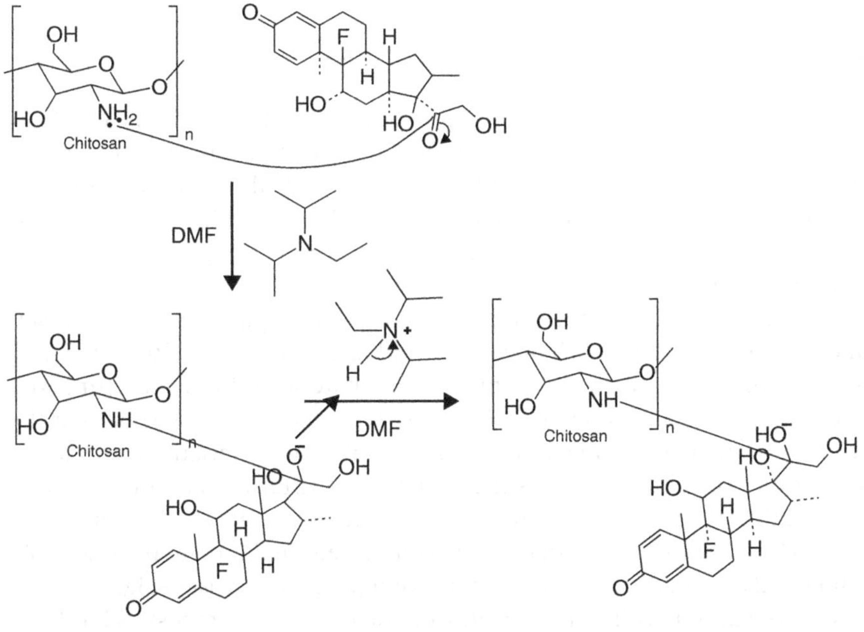

10.6 Surface immobilization of chitosan film with dexamethasone (DEX) was done by nucleophilic addition reaction. For DEX immobilization, the amino groups of the chitosan surface were reacted with carbonyl groups of DEX to form the DEX-chitosan formation. (*Source*: Adapted from Chiang *et al.*, 2012, Copyright 2012, with permission from Elsevier.)

10.4 Chitosan-based composites for bone tissue engineering

Chitosan based composites are of particular interest in creating artificial bone due to their biocompatibility, biodegradability and bone-forming ability. Several polymers and bioceramics are widely used in making composite

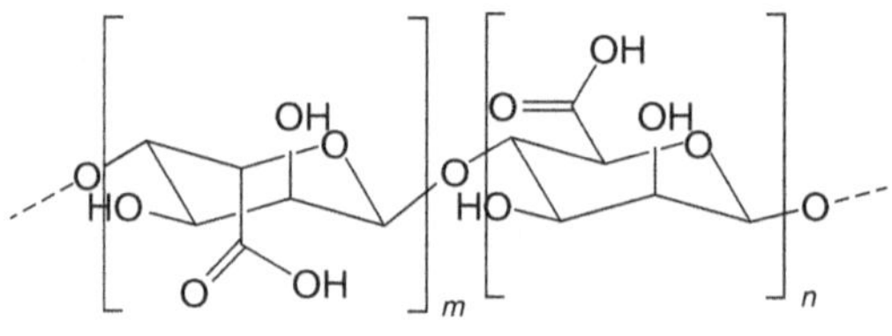

10.7 Structure of alginate.

materials with chitosan for bone tissue engineering purpose. HAp is widely used as a bioceramic since it is very similar to the natural mineral function of bone.

10.4.1 Chitosan/alginate composites

Alginate is an anionic polysaccharide found in the cell walls of algae. It can be easily modified into hydrogels due to its propensity to absorb water (Fig. 10.7). Alginate also reacts readily with the cationic nature of polymers to form a polyelectrolyte solution. Calcium chloride and oxytetracycline hydrochloride have been used as cross-linking agents to form polyelectrolytes.

In addition, several reports suggest that adding alginate to chitosan significantly improves its mechanical and biological properties. Bone-forming osteoblasts readily attach to the chitosan–alginate scaffold, proliferating well and depositing a calcified matrix (Fig. 10.8). *In vivo* studies suggest that the chitosan–alginate scaffold has a high degree of tissue compatibility. Calcium deposition occurred as early as the fourth week after implantation (Li and Zhang, 2005; Li *et al.*, 2005; Olmez *et al.*, 2007; Jeong *et al.*, 2010; Verma *et al.*, 2011). The addition of growth factor and stem cells in this composite scaffold significantly increases bone formation. An injectable chitosan–alginate with mesenchymal stem cell/bone morphogenetic protein-2 composites has been investigated, and suggests that clinically useful injectable materials may be used to generate new bone (Park *et al.*, 2005). It has also been suggested that partial protonated amine groups of chitosan react with the carboxylate groups of alginate, forming strong poly electrolytes (Li *et al.*, 2009; Tığlı and Gümüşderelioğlu, 2009).

Chitosan/alginate/sodium hyaluronate scaffolds have been prepared by the cast into film method, and studied with the addition of an Arg-Gly-Asp (RGD) containing protein to determine the potential benefits for bone tissue engineering (Hsu *et al.*, 2004).

HAp/chitosan/alginate composite scaffolds prepared by the *in situ* precipitation method have also been studied. It was found that the pore diameters of the scaffold were decreased by the addition of HAp in an agglomerated

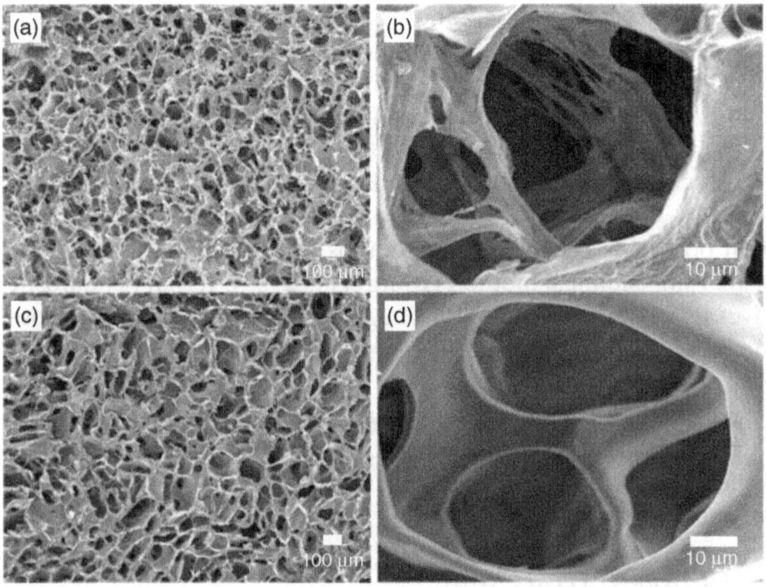

10.8 Scanning electron micrographs of the cross sections of porous scaffolds. (a) and (b) Chitosan–alginate scaffold, and (c) and (d) pure chitosan scaffold. (*Source*: Reprinted from Li *et al.*, 2005, Copyright 2009, with permission from Elsevier.)

condition. However, the compressive strength of composite scaffolds did increase in line with the HAp content (Jin *et al.*, 2008).

10.4.2 Chitosan/hydroxyapatite (HAp) composites

Chitosan with HAp has been the focus of much research into biomaterials for bone tissue engineering in recent years. Several researchers have reported that the addition of nano-sized HAp in chitosan not only increases the mechanical strength of bone, but also enhances the cellular differentiation and proliferation. The methods which have been used to prepare a nanocomposite of chitosan/nHAp are:

- the *in situ* method (Kong *et al.*, 2006)
- solvent casting (Murugan and Ramakrishna, 2004)
- simple *in situ* hybridization (Hu *et al.*, 2004)
- *in situ* mineralization (Verma *et al.*, 2008)
- calcium carbonate and calcium oxide loaded nHAp/chitosan (Zou *et al.*, 2009)
- the double diffusion method (Manjubala *et al.*, 2006)
- biocomposite (Ma *et al.*, 2011)

- the co-precipitation method (Yamaguchi *et al.*, 2003)
- hydroxypropylated chitosan and ethylene glycol functionalized nHAp (Depan *et al.*, 2011)
- homogeneous microstructure (Chen *et al.*, 2002)
- the sol-gel method (Su *et al.*, 2009)
- solvent casting and evaporation methods (Xianmiao *et al.*, 2009).

High porosity nHAp chitosan composite scaffold is prepared by a widely used *in situ* chemical method (Kong *et al.*, 2005). Ternary composite systems have also been used for bone engineering, consisting of chitosan/nHAp combined with any of the following:

- β-glycerophosphate (Huang *et al.*, 2009)
- polyamide-66 (Huang *et al.*, 2012)
- collagen (Wang *et al.*, 2009)
- starch (Jafar *et al.*, 2011)
- gelatin (Li *et al.*, 2007)
- zinc oxide (Li *et al.*, 2010)
- poly(galacturonic acid) (Verma *et al.*, 2010)
- collagen gels (Huang *et al.*, 2011)
- hyaluronic acid (Chen *et al.*, 2012)
- Cu-Zn alloy nanoparticles (nCu-Zn) (Tripathi *et al.*, 2011)
- sodium carboxymethyl cellulose (Jiang *et al.*, 2008, 2009, 2011; Liuyun *et al.*, 2008, 2009)
- poly(vinyl alcohol) (Yang *et al.*, 2008),
- pectin (Li *et al.*, 2011)
- poly(lactic acid) (Zhang *et al.*, 2012)
- magnetite (Cui *et al.*, 2008)
- genipin (Li *et al.*, 2012)
- konjac glucomannan (Zhou *et al.*, 2007)
- gelatin (nHCG) (Li *et al.*, 2009).

Chitosan/nHAp surfaces and nanostructures can significantly influence cell growth and viability. There are several key parameters which are important when studying scaffolds in the *in vitro* study, including cell proliferation, alkaline phosphatase, mineralization effect, type I collagen production and other gene markers such as osteocalcin and osteopontin. A cytocompatibility test of chitosan/nHAp using human MG63 cells revealed favorable cell viability and increased proliferation over a period of 11 days in culture (Zou *et al.*, 2009). The pre-osteoblast (MC3T3-E1) cell proliferation was about 1.5 times greater than pure chitosan after 7 days of culture and beyond. This was observed by qualitative analysis via fluorescence microscopy and quantitative study through 3-(4,5-dimethylthiazol-2-yl)-2,5-

diphenyltetrazolium bromide (MTT) assay. The addition of nHAp in the chitosan improved the mechanical properties and osteoconductivity necessary for bone formation. In terms of alkaline phosphatase (ALP) and type I collagen production, chitosan and chitosan/nHAp composite scaffolds showed similar results, but the addition of nHAp to chitosan improves osteocalcin production (Chesnutt *et al.*, 2009).

10.5 Chitosan with stem cells

Stem based therapy is a promising approach to solving several bone related issues. Different kinds of stems cells including embryonic stem cells, mesenchymal stem cells and pulp stem cells are currently used for bone tissue engineering. In addition, there has been research into the differentiation of mesenchymal stem cells into osteoblast cells. The stems can differentiate into various lineages, osteoblasts, adipocytes and chondrocytes. Applying these stem cells to bone tissue engineering requires inducing *in vitro* differentiation of these cells into bone-forming cells, osteoblasts (Seong *et al.*, 2010). In addition, scaffolds are important in treating bone defects with stem cell based therapies. An extracellular matrix (ECM) consists of self-assembled macromolecules, made up of collagens, glycoproteins, hyaluronan and proteoglycans. An ECM provides structural strength to tissues, maintaining a complex architecture around the cells and the shape of organs (Cancedda, 2009).

The search for biocompatible and biodegradable materials that can support the growth and phenotypic expression of osteoblasts is a major challenge for tissue engineering scientists. Chitosan is being investigated as an ECM for osteoblast lineages (Lahiji *et al.*, 2000). Bone formation from rat muscle-derived stem cells (rMDSCs) injected with *in situ*-forming chitosan gel is also being studied *in vivo*. Scaffolds made up of rMDSCs showed good results in terms of proliferation and cell attachment. Subcutaneous implantation of the *in situ*-forming chitosan gel demonstrated that rMDSCs containing chitosan gel induced much lower host tissue responses than did the chitosan gel alone. This is probably due to the immunosuppression of the transplanted rMDSCs (Fig. 10.9).

10.6 Conclusions

In the last three decades, chitosan has been a widely used biomaterial for bone tissue engineering. It has been used successfully to construct artificial bone due to its biocompatibility and biodegradability. The degree of deacetylation and other chemical parameters of chitosan modification influence the use of chitosan as a bone graft substitute. Sulfation, phosphorylation and carboxylation of chitosan have been used to form scaffolds. In

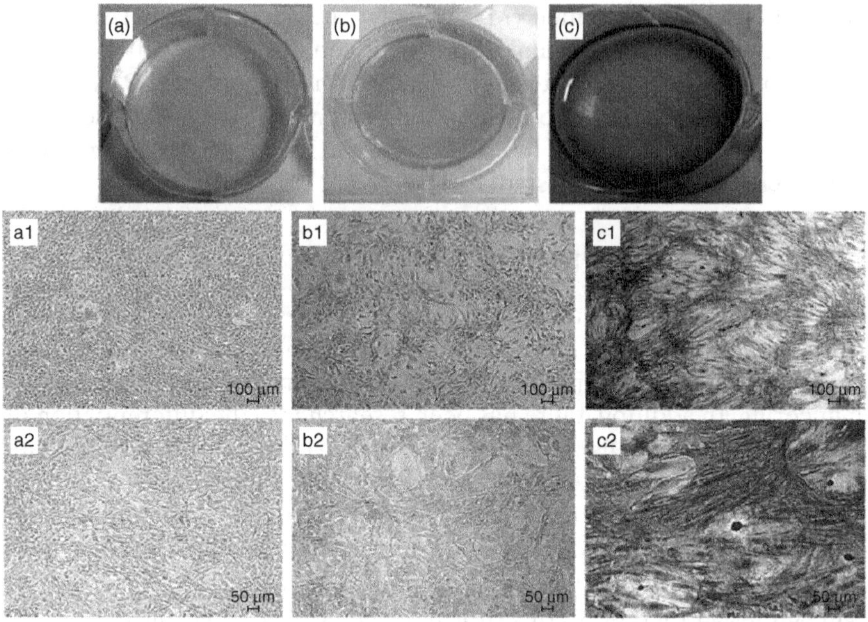

10.9 (a, a1, a2) rMDSCs isolated from pre-plate 6 imaged at passage 5. (b, b1, b2) ALP staining at 2 weeks after *in vitro* osteogenic differentiation of rMDSCs. (c, c1, c2) von Kossa staining at 4 weeks after osteogenic differentiation of rMDSCs. Magnification is (a1–c1) 40 × and (a2–c2) 100×; scale bar is (a1–c1) 100 µm and (a2–c2) 50 µm. a, b and c are 6-well plate size (BD Bioscience, Bedford, MA). (*Source*: Reprinted from Kim *et al.* 2008, copyright 2008, with permission from Elsevier.)

addition, other ceramic materials combined with chitosan have been investigated for bone tissue engineering. Chitosan composite scaffolds could act as promising biomaterials for bone tissue engineering.

10.7 Acknowledgements

This work was supported by a grant from Marine Bioprocess Research Centre of the Marine Bio 21 Center funded by the Ministry of Land, Transport and Maritime, Republic of Korea.

10.8 References

Amaral, I., Sampaio, P. and Barbosa, M. (2006). Three-dimensional culture of human osteoblastic cells in chitosan sponges: The effect of the degree of acetylation. *Journal of Biomedical Materials Research Part A,* **76** 2, 335–346.

Artan, M., Karadeniz, F., Karagozlu, M.Z., Kim, M.-M. and Kim, S.-K. (2010). Anti-HIV-1 activity of low molecular weight sulfated chitooligosaccharides. *Carbohydrate Research,* **345** 5, 656–662.

Cai, J., Yang, J., Du, Y., Fan, L., Qiu, Y., Li, J. and Kennedy, J. F. (2006). Enzymatic preparation of chitosan from the waste *Aspergillus niger* mycelium of citric acid production plant. *Carbohydrate Polymers,* **64** 2, 151–157.

Cancedda, R. (2009). Cartilage and bone extracellular matrix. *Current Pharmaceutical Design,* **15** 12, 1334–1348.

Chen, F., Wang, Z.C. and Lin, C.J. (2002). Preparation and characterization of nano-sized hydroxyapatite particles and hydroxyapatite/chitosan nano-composite for use in biomedical materials. *Materials Letters,* **57** 4, 858–861.

Chen, J., Yu, Q., Zhang, G., Yang, S., Wu, J. and Zhang, Q. (2012). Preparation and biocompatibility of nanohybrid scaffolds by *in situ* homogeneous formation of nano hydroxyapatite from biopolymer polyelectrolyte complex for bone repair applications. *Colloids and Surfaces B: Biointerfaces,* **93** 1, 100–107.

Chesnutt, B.M., Yuan, Y., Buddington, K., Haggard, W.O. and Bumgardner, J.D. (2009). Composite chitosan/nano-hydroxyapatite scaffolds induce osteocalcin production by osteoblasts *in vitro* and support bone formation *in vivo*. *Tissue Engineering Part A,* **15** 9, 2571–2579.

Chiang, Z.C., Yu, S.H., Chao, A.C. and Dong, G.C. (2012). Preparation and characterization of dexamethasone-immobilized chitosan scaffold. *Journal of Bioscience and Bioengineering,* **113** 5, 654–660.

Cui, W., Hu, Q., Wu, J., Li, B. and Shen, J. (2008). Preparation and characterization of magnetite/hydroxyapatite/chitosan nanocomposite by *in situ* compositing method. *Journal of Applied Polymer Science,* **109** 4, 2081–2088.

Depan, D., Venkata Surya, P., Girase, B. and Misra, R. (2011). Organic/inorganic hybrid network structure nanocomposite scaffolds based on grafted chitosan for tissue engineering. *Acta Biomaterialia,* **7** 5, 2163–2175.

Eaton, P., Fernandes, J.C., Pereira, E., Pintado, M.E. and Xavier Malcata, F. (2008). Atomic force microscopy study of the antibacterial effects of chitosans on *Escherichia coli* and *Staphylococcus aureus*. *Ultramicroscopy,* **108** 10, 1128–1134.

Fakhry, A., Schneider, G.B., Zaharias, R. and Şenel, S. (2004). Chitosan supports the initial attachment and spreading of osteoblasts preferentially over fibroblasts. *Biomaterials,* **25** 11, 2075–2079.

Fernandes, J.C., Eaton, P., Gomes, A.M., Pintado, M.E. and Xavier Malcata, F. (2009). Study of the antibacterial effects of chitosans on *Bacillus cereus* (and its spores) by atomic force microscopy imaging and nanoindentation. *Ultramicroscopy,* **109** 8, 854–860.

Fernandes, J.C., Eaton, P., Nascimento, H., Gião, M. S., Ramos, Ó. S., Belo, L., Santos-Silva, A., Pintado, M. E. and Malcata, F. X. (2010). Antioxidant activity of chitooligosaccharides upon two biological systems: Erythrocytes and bacteriophages. *Carbohydrate Polymers,* **79** 4, 1101–1106.

Fernandes, J.C., Tavaria, F.K., Soares, J.C., Ramos, Ó. S., João Monteiro, M., Pintado, M. E. and Xavier Malcata, F. (2008). Antimicrobial effects of chitosans and chitooligosaccharides, upon *Staphylococcus aureus* and *Escherichia coli*, in food model systems. *Food Microbiology,* **25** 7, 922–928.

Glimcher, M.J. (2006). Bone: nature of the calcium phosphate crystals and cellular, structural, and physical chemical mechanisms in their formation. *Reviews in Mineralogy and Geochemistry,* **64** 1, 223–282.

Hsu, S., Whu, S.W., Hsieh, S.C., Tsai, C.L., Chen, D.C. and Tan, T.S. (2004). Evaluation of chitosan-alginate-hyaluronate complexes modified by an RGD-containing protein as tissue-engineering scaffolds for cartilage regeneration. *Artificial Organs,* **28** 8, 693–703.

Hu, Q., Li, B., Wang, M. and Shen, J. (2004). Preparation and characterization of biodegradable chitosan/hydroxyapatite nanocomposite rods via *in situ* hybridization: a potential material as internal fixation of bone fracture. *Biomaterials,* **25** 5, 779–785.

Huang, D., Zuo, Y., Zou, Q.Wang, Y., Gao, S., Wang, X., Liu, H. and Li, Y. (2012). Reinforced nanohydroxyapatite/polyamide66 scaffolds by chitosan coating for bone tissue engineering. *Journal of Biomedical Materials Research Part B: Applied Biomaterials,* **100** 1, 51–57.

Huang, Z., Tian, J., Yu, B., Xu, Y. and Feng, Q. (2009). A bone-like nano-hydroxyapatite/collagen loaded injectable scaffold. *Biomedical Materials,* **4** 055005.

Huang, Z., Yu, B., Feng, Q., Li, S., Chen, Y. and Luo, L. (2011). *In situ*-forming chitosan/nano-hydroxyapatite/collagen gel for the delivery of bone marrow mesenchymal stem cells. *Carbohydrate Polymers,* **85** 261–267.

Jafar, A., Mostafa, R.T., Esmaeil, B., Saeed, H.K. and Rahim, J. (2011). Mechanical properties of chitosan-starch composite filled hydroxyapatite micro-and nanopowders. *Journal of Nanomaterials,* **2011** Article ID 391596 5 pages.

Jayakumar, R., Nagahama, H., Furuike, T. and Tamura, H. (2008). Synthesis of phosphorylated chitosan by novel method and its characterization. *International Journal of Biological Macromolecules,* **42** 4, 335–339.

Jayakumar, R., Reis, R. and Mano, J. (2006). Phosphorous containing chitosan beads for controlled oral drug delivery. *Journal of Bioactive and Compatible Polymers,* **21** 4, 327–340.

Jayakumar, R., Selvamurugan, N., Nair, S., Tokura, S. and Tamura, H. (2008). Preparative methods of phosphorylated chitin and chitosan – An overview. *International Journal of Biological Macromolecules,* **43** 3, 221–225.

Je, J.-Y., Park, P.-J. and Kim, S.-K. (2004). Free radical scavenging properties of hetero-chitooligosaccharides using an ESR spectroscopy. *Food and Chemical Toxicology,* **42** 3, 381–387.

Jeon, Y.-J. and Kim, S.-K. (2000). Production of chitooligosaccharides using an ultrafiltration membrane reactor and their antibacterial activity. *Carbohydrate Polymers,* **41** 2, 133–141.

Jeon, Y.-J., Park, P.-J. and Kim, S.-K. (2001). Antimicrobial effect of chitooligosaccharides produced by bioreactor. *Carbohydrate Polymers,* **44** 1, 71–76.

Jeong, S.I., Krebs, M.D., Bonino, C.A., Samorezov, J.E., Khan, S.A. and Alsberg, E. (2010). Electrospun chitosan–alginate nanofibers with *in situ* polyelectrolyte complexation for use as tissue engineering scaffolds. *Tissue Engineering Part A,* **17** 1–2, 59–70.

Jiang, H., Zuo, Y., Cheng, L., Wang, H., Gu, A. and Li, Y. (2011). A homogenous CS/NaCMC/n-HA polyelectrolyte complex membrane prepared by gradual electrostatic assembling. *Journal of Materials Science: Materials in Medicine,* **22** 2, 289–297.

Jiang, L., Li, Y., Wang, X., Zhang, L., Wen, J. and Gong, M. (2008). Preparation and properties of nano-hydroxyapatite/chitosan/carboxymethyl cellulose composite scaffold. *Carbohydrate Polymers,* **74** 3, 680–684.

Jiang, L.Y., Li, Y.B., Zhang, L. and Wang, X.J. (2009). Preparation and characterization of a novel composite containing carboxymethyl cellulose used for bone repair. *Materials Science and Engineering: C,* **29** 1, 193–198.

Jin, H.H., Lee, C.H., Lee, W.K., Lee, J.K., Park, H.C. and Yoon, S.Y. (2008). *In-situ* formation of the hydroxyapatite/chitosan-alginate composite scaffolds. *Materials Letters,* **62** 10, 1630–1633.

Kim, K.S., Lee, J.H., Ahn, H.H., Lee, J. Y., Khang, G., Lee, B., Lee, H. B., and Kim, M. S. (2008). The osteogenic differentiation of rat muscle-derived stem cells *in vivo* within *in situ*-forming chitosan scaffolds. *Biomaterials,* **29** 33, 4420–4428.

Kim, M.-M. and Kim, S.-K. (2006). Chitooligosaccharides inhibit activation and expression of matrix metalloproteinase-2 in human dermal fibroblasts. *FEBS Letters,* **580** 11, 2661–2666.

Kim, S.-K. and Rajapakse, N. (2005). Enzymatic production and biological activities of chitosan oligosaccharides (COS): A review. *Carbohydrate Polymers,* **62** 4, 357–368.

Kong, L., Gao, Y., Cao, W., Gong, Y., Zhao, N. and Zhang, X. (2005). Preparation and characterization of nano-hydroxyapatite/chitosan composite scaffolds. *Journal of Biomedical Materials Research Part A,* **75** 2, 275–282.

Kong, L., Gao, Y., Lu, G., Gong, Y., Zhao, N. and Zhang, X. (2006). A study on the bioactivity of chitosan/nano-hydroxyapatite composite scaffolds for bone tissue engineering. *European Polymer Journal,* **42** 12, 3171–3179.

Lahiji, A., Sohrabi, A., Hungerford, D.S. and Frondoza, C.G. (2000). Chitosan supports the expression of extracellular matrix proteins in human osteoblasts and chondrocytes. *Journal of Biomedical Materials Research,* **51** 4, 586–595.

Li, J., Chen, Y.P., Yin, Y., Yao, F. and Yao, K. (2007). Modulation of nano-hydroxyapatite size via formation on chitosan–gelatin network film *in situ*. *Biomaterials,* **28** 5, 781–790.

Li, J., Dou, Y., Yang, J., Yin, Y., Zhang, H., Yao, F., Wang, H. and Yao, K. (2009a). Surface characterization and biocompatibility of micro-and nano-hydroxyapatite/chitosan-gelatin network films. *Materials Science and Engineering: C,* **29** 4, 1207–1215.

Li, J., Sun, H., Sun, D., Yao, Y., Yao, F. and Yao, K. (2011). Biomimetic multicomponent polysaccharide/nano-hydroxyapatite composites for bone tissue engineering. *Carbohydrate Polymers,* **85** 885–894.

Li, X., Xie, H., Lin, J., Xie, W. and Ma, X. (2009b). Characterization and biodegradation of chitosan–alginate polyelectrolyte complexes. *Polymer Degradation and Stability,* **94** 1, 1–6.

Li, X.Y., Nan, K.H., Shi, S. and Chen, H. (2012). Preparation and characterization of nano-hydroxyapatite/chitosan cross-linking composite membrane intended for tissue engineering. *International Journal of Biological Macromolecules,* **50** 1, 43–49.

Li, Z., Ramay, H.R., Hauch, K.D., Xiao, D. and Zhang, M. (2005). Chitosan–alginate hybrid scaffolds for bone tissue engineering. *Biomaterials,* **26** 18, 3919–3928.

Li, Z., Yubao, L., Yi, Z., Lan, W. and Jansen, J.A. (2010). *In vitro* and *in vivo* evaluation on the bioactivity of ZnO containing nano-hydroxyapatite/chitosan cement. *Journal of Biomedical Materials Research Part A,* **93** 1, 269–279.

Li, Z. and Zhang, M. (2005). Chitosan–alginate as scaffolding material for cartilage tissue engineering. *Journal of Biomedical Materials Research Part A,* **75** 2, 485–493.

Liu, B., Liu, W., Han, B. and Sun, Y. (2007). Antidiabetic effects of chitooligosaccharides on pancreatic islet cells in streptozotocin-induced diabetic rats. *World Journal of Gastroenterology,* **13** 5, 725.

Liu, D., Hsieh, J., Fan, X., Yang, J. and Chung, T. (2007). Synthesis, characterization and drug delivery behaviors of new PCP polymeric micelles. *Carbohydrate Polymers,* **68** 3, 544–554.

Liuyun, J., Yubao, L. and Chengdong, X. (2009). Preparation and biological properties of a novel composite scaffold of nano-hydroxyapatite/chitosan/carboxymethyl cellulose for bone tissue engineering. *Journal of Biomedical Science,* **16** 65.

Liuyun, J., Yubao, L., Li, Z. and Jianguo, L. (2008). Preparation and properties of a novel bone repair composite: nano-hydroxyapatite/chitosan/carboxymethyl cellulose. *Journal of Materials Science: Materials in Medicine,* 19 3, 981–987.

Ma, X., Wang, Y., Guo, H. and Wang, J. (2011). Nano-hydroxyapatite/chitosan sponge-like biocomposite for repairing of rat calvarial critical-sized bone defect. *Journal of Bioactive and Compatible Polymers,* **26** 4, 335–346.

Madhavan, P. and Nair, K. (1974). Utilization of prawn waste: Isolation of chitin and its conversion to chitosan. *Fishery Technology,* **11** 1, 50–53.

Manjubala, I., Scheler, S., Bössert, J. and Jandt, K.D. (2006). Mineralisation of chitosan scaffolds with nano-apatite formation by double diffusion technique. *Acta Biomaterialia,* **2** 1, 75–84.

Mathews, S., Gupta, P., Bhonde, R. and Totey, S. (2011). Chitosan enhances mineralization during osteoblast differentiation of human bone marrow-derived mesenchymal stem cells, by upregulating the associated genes. *Cell Proliferation,* **44** 6, 537–549.

Murugan, R. and Ramakrishna, S. (2004). Bioresorbable composite bone paste using polysaccharide based nano hydroxyapatite. *Biomaterials,* **25** 17, 3829–3835.

Nishi, N., Ebina, A., Nishimura, S., Tsutsumi, A., Hasegawa, O. and Tokura, S. (1986). Highly phosphorylated derivatives of chitin, partially deacetylated chitin and chitosan as new functional polymers: preparation and characterization. *International Journal of Biological Macromolecules,* **8** 5, 311–317.

No, H.K. and Meyers, S.P. (1995). Preparation and characterization of chitin and chitosan—A review. *Journal of Aquatic Food Product Technology,* **4** 2, 27–52.

Olmez, S., Korkusuz, P., Bilgili, H. and Senel, S. (2007). Chitosan and alginate scaffolds for bone tissue regeneration. *Die Pharmazie-An International Journal of Pharmaceutical Sciences,* **62** 6, 423–431.

Park, D.J., Choi, B.H., Zhu, S.J., Huh, J.Y., Kim, B.Y. and Lee, S.H. (2005). Injectable bone using chitosan-alginate gel/mesenchymal stem cells/BMP-2 composites. *Journal of Cranio-Maxillofacial Surgery,* **33** 1, 50–54.

Prabaharan, M. (2008). Review paper: chitosan derivatives as promising materials for controlled drug delivery. *Journal of Biomaterials Applications,* **23** 1, 5–36.

Rajapakse, N., Kim, M.-M., Mendis, E., Huang, R. and Kim, S.-K. (2006). Carboxylated chitooligosaccharides (CCOS) inhibit MMP-9 expression in human fibrosarcoma cells via down-regulation of AP-1. *Biochimica et Biophysica Acta (BBA) – General Subjects,* **1760** 12, 1780–1788.

Sashiwa, H. and Aiba, S. (2004). Chemically modified chitin and chitosan as biomaterials. *Progress in Polymer Science,* **29** 9, 887–908.

Seong, J.M., Kim, B.C., Park, J.H., Kwon, I.K., Mantalaris, A. and Hwang, Y.S. (2010). Stem cells in bone tissue engineering. *Biomedical Materials,* **5** 6, 062001.

Shahidi, F. and Abuzaytoun, R. (2005). Chitin, chitosan, and co-products: chemistry, production, applications, and health effects. *Advances in Food and Nutrition Research,* **49** 93–135.

Su, W., Sun, K. and Chu, C. (2009). The behavior of rhBMP2 gene transfecting cells on nanometer hydroxyapatite/chitosan scaffold. *Journal of Medical and Biological Engineering,* **29** 290–295.

Tığlı, R.S. and Gümüşderelioğlu, M. (2009). Evaluation of alginate-chitosan semi IPNs as cartilage scaffolds. *Journal of Materials Science: Materials in Medicine,* **20** 3, 699–709.

Tripathi, A., Saravanan, S., Pattnaik, S., Moorthi, A., Partridge, N.C. and Selvamurugan, N. (2011). Bio-composite scaffolds containing chitosan/nano-hydroxyapatite/nano-copper-zinc for bone tissue engineering. *International Journal of Biological Macromolecules,* **50** 1, 294–299.

Van Ta, Q., Kim, M.-M. and Kim, S.-K. (2006). Inhibitory effect of chitooligosaccharides on matrix metalloproteinase-9 in human fibrosarcoma cells (HT1080). *Marine Biotechnology,* **8** 6, 593–599.

Verma, D., Desai, M.S., Kulkarni, N. and Langrana, N. (2011). Characterization of surface charge and mechanical properties of chitosan/alginate based biomaterials. *Materials Science and Engineering: C,* **31** 8, 1741–1747.

Verma, D., Katti, K.S. and Katti, D.R. (2010). Osteoblast adhesion, proliferation and growth on polyelectrolyte complex–hydroxyapatite nanocomposites. *Philosophical Transactions of the Royal Society A: Mathematical, Physical and Engineering Sciences,* **368** 1917, 2083–2097.

Verma, D., Katti, K.S., Katti, D.R. and Mohanty, B. (2008). Mechanical response and multilevel structure of biomimetic hydroxyapatite/polygalacturonic/chitosan nanocomposites. *Materials Science and Engineering: C,* **28** 3, 399–405.

Wang, X., Ma, J., Wang, Y. and He, B. (2001). Structural characterization of phosphorylated chitosan and their applications as effective additives of calcium phosphate cements. *Biomaterials,* **22** 16, 2247–2255.

Wang, X., Wang, X., Tan, Y., Zhang, B., Gu, Z. and Li, X. (2009). Synthesis and evaluation of collagen–chitosan–hydroxyapatite nanocomposites for bone grafting. *Journal of Biomedical Materials Research Part A,* **89** 4, 1079–1087.

Xianmiao, C., Yubao, L., Yi, Z., Li, Z., Jidong, L. and Huanan, W. (2009). Properties and *in vitro* biological evaluation of nano-hydroxyapatite/chitosan membranes for bone guided regeneration. *Materials Science and Engineering: C,* **29** 1, 29–35.

Yamaguchi, I., Iizuka, S., Osaka, A., Monma, H. and Tanaka, J. (2003). The effect of citric acid addition on chitosan/hydroxyapatite composites. *Colloids and Surfaces A: Physicochemical and Engineering Aspects,* **214** 1–3, 111–118.

Yang, D., Jin, Y., Ma, G., Chen, X., Lu, F. and Nie, J. (2008). Fabrication and characterization of chitosan/PVA with hydroxyapatite biocomposite nanoscaffolds. *Journal of Applied Polymer Science,* **110** 6, 3328–3335.

Yang, E.-J., Kim, J.-G., Kim, J.-Y., Kim, S., Lee, N. and Hyun, C.-G. (2010). Anti-inflammatory effect of chitosan oligosaccharides in RAW 264.7 cells. *Central European Journal of Biology,* **5** 1, 95–102.

Zhang, C., Zhang, C., Wang, J., Lu, C., Zhuang, Z., Wang, X. and Fang, Q. (2012). Fabrication and *in vitro* investigation of nanohydroxyapatite, chitosan, poly (L-lactic acid) ternary biocomposite. *Journal of Applied Polymer Science,* **127** 3, 2152–2159. DOI: 10.1002/app.37795.

Zhang, J., Xia, W., Liu, P., Cheng, Q., Tahi, T., Gu, W. and Li, B. (2010). Chitosan modification and pharmaceutical/biomedical applications. *Marine Drugs,* **8** 7, 1962–1987.

Zhou, G., Li, Y., Zhang, L., Zuo, Y. and Jansen, J.A. (2007). Preparation and characterization of nano-hydroxyapatite/chitosan/konjac glucomannan composite. *Journal of Biomedical Materials Research Part A,* **83** 4, 931–939.

Zhou, H., Qian, J., Wang, J., Yao, W., Liu, C., Chen, J. and Cao, X. (2009). Enhanced bioactivity of bone morphogenetic protein-2 with low dose of 2-N, 6-O-sulfated chitosan *in vitro* and *in vivo*. *Biomaterials,* **30** 9, 1715–1724.

Zou, Q., Li, Y., Zhang, L., Zuo, Y., Li, J. and Li, X. (2009). Characterization and cytocompatibility of nano-hydroxyapatite/chitosan bone cement with the addition of calcium salts. *Journal of Biomedical Materials Research Part B: Applied Biomaterials,* **90** 1, 156–164.

11

Inorganic polymer composites for bone regeneration and repair

L. GRØNDAHL, K. S. JACK and C. S. GOONASEKERA,
The University of Queensland, Australia

DOI: 10.1533/9780857099037.3.261

Abstract: In the fabrication of inorganic/polymer composites for bone regeneration and repair, the selection of components for production of a construct with adequate mechanical integrity and bioactivity is crucial. Enhanced properties can be achieved by optimising the dispersion and interfacial bond strength, and ensuring surface presentation of the bioactive inorganic phase in the polymer matrix. In nano-composites, surface modification of the inorganic phase is a promising method for achieving enhanced dispersion. This chapter discusses the development of polymer composites with micron- and nano-sized fillers and both solid and porous constructs. It also covers methods for the surface modification of the filler and shortcomings in the techniques available for the characterisation of these materials.

Key words: inorganic/polymer composites, nano-composite, composite scaffold, bioactive glass, hydroxyapatite (HAP), mechanical properties, bioactivity.

Note: This chapter is a revised and updated version of Chapter 5 'Composite materials for bone repair' by L. Grøndahl and K. S. Jack, originally published in *Biomedical composites*, ed. Luigi Ambrosio, Woodhead Publishing Limited, 2010, ISBN: 978-1-84569-436-4.

11.1 Introduction

A vast range of composite materials exists which are intended for use in bone applications. These differ with respect to the components chosen and the way in which these components are assembled and processed. The focus in this chapter will be on biocomposites consisting of an inorganic dispersed phase in a polymer matrix. This chapter will not discuss organic/inorganic fibre composites, hydrogel composites, polymer–polymer composites, multiphase ceramics, approaches to *in situ* formation of nano-composite materials, materials with composite coatings or cell–material composites. The reader is referred to the other chapters in this book as well as to the additional reading material listed at the end of this chapter for further details on these types of materials.

This chapter starts with a discussion of component selection and general design considerations for the production of inorganic/polymer composites for bone regeneration and repair. Following this, fabrication of particulate composites including the pioneering work of William Bonfield on particulate hydroxyapatite (HAP)/high density polyethylene (HDPE) composite materials will be presented. The following section will discuss the use of nano-sized inorganic particles and will address the issue of achieving a dispersed, well-bonded reinforcing phase in nano-composites. A section on the production of three-dimensional (3D), porous, composite scaffolds will follow and the chapter will conclude with a description of future trends and concluding remarks. The major focus of this chapter is on calcium phosphate based composites and, while a few examples of composites comprising bioactive glass fillers will be discussed, the reader is referred to other chapters of this book for further details on these composites.

11.2 Component selection and general design considerations

In selecting the components for a composite material the key requirements that exist for any biomaterial must, of course, be fulfilled. Thus, the components must be bio-acceptable, non-immunogenic, non-toxic and sterilisable. In addition, the cost of production, the scalability of the fabrication process and the ability to fabricate accurately sized and shaped devices are also of importance when choosing materials and processes. Finally, the polymer in the composite must have thermal properties that allow for processing and mechanical properties that approach those of the native tissue which it is to replace, since mechanical compatibility between the biocomposite and bone will encourage bone growth in and around the implant (Bonfield, 1987; Kohane and Langer, 2008). Some of the most extensively studied polymers for materials covered in this chapter are listed in Table 11.1, together with some of their physical properties including degradation mechanisms for use in bone applications. These polymers can be subdivided into those that occur in nature and those that are prepared synthetically. It is generally accepted that natural polymers including polysaccharides (e.g., chitin and chitosan) and polypeptides (e.g., collagen) possess better biocompatibility and that the so-called 'bottom-up' approach can be used to fabricate nano-composites of these types of materials where the fillers are at discrete (i.e., well-dispersed) locations (circumventing aggregation of the filler particles). However, their relatively poor mechanical properties in wet conditions and very rapid degradation rates limit their range of applications.

The fabrication of composites of these polymers is generally carried out with *in situ* precipitation of the inorganic filler and does not fall within the scope of the current chapter. The poly(β-hydroxyalkanoates), however, are naturally occurring polymers displaying good biocompatibility, suitable mechanical properties and slow degradation rates, thus making them potential materials for bone applications (Chen and Wu, 2005). Synthetic polymers have the advantage that they can be tailored to give a wide range of properties and more predictable uniformity, and they are generally free from concerns of immunogenicity, which can be a concern for naturally derived materials, for example, bacterially produced polyhydroxybutyrate (PHB) and poly(hydroxybutyrate-co-hydroxy valerate) (PHBV) (Wu *et al.*, 2006). However, it should be noted that the synthetically prepared polymers may contain low levels of chemical impurities (e.g. trace amounts of catalyst or monomer) which may lead to alternative challenges in terms of both bio- and regulatory- acceptability. Additionally, the polymers listed in Table 11.1 can be categorised into biostable (i.e., those which do not degrade *in vivo*), and biodegradable types. Whereas the use of biostable polymers avoids the issue of matching the implant degradation rate with tissue regeneration, the biodegradable ones have the potential to produce an implant that, with time, is substituted by the living tissue. The class of synthetic biodegradable polymers most widely studied for bone applications are the polyesters poly(lactic acid) (PLA), poly(glycolic acid) (PGA), poly(lactic-co-glycolic acid) (PLGA) and poly(ε-caprolactone) (PCL), all of which have their own advantages and disadvantages when used in bone applications (see Table 11.1). Biodegradable polymers may degrade either via a bulk or a surface erosion mechanism which affects the long-term stability of the material and it is generally accepted that surface-eroding polymers are preferred for use as scaffolds, as bulk-eroding polymers can display abrupt decreases in mechanical strength (Rezwan *et al.*, 2006; Chen *et al.*, 2013). An important consideration during the design of materials, therefore, is knowledge of both the mechanism and rate of degradation of the selected polymer.

Since most of the polymers possess neither the required strength to match the mechanical properties of bone nor the required bioactivity, an inorganic component is typically introduced to improve the overall mechanical stiffness and strength, as well as the bioactivity of the resulting composite material. The inorganic materials commonly studied for biocomposite fabrication, which are the focus of this chapter, will, therefore, typically be chosen to have relatively large moduli and strengths as well as proven bioactivities. A number of physical characteristics of the inorganic material can affect their bioactivity and biostability and as a result the bioactivities and -stabilities of composites in which they are contained. These characteristics include density, phase purity and

Table 11.1 A selection of polymers reported for biocomposite fabrication

Polymer	Mechanical properties	Degradation	Properties	References
HDPE	E = 0.88 GPa; UTS = 35 MPa	Biostable	Bioinert, ductile; allowing for high volume % of inorganic filler to be incorporated	Murugan and Ramakrishna, 2005
PHB	E = 2.5 GPa; UTS = 36 MPa, highly crystalline	Bioerodible	High strength but brittle, slow degradation time	Murugan and Ramakrishna, 2005
PHBV	E = 2.5–0.5 GPa; UTS = 70–20 MPa (depending on 3HV content)	Bioerodible	Less crystalline than PHB – can be modified by 3HV content, improved ductility compared to PHB, slow degradation time	Misra *et al.*, 2006
PLA	(P_LLA) or amorphous (P_{DL}LA), E = 2.7 GPa (P_LLA), 1.9 GPa (P_{DL}LA); UTS = 50 MPa ((P_LLA), 29 MPa (P_{DL}LA)	Biodegradable >24 months (P_LLA)	Semicrystalline, slow degradation time	Murugan and Ramakrishna, 2005; Rezwan *et al.*, 2006
PGA		Biodegradable, degradation time (6–12 months)	Highly crystalline, relatively short *in vivo* stability	Rezwan *et al.*, 2006
PLGA	E = 1.4–2.8 GPa; UTS = 40–55 MPa (depending on composition)	Biodegradable, degradation rate faster than either homopolymer	Tuneable degradation rate (1–12 months)	Rezwan *et al.*, 2006
PCL	E = 0.4 GPa; UTS = 10 MPa	Bioerodible, slow degradation rate (>24 months)	Semicrystalline, high tissue compatibility	Murugan and Ramakrishna, 2005

E: Young's modulus; UTS: ultimate tensile strength; HDPE: high density polyethylene; PHB: poly(hydroxy butyrate); PHBV: poly(hydroxy butyrate-co-hydroxy valerate); PLA: poly(lactic acid); PGA: poly(glycolic acid); PLGA: poly(lactic-co-glycolic acid); PCL: poly(caprolactone).

Table 11.2 Inorganic materials reported for biocomposite fabrication

Material	Key properties	Key advantages
HAP	Osteoconductive, bioactive; *E* = 80–110 GPa; UTS = 50 MPa	Easily processed into nano-sized crystals, small crystals (<10 μm) are bioresorbable
β-TCP	Bioactive, bioresorbable	Faster degradation rate than HAP
Mixtures of HAP and β-TCP (referred to as biphasic calcium phosphate (BCP) in the literature)	Osteoconductive, bioresorbable	The use of different ratios of HAP and β-TCP properties such as degradation rate and mechanical strength to be controlled
Bioglass® (BG)	Bioactive, binds to both hard and soft tissue, amorphous, non-resorbable; *E* = 35 GPa; UTS = 42 MPa	More bioactive than HAP
A-W glass-ceramic	Bioactive, amorphous with crystalline nuclei, non-resorbable; *E* = 118 GPa; UTS = 215 MPa	More bioactive than HAP

E: Young's modulus; UTS: ultimate tensile strength.
Source: This information was adapted from Daculsi *et al.* (2003), Rezwan *et al.* (2006) and Vallet-Regi (2001).

crystallinity, as well as the overall size and shape of crystallites and/or primary particles (Vallet-Regi, 2001). In Table 11.2 the most commonly used inorganic materials for bone applications are listed together with their key advantages for use in bone biocomposites. In addition to single-constituent fillers, mixtures of beta-tri calcium phosphate (β-TCP) and HAP are being used to form biphasic calcium phosphate (BCP) composites (Macedo *et al.*, 2012). The relatively high solubility of β-TCP compared to HAP at physiological conditions (Daculsi *et al.*, 2003; Schaefer *et al.*, 2011) allows composites made using BCP to have somewhat tailorable properties. It is worth noting that while the calcium phosphates used in composites for bone repair are of the biodegradable type, the micron-sized bioglasses and glass-ceramics are biostable. Glasses and glass-ceramics with a range of specific compositions from the Na_2O-CaO-P_2O_5-SiO_2 system have the capacity to bond to bone, and hence are termed bioactive. The glasses are amorphous single-phase materials and promote the rapid bonding of bone to the surface of the implant while the bioactive glass-ceramics (BGCs) are multi-phase materials, with relatively good strength and toughness, however, the bonding of the native bone to the surface does not occur as rapidly as for the glasses (Vallet-Regi, 2001). The calcium phosphate minerals such as HAP and β-TCP are obvious choices as resorbable ceramics in bone repair since HAP dominates the mineral phase in bone. Fundamental studies and clinical applications have demonstrated that calcium phosphate biomaterials are biocompatible and osteoconductive (Langstaff *et al.*, 1999; Bohner, 2000; Vallet-Regi, 2001). When implanted *in vivo*, these materials are non-toxic and bond directly to bone without any intervening connective tissue layer. In some composite materials the inorganic component can also provide a buffering capacity when used in conjunction with biodegradable polymers that produce acidic degradation products (e.g., PLGA) (Chen *et al.*, 2010b; Liang *et al.*, 2010; Wang *et al.*, 2010a) thereby avoiding locally increased acidity and subsequent adverse immune response. Finally, whereas it has long been known how to produce and stabilise HAP on the nano-scale (i.e., with at least one dimension < 100 nm), the fabrication of nano-sized bioglass commenced more recently (Brunner *et al.*, 2006; Kim *et al.*, 2006; Hong *et al.*, 2008, 2009a, 2009b; Lee *et al.*, 2008; Jo *et al.*, 2009; Lei *et al.*, 2013).

It is evident that a number of component choices exist when designing a biocomposite and it clearly depends on the intended application (i.e., load-bearing or not, biostable or degradable) as to the selections of components. One common goal in biocomposite fabrication is to achieve reinforcement of a relatively low-modulus polymer with a high-modulus, high-strength reinforcing inorganic material utilising the plastic flow of the polymeric material under stress to transfer the load to the reinforcing phase. This can result in a composite with a higher strength and modulus than the polymer alone and better toughness and processability than the pure inorganic material. The ability to achieve this goal is highly dependent on the relative volume fraction of each of the phases that can be obtained whilst ensuring that the inorganic phase is well dispersed in (as represented in Fig. 11.1A) and well bonded to the polymer matrix. The interfacial bond strength thus has to be sufficient for the load to be transferred from the matrix to the reinforcing phase if the composite is to be stronger than the polymeric matrix material. In addition to the importance of creating a large interfacial bonding strength it is also greatly beneficial to have a large interfacial area between the phases. In general, a large interfacial area provides greater transfer of load from the matrix to the reinforcing phase which leads to an improvement in mechanical properties. Moreover, this area is dependent on the shape and size of the reinforcing particle and, hence, to optimise this aspect of composite design much effort has involved the inclusion of nano-sized particles. Finally, in addition to improving the strength of the composite material, high bioactivity is necessary and this can be achieved through a high surface presentation of the bioactive inorganic component (Fig. 11.1B).

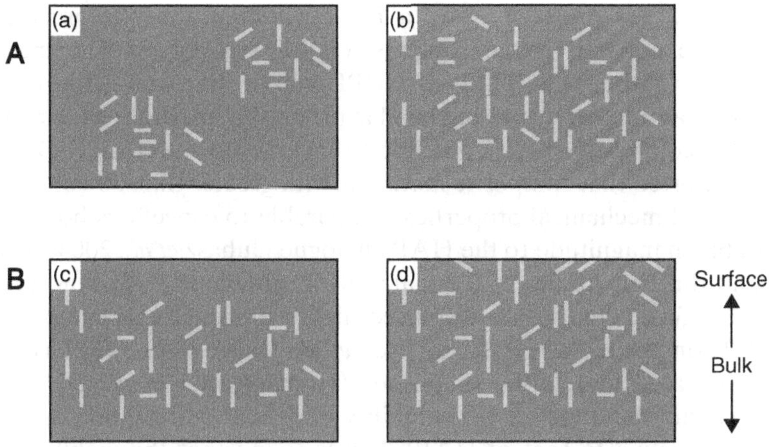

11.1 (A) Top view of composite material where grey background illustrates the polymer matrix and the light grey lines signify reinforcing crystals; (a) poorly dispersed crystals in the polymer matrix (b) well-dispersed crystals and thus enhanced mechanical properties. B: Side view of composite material where grey background illustrates the polymer matrix and the light grey lines the reinforcing crystals. (c) poor surface presentation, (d) good surface presentation and thus improved bioactivity.

11.3 Fabrication of particulate composites

The use of a bioactive inorganic filler in composite materials for bone biomaterials was pioneered by Bonfield and co-workers in the early 1980s (Bonfield *et al.*, 1981) with the production of particulate HAP/HDPE composites. Due to the ductile nature of HDPE, composites with up to 45 vol% (73 wt%) HAP could be produced using the fabrication process of compounding, powdering and compression moulding (Wang *et al.*, 1994) Uniform distribution of the HAP particles were observed on the surface of high volume percent HAP composites and the HAP particle size was unchanged from that of the HAP particles in the powder used. The lack of agglomeration of the HAP particles was attributed to the high shear forces generated during the compounding process (Wang and Bonfield, 1996). The Young's modulus and tensile strength of the composite materials increased with increasing filler content reaching values of 5.5 GPa and 19 MPa, respectively, for a 45 vol% HAP/HDPE composite (Wang *et al.*, 1998). *In vitro* osteoblast studies revealed attachment of the cells to the HAP islands followed by spreading across the implant surface (Huang *et al.*, 1997). *In vivo* studies showed an excellent bone–biomaterial interface attributed to both the bioactive nature of the implant and the mechanical matching of the implant with bone (Bonfield and Luklinske, 1991).

Another example of using a bioactive filler for reinforcement of a biostable polymer is the Bioglass®/HDPE composites. The inclusion of Bioglass® rather than HAP in HDPE composites was made to improve the overall bioactivity. Using similar processing methods as for the HAP/HDPE composites it was possible to include up to 50 vol% Bioglass®. It was reported that composite materials with greater than 30 vol% Bioglass® possessed mechanical properties comparable to cancellous bone and comparable in magnitude to the HAP analogue (Juhasz *et al.*, 2004). It was also found through studies of the materials in simulated body fluid (SBF) that the Bioglass® composites displayed higher bioactivity as compared to the HAP composites, as predicted (Rea *et al.*, 2004). Beta-TCP has also been incorporated in HDPE composites (Homaeigohar *et al.*, 2005, 2008) to investigate the effect of the solubility of β-TCP on resulting osteoconductivity (relative to that of HAP). It was found that the viability of bone-derived fibroblasts was lower for the composites with the higher volume fraction of β-TCP (40%), than for the control polyethylene samples and the tissue culture polystyrene and was attributed to the ion concentration at the surface reaching toxic levels, which has not been reported for HAP composites (Homaeigohar *et al.*, 2008). Additionally, composites of HAP with biostable polymers such as poly(1,4-phenylene ether-ether-sulphone) (PSU) (Wang, 2003; Chlopek *et al.*, 2006), polypropylene (Liu and Wang, 2007; Liao *et al.*, 2013), high-impact polystyrene (HIPS) (Gong *et al.*, 2004), poly(etheretherketone) (Abu Baker *et al.*, 2003; Ma *et al.*, 2013) and polyamide66 (Wang *et al.*, 2009; Zhou *et al.*, 2012) have also been reported. The polymers are chosen for investigation because of their superior mechanical properties (stiffness, strength or toughness) compared with HDPE and, in the case of PSU, its stability with respect to oxidation, hydrolysis and radiation, which are typical processes used in the sterilisation of medical devices. It is reported that the incorporation of HAP leads to improvements in the mechanical properties of the resulting composites, provided that sufficient dispersion of filler particles can be achieved. Indeed, in the work of Gong *et al.* (2004), the authors have shown that the degree of dispersion and the interfacial adhesion of micron-sized HAP particles in a HIPS matrix could be enhanced through an *in situ* polymerisation of styrene on the surface of the HAP particles. Furthermore, Chlopek *et al.* (2006) have investigated the temporal variation of mechanical properties of a 15 wt% HAP/PSU composite under *in vitro* conditions via creep measurement at various applied stresses. They found that the HAP/PSU composite showed long-term stability only when the applied stress was less than 20% of the initial (static) strength of the material, suggesting stringent limitations on the applicable sites of implantation of these materials. These findings are important as they highlight the need for the measurement of dynamic mechanical properties *in vitro*, as opposed to static measurements, when assessing the performance of biomaterials. When comparing the overall performance of the biostable composites, however, there appears to be no definitive conclusions in the literature as to the general superiority of any of these composites over the others, including HDPE.

The use of biodegradable polymers in the fabrication of particulate reinforced composites has predominantly involved the use of the polyhydroxyalkanoates (PHAs), for example PHB and PHBV (Chen and Wu, 2005), as well as synthetic polyesters. However, it is not within the scope of this chapter to comprehensively cover particulate reinforced composites of both classes of polymers and the reader is referred to the reading list in Section 11.7. The sections on nano-composites and composite scaffolds will, however, include studies on a much broader range of polymers (Sections 11.4 and 11.5).

Hydroxyapatite particle reinforcement of PHB and PHBV produced by dry mixing and melt processing resulted in composites displaying compressive strength of 62 MPa (30 wt% HAP) (Galego *et al.*, 2000). *In vitro* testing in SBF demonstrated high bioactivity as evident from formation of a bone-like apatite layer on the material surface with the bioactivity increasing with filler content (10–30 vol% HAP) (Ni and Wang, 2002). Coskun *et al.* (2005) studied the inclusion of HAP rods (2–4 × 20–30 μm) as filler in PHB and PHBV composites produced by blending and injection moulding. Scanning electron microscopy (SEM) images indicated highly embedded and homogeneously distributed rods; however, based on the mechanical properties it was concluded that the composite did not possess a strongly bonded interface. Indeed, the Young's modulus decreased with inclusion of filler (15 wt%) whereas the tensile strength increased for PHB but decreased for PHBV composites. *In vivo* implantation into bone tissue of particulate reinforced HAP/PHBV or β-TCP/PHBV composites have shown good bone integration and formation of a layer of bone-like apatite at the implant interface after implantation (Luklinska and Bonfield, 1997; Luklinska and Schluckwerder, 2003). Notably, none of these studies compared the bioactivity to that of the PHBV polymer alone. However, a separate *in vivo* study on HAP/PHB particulate composites failed to detect improved bioactivity when compared to PHB (Doyle *et al.*, 1991) and this was attributed to the lack of surface presentation of HAP.

Composite Bioglass®/PHB films (5 and 20 wt%) were produced by solvent casting from a chloroform suspension of 5 μm Bioglass® particles which were dispersed by sonication (Misra *et al.*, 2007). It was found that the presence of Bioglass® reduced the crystallinity of the PHB polymer. In addition, it was proposed that poor wetting of the particles caused poor interfacial bonding which, in turn, resulted in lowered Young's moduli for the composites compared to the pure polymer. Although the surface presentation of the Bioglass® particles was only evaluated using microscopy, it was concluded that it was low and that the particles were present mainly in the bulk of the film. This explained the relatively low bioactivity as evaluated *in vitro* by immersion in SBF. However, due to the lack of including a pure PHB control in their reported work, it is not clear if the filler improved the bioactivity to any extent. It is also evident that the porosity of the bioglass filler appears to influence bioactivity. Wu *et al.* (2009) have shown that PLGA films containing mesoporous bioactive glass (MBG) particles, with a molar ratio of Si/Ca/P = 80/15/5, had higher bioactivity (both in terms of mineralisation and cellular activity) than films containing the same mass of non-porous BG. In addition, the MBG composites showed enhanced *in vitro* degradation and significantly greater tensile strengths and moduli at all mass percentages (10–30 wt%) compared to the BG composite. The attachment, spreading and proliferation of human osteoblastic cells was comparable for both types of composites and significantly greater than for PLGA alone. However, cell differentiation was significantly greater on the 30 wt% containing MBG composite compared to the 30 wt% BG composite. It is unclear if this is related to differences in the total surface area or the surface presentation of MBG and BG in the composites.

11.4 Fabrication of nano-composites

In micron-scale composite fabrication agglomeration of particles can, to a large extent, be avoided by using blending, solvent processing, extrusion and/or sonication. However, these methods are generally insufficient when working with nano-sized crystals. Indeed, one of the main challenges in nanotechnology is to achieve good dispersion of the individual nano-sized (primary) particles which, due to their high surface area, have a strong tendency to agglomerate. Agglomeration is undesirable as it reduces the interfacial area and, in addition, agglomerates formed in the final material may act as stress initiators and stress concentrators, triggering material failure.

According to Wang and Bonfield (2001), a composite material undergoes mechanical failure due to a series of steps. First, stress concentration occurs during tensile stressing, which leads to dewetting of the interface and the formation of cavities at the particle sites. With further stressing these cavities coalesce and grow in the tensile stress direction, leading to subsequent matrix deformation and failure of the composite. With the increase of the volume of the filler, the stress needed for growth and coalescence of the cavities decreases causing the transition of the behaviour of the composite to change from ductile to brittle.

There are a number of examples of composite preparation from a dispersion of the nano-sized HAP (nHAP) particles in an organic solvent which can also be used to dissolve the polymer of choice. An important consideration here is the colloidal stability of the nHAP particles in the solvent which have led to the choice of solvents such as dimethylformamide (DMF) (Deng *et al.*, 2001; Hao *et al.*, 2002) which offers relatively good stability of nHAP particles (Rai *et al.*, 2008a). Co-precipitation of nHAP particles and a polymer from a combined suspension allows production of a nano-composite powder which can be further processed by, e.g. melt extrusion or compression. For such composites of PCL or PLA an increase in tensile modulus was observed with increasing nHAP content, however, no change in the tensile strength was found (Deng *et al.*, 2001; Hao *et al.*, 2002). Since the particles in these composites appeared to be well dispersed when studied by microscopy, the authors suggested that the observed properties can be attributed to poor interfacial bonding between the hydrophilic nHAP particles and the hydrophobic polymers. However, work by Schaefer and Justice (2007) has shown using ultra-small-angle scattering techniques that in many apparently 'well-dispersed' materials determined by microscopy, there exists micron-scale agglomeration of the nano-sized particles which in turn reduces the overall efficiency of the filler particles to provide reinforcement of the composite.

In the production of nHAP/polymer composites a number of different approaches which involve innovative chemistry as illustrated in Fig. 11.2 have been employed not only to create well dispersed particles but also to enhance the interfacial bonding between the particles and the matrix polymer. Thus, when tensile stresses are applied the chains in the polymer matrix are forced to orientate and elongate, rather than cause dewetting and cavitation at the polymer–filler interface and this results in higher toughness and ductility of the composite. These approaches involve surface modification of the nHAP particles prior to composite fabrication and include adsorption of surfactants or polyelectrolytes, surface grafting techniques and the use of linker molecules to attach high molecular-weight molecules to the surface as summarised in Fig. 11.2.

The successful use of surfactant adsorption to nHAP was demonstrated by Kim (2007) who used oleic acid to both increase the colloidal stability of nHAP in a chloroform suspension and to mediate the interaction between the hydrophilic nHAP crystals and the hydrophobic PCL matrix. An increase in both tensile stress and elastic modulus was observed for composites prepared using oleic acid compared to those prepared using non-stabilised nHAP. The latter composites displayed an increase in the modulus but not strength, similar to that observed for the systems dispersed in DMF. Kim (2007) also investigated the biological response and found that cell proliferation was enhanced on the nano-composite produced using oleic acid; however, it could not be concluded if this was due to different surface presentation of nHAP or differences in surface roughness of the films or if indeed it was due to the nano-size of the reinforcing particles.

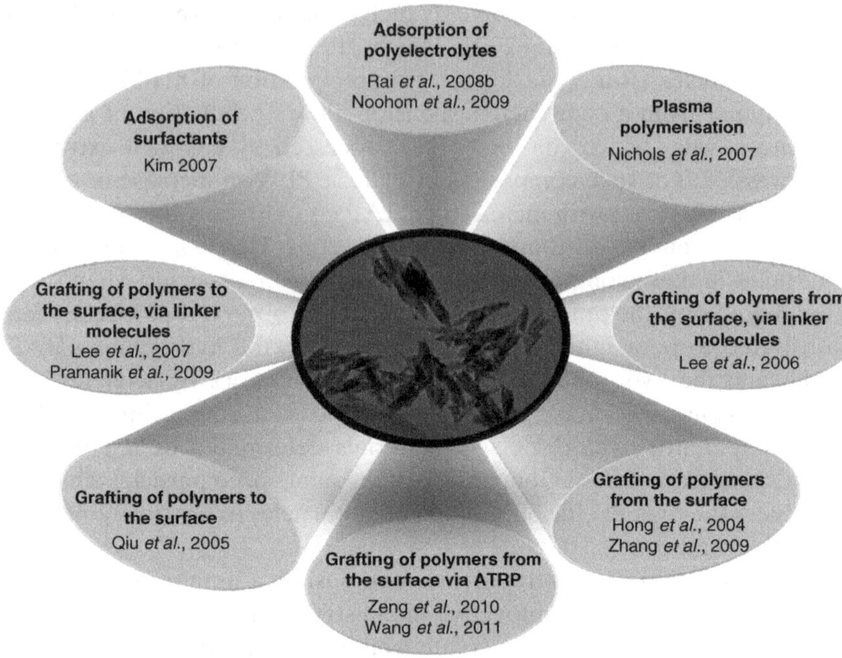

11.2 Surface modification methods of nHAP (TEM micrograph in centre) and a selection of relevant references from literature.

Adsorption of polyelectrolytes such as poly(acrylic acid) (PAA) and heparin to nHAP particles is well studied (Misra 1996; Rees *et al.*, 2002) and the resulting enhancement in colloidal stability well documented (Misra 1996; Rai *et al.*, 2008b; Noohom *et al.*, 2009). The use of PAA coated nHAP in the preparation of PHBV composite powders which were subsequently processed into disks via compression moulding or films via solvent casting (Fig. 11.3) showed an enhanced surface presentation and concomitant improved cell response compared to composites produced using micron-sized HAP particles (Cool *et al*, 2007). In addition, the mechanical properties were found to be enhanced for particle loadings of up to ca. 15 wt%. At higher loadings the composites were found to contain significant large-scale agglomeration of particles and could not be processed into structures suitable for mechanical testing (Noohom *et al.*, 2009). It was also shown that the level of mechanical improvements observed was well below the theoretical values predicted and this is most likely the result of the presence of micron-scale agglomeration as discussed by Schaefer and Justice (2007). The choice of heparin for stabilising nHAP particles (Rai *et al.*, 2008a) was based on its dual effect in enhancing colloidal stability and providing biological stimulus, that is, it was used as a model for bone specific heparan

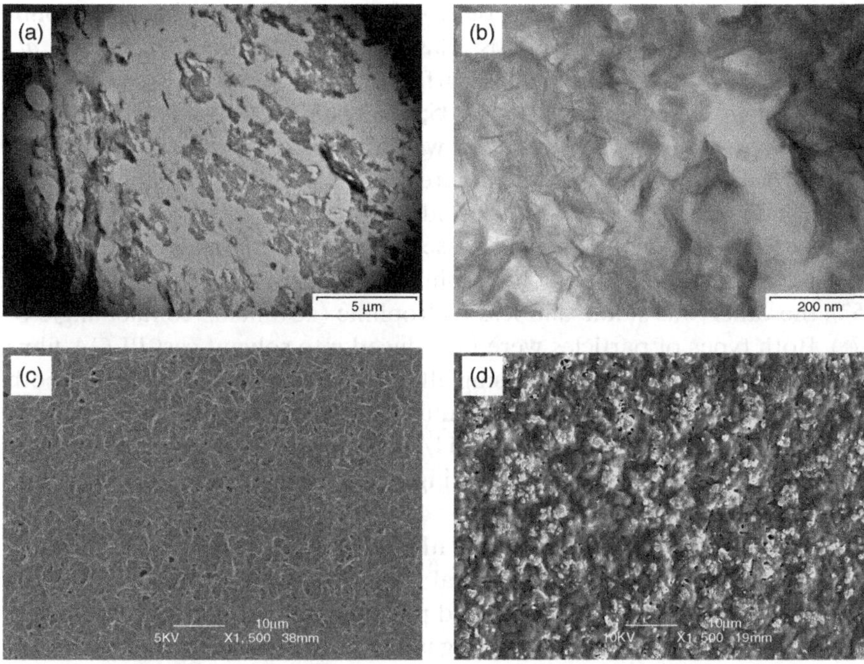

11.3 10% w/w nHAP/PHBV solvent cast films produced using PAA stabilised HAP nano crystals, **a+b**: TEM images; nHAP clusters are visibly scattered throughout the composite (**a**). However, at higher magnification (**b**) individual nHAP particles with spindle-like morphology are visible. **c+d**: SEM images; solvent cast PHBV film (**c**) and 10% nHAP/PHBV composite (**d**). Micron-sized nHAP agglomerates are visible on the surface of the nHAP/PHBV composite films. The dimensions of the scale bars are: **a** = 5 μm; **b** = 200 nm; **c** and **d** = 10 μm. (*Source*: Images supplied by Dr C K A Wu, The University of Queensland.)

sulfates (Nurcombe *et al.*, 2004). Solvent cast films produced from co-precipitated composite powder displayed enhanced wettability and nHAP surface presentation (yielding an advancing contact angle of 35° for 30 wt% composite) compared to an insignificant change for composites prepared using non-stabilised particles (Rai *et al.*, 2008a). In addition, the mechanical properties were improved for the heparin-stabilised composite films yielding a four-fold increase in both elastic modulus and tensile strength for the 30 wt% composite. Again, non-stabilised composites displayed similar mechanical properties to the pure polymer. The enhancements seen were to a large extent attributed to the heparin-stabilised nHAP particles being well dispersed in the polymer matrix with minimal agglomeration observed by electron microscopy.

Surface grafting of polymers onto nHAP particles has been predominantly achieved by two methods: plasma polymerisation (Nichols *et al.*, 2007) and chemical reaction to the surface hydroxyl groups (Liu *et al.*, 1998, 2008; Hong *et al.*, 2004; Qiu *et al.*, 2005; Lee *et al.*, 2006; Wang *et al.*, 2011b). In the first method, nHAP particles were modified by acrylic acid plasma polymerisation to introduce a tethered PAA network. By manipulating the plasma parameters (i.e., power and gas pressure) two types of coatings resulted: one was degradable and possessed a high percentage of carboxylic acid groups yielding a highly hydrophilic coat (advancing contact angle of 15°); the other was stable and less hydrophilic (advancing contact angle of 57°). Both types of particles were introduced into solvent cast PLGA films (3 wt%) with mixing achieved using ultrasonication. It was found that both types of coatings improved the modulus and yield strength of the composite films (compared to non-coated nHAP composites) and, in addition, the use of particles with degradable coatings yielded significantly better results (Nichols *et al.*, 2007).

Chemical reactions involving the nHAP surface hydroxyl groups have been achieved using the two chemical approaches of grafting-to and grafting-from. In the former, a pre-formed polymer chain is attached to surface hydroxyl groups; in the latter, grafting is initiated from the surface hydroxyl groups. The grafting-to approach in this context was pioneered by Liu *et al.* (1998) with the fabrication of polyethylene glycol (PEG)-grafted nHAP particles. Later, Qiu *et al.* (2005) used the grafting-to approach to attach poly(L-lactic acid) (PLLA) via a condensation reaction between the oligomer carboxylic acid end group and the nHAP surface hydroxyl groups. These particles formed a stable dispersion in chloroform and were subsequently used in composite formation (5–30 wt% filler) using mechanical mixing. It was found that the tensile modulus increased with increasing filler percentage over the range studied, while the tensile strength increased to 15% filler and then decreased to reach values of the pure polymer at 30% filler. Both modulus and strength were vastly improved compared to non-grafted nHAP composites and this was attributed to the improved distribution of the particles within the polymer matrix on the micron scale. This was elegantly supported using energy dispersive X-ray spectroscopy (EDX) calcium elemental distribution maps of the fracture surfaces.

The grafting-from approach for the surface modification of nHAP has been more extensively used. Some of the advantages of this technique are that it offers higher graft density and easy purification as compared to the grafting-to method. Hong *et al.* (2004) grafted PLA onto nHAP and subsequently introduced these particles into PLLA composites (1–30 wt%). Increased particle dispersion as compared to non-grafted nHAP was observed by transmission electron microscopy (TEM) and, although better tensile strength was found for composites made from grafted nHAP, no difference in tensile modulus between stabilised and non-stabilised particle composites were observed. Improved grafting-from methods were developed by Lee *et al.* (2006) in which linker molecules (e.g., lactic acid and PEG) were introduced prior to grafting. This was done to decrease steric effects and increase the nucleophilicity of the hydroxyl groups and it allowed for more efficient grafting (especially for the PEG modified nHAP particles). In this study, grafting of PCL by ring-opening polymerisation was performed. Nano-HAP/PCL composites with 10 wt% filler showed improved tensile modulus as well as tensile strength when using grafted nHAP as compared to the non-grafted system. A subsequent study evaluated the *in vitro* biocompatibility of the composites using protein adsorption and fibroblasts cell culture (Lee *et al.*, 2007). Improved biological activity was observed for composites produced using grafted nHAP compared to the non-grafted analogue; however, the surface presentation of the nHAP particles were not evaluated and, therefore, precisely what caused this improvement cannot be assessed directly from this work.

The method of atom transfer radical polymerisation (ATRP) has been used in recent years to graft material from the surface of nHAP. Zeng and co-workers grafted comb-shaped PCL-*g*-poly(hydroxyethyl methacrylate) brushes on the surface of nHAP particles by using a combination of ATRP and ring-opening polymerisation (Zeng *et al.*, 2010). ATRP was used to introduce new hydroxyl functional groups to the surface by polymerising 2-hydroxyethyl methacrylate, to which PCL was grafted by ring-opening polymerisation of ε-caprolactone. This method allowed over 20 wt% PCL to be grafted from the surface. Other studies have also used ATRP for grafting polymers from the nHAP surface. Wang and co-workers grafted poly(methyl methacrylate) (PMMA) from nHAP via surface initiated ATRP (Wang *et al.*, 2011b) achieving 49 wt% attachment. They also investigated the use of reverse ATRP to control the polydispersity of the attached chains. Though this method only led to 9 wt% grafting, in both cases, the surface of the HAP was observed to change from hydrophilic to hydrophobic after grafting, and showed improved stability to sedimentation in organic solvents. The PMMA-grafted nHAP particles, made by the reverse ATRP method, were incorporated at a 2 wt% loading into a PMMA composite and it was determined that both the compressive strengths and compressive moduli of these materials were greater than the composite containing ungrafted nHAP; however, it is unclear whether there was an improvement compared to the unfilled polymer as this was not investigated. (Wang *et al.*, 2011a)

Low molecular-weight molecules have been attached to the HAP surface as linker molecules. In a study by Pramanik *et al.* (2009), 2-carboxyethylphosphonic acid was grafted onto the nHAP surface and composites with poly(ethylene-*co*-vinyl alcohol) were fabricated by blending and compression moulding. The resulting composites had a low porosity (15–17%). Nano-HAP loadings in the range 10–60 wt% were investigated and the use of TEM imaging showed better dispersion of the surface modified nHAP than unmodified nHAP in the composite. The tensile strength of the composites was greater than the pure polymer at all loadings (and significantly higher for the modified nHAP in all cases), and increased with increasing filler content up to 50 wt%. At loadings greater than 60 wt% the tensile strengths of both composites with the modified and unmodified filler decreased.

An investigation by Berger *et al.* (2009) showed that the chemical nature of the material used for surface modifying nHAP can have a significant influence on the properties of the fabricated composite. 2-(Methacryloyloxy) ethyl phosphate (MOEP) and bis[2-(methacryloyloxy)ethyl] phosphate (BIS-MOEP) were used in varying amounts to modify the surface of nHAP before it was incorporated at 40 wt% with a cross-linkable methacrylated oligo-L-lactide macromer to form a composite. The composite containing unmodified nHAP showed a compressive strength of 66 MPa and a Young's modulus of 611 MPa. In the composites incorporating MOEP–nHAP particles, an increase in both the compressive strengths and Young's moduli was observed with increasing MOEP content on the nHAP particles. For example, when 10 wt% MOEP was grafted to nHAP, the compressive strength and modulus were 144 and 1419 MPa, respectively. In contrast to this, the compressive strengths of 0.5–2 wt% grafted BIS-MOEP–nHAP composites were lower than the composite fabricated from unmodified nHAP, but increased for higher graft percentages. A maximum compressive strength and Young's modulus of 155 and 1657 MPa, respectively, were achieved in the composite containing 5 wt% grafted BIS-MOEP–nHAP. Further evaluation of the effect of the surface modifier on the mechanical properties was based on close examination of the shape of the plastic region of the stress–strain curves. Both the nHAP and the MOEP–nHAP composites showed ductile behaviour, with plastic deformation occurring after the ultimate compressive strength was reached. However, the BIS-MOEP–nHAP composite did not show this plastic deformation, indicating that they were brittle.

Since the development of Bioglass® on the nano-scale, studies using these nano-fillers in biocomposites for bone biomaterials applications have emerged. Misra *et al.* (2008) compared the inclusion of micron- and nano-sized Bioglass® in PHB composite films prepared by solvent casting. It was found that the micron-sized Bioglass® reinforced composites (10–30 wt% filler) displayed a reduction in Young's modulus compared to the pure polymer, while the introduction of the nano-sized filler improved the Young's modulus. However, the improvement was greatest for the low filler content (10 wt%) and decreased with increasing amount of filler (up to 30 wt%). The nano-sized filler was found to be present at the material surface (assessed by SEM and contact angle measurement) and this resulted in higher protein adsorption and bioactivity (assessed by SBF) compared to pure polymer and composites with micron-sized filler. In a study by Liu *et al.* (2008) the grafting-to approach was used to modify nano-sized Bioglass® particles with PLLA. These particles were subsequently introduced into PLLA disks prepared by hot pressing. An increase in tensile modulus with filler content (4–20 wt%) was observed for both grafted and non-grafted Bioglass® composites, however, only grafted Bioglass® composites showed improvement in tensile strength and this only to a filler content of 6 wt%.

The fabrication of BG nano-fibres (of composition $70SiO_2.25CaO.5P_2O_5$) and their use in a range of nano-composites have been reported. Kim and co-workers have fabricated such nano-composites with PCL (Lee *et al.*, 2008) and PLA (Kim *et al.*, 2008). The BG nano-fibres were made from a process of electrospinning based on a sol-gel precursor (Kim *et al.*, 2006) and their inclusion in composites has been observed to lead to improved mineralisation in SBF and osteoblast attachment and proliferation, compared to the polymers alone. The significance of the aspect ratio of the BG fillers on mechanical properties and bioactivity were also investigated (Jo *et al.*, 2009). A composite of PCL and BG nano-fibres was compared to a composite of PCL containing micron-sized BG particles and it was determined that at 20 wt% incorporation, the nano-fibre containing composites showed better *in vitro* and *in vivo* biocompatibility, as well as a higher Young's modulus (though there was no significant difference in the tensile strength) than the composite containing micron-sized particles. HAP has also been recently synthesised in nanowire form (aspect ratio ~100). Solvent cast composite films of PCL and nanowire HAP (10–50 wt%) were investigated by Costa *et al.* (2012). Observation with SEM-EDX showed good dispersion and homogeneous distribution of the nanowires. The Young's modulus and compressive modulus of the 50 wt% HAP composites was 665 MPa and 487 MPa, respectively, a significant increase from 193 and 230 MPa of the PCL films.

As can be seen from the examples given in this section, the use of intelligent manipulation of the nano-sized particles can lead to improvements in mechanical properties of nano-composite materials. Though the surface modification of the filler material has been observed to lead to vast improvements in mechanical properties, it is still limited in its success at higher filler loadings. In the above examples, only the approach of using the phosphonic acid linker was successful in increasing the strength up to 50 wt% filler content. Heparin adsorption was successful up to 30 wt% filler content (the highest tested in this work), with most of the remaining strategies being effective to a filler content of 15% or less. Given these limitations, there are an increasing number of studies focussing on controlling the morphology of the filler particles and these approaches appear to more successfully lead to improvements in mechanical properties up to loadings of 50 wt%. There is, however, no clear evidence that it is due to their nanoscale size that these reinforcing particles improve the biological response of the composites. In all cases, it can indeed equally well be attributed to secondary effects such as either surface presentation of the filler or processing effects such as altered surface roughness. These secondary effects are of course brought about by their nano-size, and thus it can be concluded that the use of nano-sized particles is generally highly beneficial for such biomaterials.

11.5 Composite scaffolds

The fabrication of 3D porous structures with controlled and tuneable architectures remains an important challenge in the field of tissue engineering and has been the subject of many reviews (Rezwan *et al.*, 2006; Hutmacher *et al.*, 2007; Kohane and Langer, 2008; Yunos *et al*, 2008; Holzwarth and Ma, 2011; Ng *et al.*, 2012). In general, such scaffolds are engineered to enhance the rate of tissue repair by providing a large surface area for promoting vascularisation and the integration of the regenerating tissue within the scaffold. This is often coupled with the delivery of, for example, stem cells or growth factors. Furthermore, scaffolds for tissue engineering applications are designed such that they will fully degrade once the native tissue has reformed. For applications in bone regeneration a number of key parameters, in addition to the general requirements for bulk composite materials discussed in Section 11.2, need to be addressed when choosing both the materials for and the method of fabrication of the scaffolds. As discussed below, the resulting scaffolds should have optimal porosity as well as an optimal distribution of pore sizes and shapes, and their mechanical properties and degradation rates need to be considered.

The ability to form dense 3D networks of interconnected pores of a suitable size to allow for vascularisation and the penetration of osteogenic cells is a key requirement. The optimal pore size required for successful tissue ingrowth and regeneration of bone tissue is currently a matter of much debate. It has been proposed that a minimal pore size of 100 μm is necessary for cell permeation and vascularisation (Nam and Park, 1999; Rezwan *et al.*, 2006). Furthermore, a number of studies suggest that a pore size in the range of 200–400 μm is optimal for vascularisation and bone tissue ingrowth (Robinson *et al.*, 1995; Brekke and Toth, 1998; Burg *et al.*, 2000; Yang *et al.*, 2001; Leong *et al.*, 2003; Oh *et al.*, 2007) while it has been proposed that pore sizes of greater than 300 μm and maximal porosities of up to 90% are best suited for successful regeneration of bone tissue (Karageorgiou and Kaplan, 2005). It was observed that, while lower porosity enhanced osteogenesis *in vitro* (Itoh *et al.*, 2004; Takahashi and Tabata, 2004), greater ingrowth of bone was achieved *in vivo* under conditions of higher porosity and larger pore sizes (Kruyt *et al.*, 2003; Roy *et al.*, 2003). In addition, Oh *et al.* (2007) have demonstrated a difference in the growth and proliferation of osteoblasts, chondrytes and fibroblasts as a function of the pore size in scaffolds of PCL when comparing *in vitro* and *in vivo* models. The above studies indicate that differences are observed under *in vitro* and *in vivo* conditions and, therefore, results from these studies must be considered with care. In addition, caution is warranted as pore shape and tortuosity are also likely to affect the biological response.

The desire for high porosities must be tempered by the need to maintain suitable mechanical properties of the scaffold throughout its lifetime (Karageorgiou and Kaplan, 2005; Bonfield, 2006). The maximum porosity that can be used is, therefore, highly dependent on the mechanical properties of materials used, their degradation properties and their intended implantation site. Scaffolds with porosities of between 70% and 95% are typically reported. The development of materials with mechanical properties which are sufficient to be at least self-supporting over the intended lifetime of a scaffold and, in some cases, match those of native bone (e.g., in the case of load-bearing implants) is another key challenge. This requirement is more restrictive on the choice of materials when compared to the bulk materials as the stiffness and strength are greatly reduced due to porosity and in addition must be maintained *in vivo* (i.e., in wet conditions). Moreover, for these degradable scaffolds the rates of degradation of the scaffold must be balanced with the rates of tissue regeneration (and any decline in mechanical properties) to maintain the integrity of the scaffold throughout the remodelling phase. In addition, the mechanism of polymer degradation (i.e., surface or bulk eroding) may be dependent on the thickness of the material. Hence a polymer that degrades by a surface erosion mechanism in a thick specimen may degrade via bulk erosion when fabricated as a thin specimen (Burkersroda *et al.*, 2002). Given the typically large surface areas and thin walls engineered into tissue scaffolds, the issue of degradation rate and mechanism need to be determined for the scaffolds and cannot be simply extrapolated from the behaviour of bulk polymers.

A range of methods exists for the fabrication of porous polymeric and composite scaffolds which include solvent casting and salt/particulate leaching (Hou *et al.*, 2003), electrospinning (Phipps *et al.*, 2012), microsphere templating (Linnes *et al.*, 2007), photolithography (Bryant *et al.*, 2007), sintering (Gross and Rodriguez-Lorenzo, 2004), thermally induced phase separation (TIPS) (Cao *et al.*, 2006; Wang *et al.*, 2010b), rapid prototyping (Russias *et al.*, 2007; Jiang *et al.*, 2012), solid free form fabrication (Hutmacher and Cool, 2007), supercritical CO_2 induced phase separation (Ding *et al.*, 2012; Karakeçili and Arıkan, 2012), combinations of these methods (Gupta *et al.*, 2009; Francis *et al.*, 2010; Shin *et al.*, 2010; Vaquette and Cooper-White, 2013) and *in situ* formation of HAP crystals in the composite scaffold (Chen *et al.*, 2010a; Fabbri *et al.*, 2010). It is beyond the scope of this chapter to describe in detail the methods and their relative merits; instead the reader is directed toward reviews in the literature (e.g., Rezwan *et al.*, 2006; Puppi *et al.*, 2010; Sun *et al.*, 2011; Ng *et al.*, 2012). It should, however, be noted here that although all methods allow for some control over the pore size, scaffolds prepared by salt/particulate leaching always have high porosity which greatly affects their mechanical properties while, for example, solid free form fabrication allows for much lower porosity which produces scaffolds that are in general more suitable for bone tissue engineering. Moreover, techniques such as rapid prototyping and microsphere templating allow for the detailed control of the distribution of pore sizes and shapes and can therefore be used to investigate the effect of these parameters on scaffold performance.

The choices of component materials for composite scaffolds are primarily governed by the same design considerations discussed in Section 11.2 and, hence, the vast majority of composite scaffolds reported for bone repair contain HAP, β-TCP or bioactive glasses as the reinforcing phase. Composites of these inorganic particles with the biodegradable polyesters (PCL, PLA, PLGA) have been most extensively investigated (Ma *et al.*, 2001; Wei and Ma, 2004; Verma, 2006; Kim *et al.*, 2006a, b; Hutmacher, *et al.*, 2007; Lei *et al.*, 2007; Russias *et al.*, 2007; Charles-Harris *et al.*, 2008; Hong *et al.*, 2008; Minamiguchi *et al.*, 2008; Wang *et al.*, 2010b; Scaglione *et al.*, 2012). In addition, the fabrication and characterisation (physical, *in vitro* and *in vivo*) of composite scaffolds of PHAs have also received much attention in the literature (Sultana and Wang, 2008; Jack *et al.*, 2009; Li and Chang, 2004, 2005; Wang *et al.*, 2004, 2005; Wang, 2006; Hajiali *et al.*, 2010; Misra *et al.*, 2010; Sultana and Khan, 2012).

In the vast majority of these works the authors have demonstrated that the incorporation of inorganic particles within the polymer matrix leads to significant reinforcement of the composite scaffold, though it seems that there is a critical particle concentration over which this is true. In the case of nHAP, this concentration is most commonly between 5 and 50 wt% (Wei and Ma, 2004; Boissard *et al.*, 2009; Cui *et al.*, 2009; Wang *et al.*, 2010b), regardless of whether surface modification strategies have been applied and the fabrication technique used. For example Wei and Ma (2004) have shown that the incorporation of HAP particles led to an increase in the compressive modulus from ca. 4 to 8 MPa when 30–50 wt% of HAP was added to a PLLA scaffold fabricated by TIPS. However, only a small difference in the modulus of the composites was shown when comparing the incorporation of 50 wt% nano-sized HAP with micron-sized HAP. It has been demonstrated that the addition of as little as 2 wt% of nHAP particles to a porous scaffold of PHBV produced by the TIPS method resulted in a significant increase in both the compressive modulus (1.8–4.5 MPa) and strength (0.8–2.5 MPa)

(Jack *et al.*, 2009). A three-fold increase in compressive modulus upon addition of 20 wt% of nHAP to PHBV TIPS scaffolds was also reported by Sultana and Wang (2007). In addition, in the work of Jack *et al.* (2009) the compressive modulus continued to show an increase to ca. 10 MPa after immersion of the composite scaffold in SBF for two weeks while no increase was observed in a scaffold of pure PHBV during this time. The addition of BGC to PLLA to produce composite scaffolds has also been investigated. For example, Hong and co-workers (Hong *et al.*, 2008, 2009a, b) have fabricated scaffolds with a range of compositions of BGC/PLLA (up to 30 wt% of BGC) by TIPS. They have shown that the addition of up to 30 wt% of BGC led to an increase in the compressive modulus (from ca. 5.5 to 8 MPa) and strength (from ca 0.28 to 0.35 MPa) (Hong *et al.*, 2008). Moreover, in all of these works the addition of the reinforcing particles were found to have little effect on the pore size or the degree of porosity of the scaffolds, which were in the range of (88–95%). More recently, there have been some reports that do indicate a dependence of these properties with increasing content of nHAP (i.e., a shift from 94% to 75% porosity and 97 to 83 μm in the mean pore size between the unloaded polymer and a 30 wt% nHAP composite) (Boissard *et al.*, 2009) and BG particles (from 67% to 85% porosity and 180 to 136 μm in the mean pore size for unfilled polymer and a 70 wt% bioactive glass composite) (Gentile *et al.*, 2012). Some investigations have also been made into the effect of nHAP inclusion on the crystallinity of the matrix polymer (Wang *et al.*, 2010b; Han *et al.*, 2013); however, this too does not show a simple correlation with filler content as it is affected by the fabrication technique used and physicochemical properties of the polymer in question.

In general, for the highly porous biodegradable composite scaffolds reviewed in this work the maximum obtainable values for the elastic moduli were of the order of a few to a few tens of megapascals. One exception is the work of Russias *et al.* (2007) in which highly ordered scaffolds (75% porosity) containing 70 wt% HAP in PCL and PLA were prepared by a printing method and the resulting Young's modulus was reported to be as high as 150 MPa when measured parallel to the printing plane. In general, these values are within that of typical cancellous bone (50–500 MPa) and orders of magnitude less than that of cortical bone (7–30 GPa), but are large enough to provided suitable mechanical integrity to the scaffolds to allow them to be self-supporting.

Arguably of greater importance, given that the current levels of reinforcement result in mechanical properties significantly less than those observed in bone, is that the incorporation of bioactive fillers has been shown to provide significant advantages to the bioactivity of the scaffolds both *in vivo* and *in vitro*. The presence of HAP and BGC particles in porous scaffolds have extensively been shown to enhance the rate of mineralisation onto the surface of the scaffold in SBF and *in vivo* when incorporated into a range of polymeric matrices; this potentially provides materials with a higher degree of osteogenic potential in addition to enhanced mechanical properties (Wei and Ma, 2004; Kim *et al.*, 2006a, b; Hong *et al.*, 2008; Jack *et al.*, 2009; Gentile *et al.*, 2012). It is also widely reported that the incorporation of bioactive fillers results in materials that show higher levels of cell attachment, differentiation, proliferation and penetration into the scaffold (Ma *et al.*, 2001; Wang *et al.*, 2005; Kim 2006a, b; Charles-Harris *et al.*, 2008; Minamiguchi *et al.*, 2008; Gentile *et al.*, 2012; Han *et al.*, 2013; Jaiswal *et al.*, 2013) and enhanced formation of bone *in vivo* (Cao and Kuboyama, 2010; Cao *et al.*, 2012). In addition, it has been demonstrated that the presentation of the HAP at the surface of the scaffold is of key importance to the full realisation of these improvements (Wang *et al.*, 2005; Jack *et al.*, 2009; Gupta *et al.*, 2009; Abdal-hay *et al.*, 2013). Finally, the incorporation of biodegradable filler particles can provide some method for modifying the rates of degradation of the scaffold and can also act as a buffer against undesirable degradation products (Li and Chang, 2004, 2005; Chen *et al.*, 2010b; Liang *et al.*, 2010; Wang *et al.*, 2010a).

An aspect in the fabrication of composite scaffolds that is often overlooked is the effect of processing on the filler material. Recently, Abdollahi and co-workers carried out a study to investigate whether the surface properties of Bioglass® changed when incorporated into poly-D,L-lactic acid composites (Abdollahi *et al.*, 2013). They identified, through the use of X-ray photoelectron and infra-red spectroscopy, that the interaction of the Bioglass® particles with the polymeric matrix within the composite, as well as interactions with the processing liquids during composite fabrication, altered the surface chemistry and, hence, the resulting bioactivity of Bioglass®.

In the bulk composites it has been shown that the stabilisation of primary particles (i.e., prevention of agglomeration), particularly at higher particle content, is of vital importance to maintaining effective reinforcement (Rai *et al.*, 2008a; Schaefer and Justice, 2007; Noohom *et al.*, 2009). Investigating these questions in porous composite scaffolds is, however, more challenging, in part because the analytical tools available at present do not have the necessary resolution and dynamic range required to simultaneously resolve individual particles and porosity in three dimensions *in situ*. Many investigators have used tools such as SEM-EDX to determine that HAP is distributed throughout the composite (Shen *et al.*, 2011; Costa *et al.*, 2012); however, it does not have the resolution to determine whether the particles are non-aggregated as illustrated for our nHAP/PCL scaffolds in Plate IVa and b (in the colour section between pages 186 and 187). Others have investigated methods such as specific surface area analysis (Han *et al.*, 2013) to determine surface presentation; however, any observed increase in interfacial area may equally be due to surface roughening caused during processing, and not necessarily the particles themselves being on the surface. Direct methods of imaging particles with nano-scale resolution include TEM and synchrotron-based X-ray fluorescence imaging, but these methods have problems of artefact introduction during sample preparation and are inherently two-dimensional. An illustration from our work of the information that can be gathered from TEM is given in Plate IVc. Indirect methods that may be used to give information on the nano-scale include time-of-flight secondary-ion mass spectrometry (ToF-SIMS) surface imaging, small-angle scattering methods, atomic force microscopy (AFM) and nano-computed tomography (nano-CT). However, each of these methods has its own inherent limitations; for example, a lack of three-dimensional information in the case of ToF-SIMS and AFM and the complexity of modelling required to interpret the Fourier space images afforded by scattering methods. It is also noted that nano-CT has demonstrated some applicability for characterising nano-composite scaffolds (Salmon and Sasov, 2007); however, it is still a novel technique and, hence, still to be fully developed.

11.6 Conclusions and future trends

In general, the majority of studies indicate that the incorporation of a suitable inorganic filler results in an increase in mechanical properties such as compressive strength and Young's modulus, as well as bioactivity compared to polymer alone. This is true of micron-sized particulate composites (Jo *et al.*, 2009), nano-composites (Pramanik *et al.*, 2009; Costa *et al.*, 2012; Abdal-hay *et al.*, 2013) and porous composite scaffolds (Han *et al.*, 2013). However, the fabrication method does not seem to play a major role in how the filler improves these properties. An important aspect in the fabrication of nano-composites is the ability to obtain the most complete dispersion possible of the primary particles and to maintain this dispersion throughout the lifetime of the composite, if the full benefits of the nano-scale reinforcing particles are to be realised. More knowledge and understanding of methods for achieving these higher degrees of dispersion are still required even in bulk materials. In the case of HAP, the physicochemical nature of the material itself makes improvements in both mechanical strength and bioactivity of the composite highly likely and the real question is to what degree such improvements are influenced by the crystal size, dispersion and/or surface presentation of the HAP in the composite. It is generally accepted that the increased surface area-to-mass ratio of the nano-particles in combination with any higher degree of surface presentation results in higher levels of bioactivity, for example, protein adsorption, *in vitro* mineralisation, and cell proliferation and differentiation (Wei and Ma, 2004; Sato and Webster, 2004; Kim *et al.*, 2006a, b; Kim and Fisher, 2007). Although the question of efficacy of nano- versus micron-sized particles has also been discussed for bone scaffolds, most typically when comparing HAP filled polymers, (see e.g., Wei and Ma, 2004; Kim and Fisher, 2007), there appears to be an absence of systematic studies of the effect of the filler particle size on the mechanical properties or bioactivity. Finally, the development of better analytical methods to fully characterise and quantify the dispersion and surface presentation in 3D will facilitate a greater understanding of the interplay between the methods of stabilisation and scaffold fabrication on the *in vitro* and *in vivo* response of the scaffolds.

11.7 Sources of further information and advice

As pointed out throughout this chapter, not all aspects of biocomposites for bone regeneration could be extensively covered. The reader is therefore referred to the following key review papers and books for further details.

Topics in Applied Physics: Molecular Building Blocks for Nanotechnology – From Diamondoids to Nanoscale Materials and Applications, Ed.: G. Ali Mansoori, Thomas, F. George, Lahsen Assoufid, Guoping Zhang. Chapter 13. *Nanoengineering Biomimetic Bone-Building Blocks* by B. Murugan and S. Ramakrishna, Springer, 2007.

Burg, K J L, Porter, S and Kellam, J F (2000) 'Biomaterial developments for bone tissue engineering' *Biomaterials*, **21**, 2347–2359.

Seal, B L, Otero, T C and Panitch, A (2001) 'Polymeric biomaterials for tissue and organ regeneration' *Materials Science and Engineering R*, **34**, 147–230.

Rahaman, M N and Mao, J J (2005) 'Stem cell-based composite tissue constructs for regenerative medicine' *Biotechnology and Bioengineering*, **91**, 261–284.

Allo, B A, Costa, D O, Dixon, S J, Mequanint, K and Rizkalla, A S (2012) 'Bioactive and biodegradable nano-composites and hybrid biomaterials for bone regeneration' *Journal of Functional Biomaterials*, **3** (2), 432–463.

11.8 References

Abdal-hay, A, Sheikh, F A and Lim, J K (2013) 'Air jet spinning of hydroxyapatite/poly(lactic acid) hybrid nanocomposite membrane mats for bone tissue engineering' *Colloids Surf, B*, **102**, 635–643.

Abdollahi, S, Ma, A C C and Cerruti, M (2013) 'Surface transformations of bioglass 45S5 during scaffold synthesis for bone tissue engineering' *Langmuir*, **29**, 1466–1474.

Abu Bakar, M S, Cheang, P and Khor, K A (2003) 'Mechanical properties of injection molded hydroxyapatite-polyetheretherketone biocomposites' *Comp Sci Tech*, **63**, 421–425.

Berger, S, Muller, E and Schnabelrauch, M (2009) 'Influence of methacrylate-containing surface modifiers on the mechanical properties of nano-hydroxyapatite/polylactide network composites' *Mater Lett*, **63**, 2714–2717.

Bohner, M (2000) 'Calcium orthophosphates in medicine: from ceramics to calcium phosphate cements' *Injury*, **31** (Suppl. 4) S37–S47.

Boissard, C I R, Bourban, P E, Tami, A E, Alini, M and Eglin, D (2009) 'Nanohydroxyapatite/poly(ester urethane) scaffold for bone tissue engineering' *Acta Biomater,* **5**, 3316–3327.

Bonfield, W (2006) 'Designing porous scaffolds for tissue engineering' *Phil Trans R Soc,* **364**, 227–232.

Bonfield, W (1987) 'Materials for the replacement of osteoarthritic hip joints' *Metals Mater,* **3**, 712–716.

Bonfield, W, Grynpas, M D, Tully, A E, Bowman, J and Abram, J (1981) 'Hydroxyapatite reinforced polyethylene – a mechanically compatible implant material for bone replacement' *Biomaterials,* **2**, 185–186.

Bonfield, W and Luklinska, A B (1991) *The Bone-Biomaterial Interface*, Ed J E Davies, Uni Toronto Press 89093.

Brekke, J H and Toth, J M (1998) 'Principles of tissue engineering applied to programmable osteogenesis' *J Biomed Mater Res,* **43**, 380–398.

Brunner, T J, Grass, R N and Stark, W J (2006) 'Glass and bioglass nanopowders by flame synthesis' *Chem Comm,* 1384–1386.

Bryant, S J, Cuy, J L, Hauch, K D and Ratner, B D (2007) 'Photo-patterning of porous hydrogels for tissue engineering' *Biomaterials,* **28**, 2978–2986.

Burg, K J, Porter, S and Kellam, J F (2000) 'Biomaterial developments for bone tissue engineering' *Biomaterials,* **21**, 2347–2359.

Burkersroda, F, Schedl, L and Göpferich, A (2002) 'Why degradable polymers undergo surface erosion or bulk erosion' *Biomaterials,* **23**, 4221–4231.

Cao, H and Kuboyama, N (2010) 'A biodegradable porous composite scaffold of PGA/β-TCP for bone tissue engineering' *Bone,* **46**, 386–395.

Cao, L, Duan, P G, Wang, H R, Li, X L, Yuan, F L, Fan, Z Y, Li, S M and Dong, J (2012) 'Degradation and osteogenic potential of a novel poly(lactic acid)/nano-sized b-tricalcium phosphate scaffold' *Int J Nanomedicine,* **7**, 5881–5888.

Cao, Y, Croll, T I, O'Connor, A J, Stevens, G W and Cooper-White, J J (2006) 'Systematic selection of solvents for the fabrication of 3D combined macro- and microporous polymeric scaffolds for soft tissue engineering' *J Biomater Sci: Polym Ed,* **17**, 369–402.

Charles-Harris, M, Koch, M A, Navarro, M, Lacroix, D, Engel, E and Planell, J A (2008) 'A PLA/calcium phosphate degradable composite material for bone tissue engineering: an *in vitro* study' *J Mater Sci: Mater Med,* **19**, 1503–1513.

Chen, G-Q and Wu, Q (2005) 'The application of polyhydroxyalkanoates as tissue engineering materials' *Biomaterials,* **26**, 6565–6578.

Chen, J, Li, X, Cui, W, Xie, C, Zou, J and Zou, B (2010a) '*In situ* grown fibrous composites of poly(dl-lactide) and hydroxyapatite as potential tissue engineering scaffolds' *Polymer,* **51**, 6268–6277.

Chen, Q, Jin, L, Cook, W D, Mohn, D, Lagerqvist, E L, Elliott, D A, Haynes, J M, Boyd, N, Stark, W J, Pouton, C W, Stanley, E G and Elefanty, A G (2010b) 'Elastomeric nanocomposites as cell delivery vehicles and cardiac support devices' *Soft Matter,* **6**, 4715–4726.

Chen, Q, Liang, S and Thouas, G A (2013) 'Elastomeric biomaterials for tissue engineering' *Prog Polym Sci,* **38**, 584–671.

Chlopek, J, Rosol, P and Morawska-Chochol, A (2006) 'Durability of polymer-ceramics composite implants determined in creep tests' *Comp Sci Tech,* **66**, 1615–1622.

Cool, S M, Kenny, B, Wu, A, Nurcombe, V, Trau, M, Cassady, A I and Grøndahl, L (2007) 'Poly(3-hydroxybutyrate-co-3-hydroxyvalerate) composite biomaterials for bone tissue regeneration: *in vitro* performance assessed by osteoblast proliferation, osteoclast adhesion and resorption, and macrophage pro-inflammatory response' *J Biomed Mater Res,* **82A**, 599–610.

Coskun, S, Korkusuz, F and Hasirci, V (2005) 'Hydroxyapatite reinforced poly(3-hydroxybutyrate) and poly(3-hydroxybutyrate-co-3-hydroxyvalerate) based degradable composite bone plate' *J Biomater Sci: Polym Ed*, **16**, 1485–1502.

Costa, D O, Dixon, S J and Rizkalla, A S (2012) 'One- and three-dimensional growth of hydroxyapatite nanowires during sol–gel–hydrothermal synthesis' *ACS Appl Mater Interfaces,* **4**, 1490–1499.

Cui, Y, Liu, Y, Jing, X B, Zhang, P B A and Chen, X S (2009) 'The nanocomposite scaffold of poly(lactide-co-glycolide) and hydroxyapatite surface-grafted with L-lactic acid oligomer for bone repair' *Acta Biomater,* **5**, 2680–2692.

Daculsi, G, Laboux, O, Malard, O and Weiss, P (2003) 'Current state of the art of biphasic calcium phosphate bioceramics' *J Mater Sci Mater Med,* **14**, 195–200.

Deng, X, Hao, J and Wang, C (2001) 'Preparation and mechanical properties of nano-composites of poly(D,L-lactide) with Ca-deficient hydroxyapatite nanocrystals' *Biomaterials*, **22**, 2867–2873.

Ding, Z, Liu, Z, Wei, W and Li, Z (2012) 'Preparation and characterization of PLLA composite scaffolds by $ScCO_2$-induced phase separation' *Polym Compos,* **33**, 1667–1671.

Doyle, C, Tanner, E T and Bonfield, W (1991) '*In vitro* and *in vivo* evaluation of polyhydroxybutyrate and of polyhydroxybutyrate reinforced with hydroxyapatite' *Biomaterials*, **12**, 841–847.

Fabbri, P, Bondioli, F, Messori, M, Bartoli, C, Dinucci, D and Chiellini, F (2010) 'Porous scaffolds of polycaprolactone reinforced with *in situ* generated hydroxyapatite for bone tissue engineering' *J Mater Sci: Mater Med,* **21**, 343–351.

Francis, L, Venugopal, J, Prabhakaran, M P, Thavasi, V, Marsano, E and Ramakrishna, S (2010) 'Simultaneous electrospin–electrosprayed biocomposite nanofibrous scaffolds for bone tissue regeneration' *Acta Biomater,* **6**, 4100–4109.

Galego, N, Rozsa, C, Sanchez, R, Fung, J, Vazquez, A and Tomas, J S (2000) 'Characterization and application of poly(β-hydroxyalkanoates) family as composite biomaterials' *Polym Test*, **19**, 485–92.

Gentile, P, Mattioli-Belmonte, M, Chiono, V, Ferretti, C, Baino, F, Tonda-Turo, C, Vitale-Brovarone, C, Pashkuleva, I, Reis, R L and Ciardelli, G (2012) 'Bioactive glass/polymer composite scaffolds mimicking bone tissue' *J Biomed Mater Res A,* **100**, 2654–2667.

Gong, X-H, Tang, C-K, Hu, H-C, Zhou, X-P and Xie, X-L (2004) 'Improved mechanical properties of HIPS/hydroxyapatite composites by surface modification of hydroxyapatite via *in-situ* polymerisation of styrene' *J Mater Sci: Mater Med,* **15**, 1141–1146.

Gross, K A and Rodriguez-Lorenzo, L M (2004) 'Biodegradable composite scaffolds with an interconnected spherical network for bone tissue engineering' *Biomaterials,* **25**, 4955–4962.

Gupta, D, Venugopal, J, Mitra, S, Giri Dev, V R and Ramakrishna, S (2009) 'Nanostructured biocomposite substrates by electrospinning and electrospraying for the mineralization of osteoblasts' *Biomaterials,* **30**, 2085–2094.

Hajiali, H, Karbasi, S, Hosseinalipour, M and Rezaie, H (2010) 'Preparation of a novel biodegradable nanocomposite scaffold based on poly (3-hydroxybutyrate)/bioglass nanoparticles for bone tissue engineering' *J Mater Sci: Mater Med,* **21**, 2125–2132.

Han, W, Zhao, J, Tu, M, Zeng, R, Zha, Z and Zhou, C (2013) 'Preparation and characterization of nanohydroxyapatite strengthening nanofibrous poly(L-lactide) scaffold for bone tissue engineering' *J Appl Polym Sci,* **128**, 1332–1338.

Hao, J, Yuan, M and Deng, X (2002) 'Biodegradable and biocompatible nano-composites of poly(ε-caprolactone) with hydroxyapatite nanocrystals: thermal and mechanical properties' *J Appl Polym Sci,* **86**, 676–683.

Holzwarth, J M and Ma, P X (2011) '3D nanofibrous scaffolds for tissue engineering' *J Mater Chem,* **21**, 10243–10251.

Homaeigohar, S, Shokrgozar, M A, Sadi, A Y, Khavandi, A, Javadpour, J and Hosseinalipour, M (2005) '*In vitro* evaluation of biocompatibility of beta-tricalcium phosphate-reinforced high-density polyethylene; an orthopedic composite' *J Biomed Mater Res A,* **75**, 14–22.

Homaeigohar, S S, Shokrgozar, M A, Khavandi, A and Sadi, A Y (2008) '*In vitro* biological evaluation of β–TCP/HDPE—A novel orthopedic composite: a survey using human osteoblast and fibroblast bone cells' *J Biomed Mater Res A,* **84A**, 491–499.

Hong, Z, Reis, R L and Mano, J F (2008) 'Preparation and *in vitro* characterization of scaffolds of poly(l-lactic acid) containing bioactive glass ceramic nanoparticles' *Acta Biomater,* **4**, 1297–1306.

Hong, Z, Liu, A, Chen, L, Chen, X and Jing, X (2009a) 'Preparation of bioactive glass ceramic nanoparticles by combination of sol–gel and coprecipitation method' *J Non-Cryst Solids,* **355**, 368–372.

Hong, Z, Reis, R L and Mano, J F (2009b) 'Preparation and *in vitro* characterization of novel bioactive glass ceramic nanoparticles' *J Biomed Mater Res A,* **88**, 304–313.

Hong, Z, Qiu, X, Sun, J, Deng, M, Chen, X and Jing, X (2004) 'Grafting polymerization of L-lactide on the surface of hydroxyapatite nano-crystals' *Polymer,* **45**, 6705–13.

Hou Q P, Grijpma, D W and Feijen, J (2003) 'Porous polymeric structures for tissue engineering prepared by a coagulation, compression moulding and salt leaching technique' *Biomaterials,* **24**, 1937–1947.

Huang, J, Di Silvio L, Wang, M, Tanner, K E and Bonfield, W (1997) '*In vitro* mechanical and biological assessment of hydroxyapatite-reinforced polyethylene composite' *J Mater Sci: Mater Med,* **8**, 775–779.

Hutmacher, D W and Cool, S (2007) 'Concepts of scaffold-based tissue engineering – the rationale to use solid free-form fabrication techniques' *J Cell Mol Med,* **11**, 654–669.

Hutmacher, D W, Schantz, J T, Lam, C X F, Tan, K C and Lim T C, (2007) 'State of the art and future directions of scaffold-based bone engineering from a biomaterials perspective' *J Tissue Eng Regen Med,* **1**, 245–260.

Itoh, M, Shimazu, A, Hirata, I, Yoshida, Y, Shintani, H and Okazaki, M (2004) 'Characterization of CO_3Ap-collagen sponges using X-ray high-resolution microtomography' *Biomaterials,* **25**, 2577–2583.

Jack, K S, Velayudhan, S, Luckmann P, Trau M, Grøndahl L and Cooper-White, J (2009) 'The fabrication and characterisation of biodegradable HA/PHBV nanoparticle-polymer composite scaffolds' *Acta Biomater,* **5**, 2657–2667.

Jaiswal, A K, Chhabra, H, Soni, V P and Bellare, J R (2013) 'Enhanced mechanical strength and biocompatibility of electrospun polycaprolactone-gelatin scaffold with surface deposited nano-hydroxyapatite' *Mater Sci Eng C Mater Biol Appl,* **33**, 2376–2385.

Jiang, L, Li, Y, Wang, X, Zhang, L, Wen, J and Gong, M (2008) 'Preparation and properties of nano-hydroxyapatite/chitosan/carboxymethyl cellulose composite scaffold' *Carbohydr Polym,* **74**, 680–684.

Jiang, W, Shi, J, Li, W and Sun, K (2012) 'Morphology, wettability, and mechanical properties of polycaprolactone/hydroxyapatite composite scaffolds with interconnected pore structures fabricated by a mini-deposition system' *Polym Eng Sci,* **52**, 2396–2402.

Jo, J H, Lee, E J, Shin, D S, Kim, H E, Kim, H W, Koh, Y H and Jang, J H (2009) '*In vitro*/*in vivo* biocompatibility and mechanical properties of bioactive glass nanofiber and poly(epsilon-caprolactone) composite materials' *J Biomed Mater Res B Appl Biomater,* **91**, 213–220.

Juhasz, J A, Best S M, Brooks, R, Kawashita, M, Miyata, N, Kokubo, T, Nakamura, T and Bonfield, W (2004) 'Mechanical properties of glass-ceramic A-W-polyethylene composites: effect of filler content and particle size' *Biomaterials,* **25**, 949–955.

Karageorgiou V and Kaplan D (2005) 'Porosity of 3D biomaterial scaffolds and osteogenesis' *Biomaterials,* **26**, 5474–5491.

Kim, H-W (2007) 'Biomedical nano-composites of hydroxyapatite/polycaprolactone obtained by surfactant mediation' *J Biomed Mater Res,* **83A**, 169–177.

Kim, K and Fisher, J P (2007) 'Nanoparticle technology in bone tissue engineering' *J Drug Target,* **15**, 241–252.

Kim, S-S, Park, M S, Jeon, O, Choi, C Y and Kim, B-S (2006a) 'Poly(lactide-*co*-glycolide)/hydroxyapatite composite scaffolds for bone tissue engineering' *Biomaterials,* **27**, 1399–1409.

Kim, S-S, Park, M S, Gwak, S-J, Choi, C Y and Kim, B-S (2006b) 'Accelerated bone-like apatite growth on porous polymer/ceramic composite scaffolds *in vitro*' *Tissue Eng,* **12**, 2997–3006.

Kohane, D S and Langer, R (2008) 'Polymeric biomaterials in tissue engineering' *Paediatr Res,* **63**, 487–491.

Karakeçili, A and Arıkan, A (2012) 'Preparation of chitosan-nanohydroxyapatite composite scaffolds by a supercritical CO_2 assisted process' *Polym Compos,* **33**, 1215–1223.

Kim, H-W, Lee, H-H and Chun, G-S (2008) 'Bioactivity and osteoblast responses of novel biomedical nanocomposites of bioactive glass nanofiber filled poly(lactic acid)' *J Biomed Mater Res A,* **85A**, 651–663.

Kim, H W, Kim, H E and Knowles, J C (2006) 'Production and potential of bioactive glass nanofibers as a next-generation biomaterial' *Adv Funct Mater,* **16**, 1529–1535.

Kruyt, M C, de Bruijn, J D, Wilson, C E, Oner, F C, van Blitterswijk, C A, Verbout, A J and Dhert, W J (2003) 'Viable osteogenic cells are obligatory for tissue-engineered ectopic bone formation in goats' *Tissue Eng,* **9**, 327–336.

Langstaff, S, Sayer, M, Smith, T J N, Pugh, S M, Hesp, S A M and Thompson, W T (1999) 'Resorbable bioceramics based on stabilized calcium phosphates. Part I: rational design, sample preparation and material characterisation' *Biomaterials,* **20**, 1727–1741.

Lee, H H, Yu, H S, Jang, J H and Kim, H W (2008) 'Bioactivity improvement of poly(epsilon-caprolactone) membrane with the addition of nanofibrous bioactive glass' *Acta Biomater,* **4**, 622–629.

Lee, H J, Choi, H W, Kim, K J and Lee, S C (2006) 'Modification of hydroxyapatite nanosurfaces for enhanced colloidal stability and improved interfacial adhesion in nano-composites' *Chem Mater*, **18**, 5111–5118.

Lee, H J, Kim, S E, Choi, H W, Kim, C W, Kim, K J and Lee, S C (2007) 'The effect of surface-modified nano-hydroxyapatite on biocompatibility of poly(ε-caprolactone)/hydroxyapatite nano-composites' *Eur Polym J*, **43**, 1602–1608.

Lee, S C, Choi, H W, Lee, H J, Kim, K J, Chang, J H, Kim, S Y, Choi, J, Oh, K S and Jeong, Y K (2007) 'In-situ synthesis of reactive hydroxyapatite nano-crystals for a novel approach of surface grafting polymerization' *J Mater Chem,* **17**, 174–180.

Lei, B, Shin, K-H, Noh, D-Y, Jo, I-H, Koh, Y-H, Kim, H-E and Kim, S E (2013) 'Sol–gel derived nanoscale bioactive glass (NBG) particles reinforced poly(ε-caprolactone) composites for bone tissue engineering' *Mater Sci Eng C Mater Biol Appl,* **33**, 1102–1108.

Lei, Y, Rai, B, Ho, K H and Teoh, S H (2007) '*In vitro* degradation of novel bioactive polycaprolactone-20% tricalcium phosphate composite scaffolds for bone engineering' *Mater Sci Eng,* **27C**, 293–298.

Leong, K F, Cheah, C M and Chua, C K (2003) 'Solid freeform fabrication of three-dimensional scaffolds for engineering replacement tissues and organs' *Biomaterials,* **24**, 2363–2378.

Li, H Y and Chang, J (2005) '*In vitro* degradation of porous degradable and bioactive PHBV/wollastonite composite scaffolds' *Polym Degrad Stab,* **87**, 310–307.

Li, H and Chang, J (2004) 'Fabrication and characterization of bioactive wollastonite/PHBV composite scaffolds' *Biomaterials,* **25**, 5473–5480.

Liang, S L, Cook, W D, Thouas, G A and Chen, Q Z (2010) 'The mechanical characteristics and *in vitro* biocompatibility of poly(glycerol sebacate)-bioglass elastomeric composites' *Biomaterials,* **31**, 8516–8529.

Liao, C Z, Li, K, Wong, H M, Tong, W Y, Yeung, K W K and Tjong, S C (2013) 'Novel polypropylene biocomposites reinforced with carbon nanotubes and hydroxyapatite nanorods for bone replacements' *Mater Sci Eng C Mater Biol Appl,* **33**, 1380–1388.

Linnes, M P, Ratner, B D and Giachelli, C M (2007) 'A fibrogen-based precision microporous scaffold for tissue engineering' *Biomaterials,* **28**, 5298–5306.

Liu, A, Hong, Z, Zhang, X, Chen, X, Cui, Y, Liu, Y and Jing, X (2008) 'Surface modification of bioactive glass nanoparticles and the mechanical and biological properties of poly(L-lactide) composites' *Acta Biomater*, **4**, 1005–1015.

Liu, Y and Wang, M (2007) 'Fabrication and characteristics of hydroxyapatite reinforced polypropylene as a bone analogue biomaterial' *J Appl Polym Sci,* **106**, 2780–2790.

Liu, Q, de Wijn, J R, de Groot, K and van Blitterswijk, C A (1998) 'Surface modification of nano-apatite by grafting organic polymer' *Biomaterials*, **19**, 1067–1072.

Luklinska, Z B and Bonfield, W (1997) 'Morphology and ultrastructure of the interface between hydroxyapatite-polyhydroxybutyrate composite implant and bone' *J Mater Sci Mater Med*, **8**, 379–83.

Luklinska, Z B and Schluckwerder, H (2003) '*In vivo* response to HA-polyhydroxybutyrate/polyhydroxyvalerate composite' *J Microsc,* **211**, 121–129.

Ma, P X, Zhang, R, Xiao, G and Franceschi, R (2001) 'Engineering new bone tissue *in vitro* on highly porous poly (α-hydroxyl acids)/hydroxyapatite composite scaffolds' *J Biomed Mater Res,* **54**, 284–293.

Ma, R, Weng, L, Bao, X, Song, S and Zhang, Y (2013) '*In vivo* biocompatibility and bioactivity of *in situ* synthesized hydroxyapatite/polyetheretherketone composite materials' *J Appl Polym Sci,* **127**, 2581–2587.

Macedo, F A, Nunes, E H M, Vasconcelos, W L, Santos, R A, Sinisterra, R D and Cortes, M E (2012) 'A biodegradable porous composite scaffold of PCL/BCP containing Ang-(1–7) for bone tissue engineering' *Cerâmica,* **58**, 481–488.

Minamiguchi, S, Takechi, M, Yuasa, T, Momota, Y, Tatehara, S, Takano, H, Miyamoto, Y, Satomura, K, and Nagayama, M (2008) 'Basic research on aw-AC/PLGA composite scaffolds for bone tissue engineering' *J Mater Sci: Mater Med,* **19**, 1165–1172.

Misra, D N (1996) 'Adsorption of polyacrylic acids and their sodium salts on hydroxyapatite: effect of relative molar mass' *J Colloid Interface Sci,* **181**, 289–296.

Misra, S K, Valappil, S P, Roy, I and Boccaccini, A R (2006) 'Polyhyroxyalkanoate (PHA)/inorganic phase composites for tissue engineering applications' *Biomacromolecules,* **7**, 2249–2258.

Misra, S K, Nazhat, S N, Valappil, S P, Moshrefi-Torbati, M, Wood, R J K, Roy, I and Boccaccini, A R (2007) 'Fabrication and characterisation of biodegradable poly(3-hydroxybutyrate) composites containing bioglass' *Biomacromolecules,* **8**, 2112–2119.

Misra, S K, Mohn, D, Brunner, T J, Stark, W J, Philip, S E, Roy, I, Salih, V, Knowles, J C and Boccaccini, A R (2008) 'Comparison of nanoscale and microscale bioactive glass on the properties of P(3HB)/Bioglass® composites' *Biomaterials,* **29**, 1750–1761.

Misra, S K, Ansari, T I, Valappil, S P, Mohn, D, Philip, S E, Stark, W J, Roy, I, Knowles, J C, Salih, V and Boccaccini, A R (2010) 'Poly(3-hydroxybutyrate) multifunctional composite scaffolds for tissue engineering applications' *Biomaterials,* **31**, 2806–2815.

Murugan, R and Ramakrishna, S (2005) 'Development of nano-composites for bone grafting' *Composite Sci Tech,* **65**, 2385–2406.

Nam, Y S and Park, T G (1999) 'Biodegradable polymeric microcellular foams by modified thermally induced phase separation method' *Biomaterials,* **20**, 1783–1790.

Ng, R, Zang, R, Yang, K K, Liu, N and Yang, S-T (2012) 'Three-dimensional fibrous scaffolds with microstructures and nanotextures for tissue engineering' *RSC Adv,* **2**, 10110–10124.

Ni, J and Wang, M (2002) '*In vitro* evaluation of hydroxyapatite reinforced polyhydroxybutyrate composite' *Mater Sci Eng C,* **20**, 101–109.

Nichols, H L, Zhang, N, Zhang, J, Shi, D, Bhaduri, S and Wen, X (2007) 'Coating nanothickness degradable films on nanocrystalline hydroxyapatite particles to improve the bonding strength between nanohydroxyapatite and degradable polymer matrix' *J Biomed Mater Res,* **82A**, 373–382.

Noohom, W, Jack, K S, Martin, D and Trau, M (2009) 'Understanding the roles of nanoparticle dispersion and polymer crystallinity in controlling the mechanical properties of HA/PHBV nano-composites' *Biomed Mater,* **4**, 015003.

Nurcombe, V, Kumarasuriyar, A and Cool, S M (2004) 'The development of heparan sulfate sugars as therapeutics: versatility that couples stem cells, tissue engineering, and wound repair' *Drug Dev Res,* **62**, 303–316.

Oh, S H, Park, I K, Kim, J M and Lee, J H (2007) '*In vitro* and *in vivo* characteristics of PCL scaffolds with pore size gradient fabricated by a centrifugation method' *Biomaterials,* **28**, 1664–1671.

Phipps, M C, Clem, W C, Grunda, J M, Clines, G A and Bellis, S L (2012) 'Increasing the pore sizes of bone-mimetic electrospun scaffolds comprised of polycaprolactone, collagen I and hydroxyapatite to enhance cell infiltration' *Biomaterials,* **33**, 524–534.

Pramanik, N, Mohapatra, S, Bhargava, P and Pramanik, P (2009) 'Chemical synthesis and characterization of hydroxyapatite (HAp)-poly (ethylene co vinyl alcohol) (EVA) nanocomposite using a phosphonic acid coupling agent for orthopedic applications' *Mat Sci Eng C-Biomim Supramole Syst,* **29**, 228–236.

Puppi, D, Chiellini, F, Piras, A M and Chiellini, E (2010) 'Polymeric materials for bone and cartilage repair' *Prog Polym Sci,* **35**, 403–440.

Qiu, X, Chen, L, Hu, J, Sun, J, Hong, Z, Liu, A, Chen, X and Jing, X (2005) 'Surface-modified hydroxyapatite linked by L-lactic acid oligomer in the absence of catalyst' *J Polym Sci Part A: Polym Chem,* **43**, 5177–5185

Rai, B, Noohom, W, Kithva, P H, Grøndahl, L and Trau, M (2008a) 'Bionanohydroxyapatite/ poly(3-hydroxyutyrate-co-3-hydroxyvalerate) composites with improved particle dispersion and superior mechanical properties' *Chem Mater,* **20**, 2801–2808.

Rai, B, Grøndahl, L and Trau, M (2008b) 'Combining chemistry and biology to create stable bionanohydroxyapatite particles towards load-bearing bone applications' *Langmuir,* **24**, 7744–8849.

Rea, S M, Best, S M and Bonfield W (2004) 'Bioactivity of ceramic-polymer composites with varied composition and surface topography' *J Mater Sci Mater Med,* **15**, 997–1005.

Rees, S G, Wassell, D T H and Embery, G (2002) 'Interaction of glucuronic acid and iduronic acid-rich glycosaminoglycans and their modified forms with hydroxyapatite' *Biomaterials,* **23**, 481–489.

Rezwan, K, Chen, Q Z, Blaker, J J and Boccaccini, A R (2006) 'Biodegradable and bioactive porous polymer/inorganic composite scaffolds for bone tissue engineering' *Biomaterials,* **27**, 3413–3431.

Robinson, B P, Hollinger, J O, Szachowicz, E H and Brekke, J (1995) 'Calvarial bone repair with porous D,L-polylactide' *Otolaryngol Head Neck Surg,* **112**, 707–713.

Roy, T D, Simon, J L, Ricci, J L, Rekow, E D, Thompson, V P and Parsons, J R (2003) 'Performance of degradable composite bone repair products made via three-dimensional fabrication techniques' *J Biomed Mater Res A,* **66**, 283–291.

Russias, J, Saiz, E, Deville, S, Gryn, K, Liu, G, Nalla, R K and Tomsia, A P (2007) 'Fabrication and *in vitro* characterization of three-dimensional organic/inorganic scaffolds by robocasting' *J Biomed Mater Res,* **83A**, 434–445.

Salmon, P and Sasov, A (2007), *Application of Nano-CT and High-Resolution Micro-CT to Study Bone Quality and Ultrastructure, Scaffold Biomaterials and Vascular Networks – From Advanced Bioimaging Technologies in Assessment of the Quality of Bone and Scaffold Materials*, Ed.: L. Qin, H. Genant, J. Griffith and K. Leung, Springer Berlin Heidelberg: 323–331.

Sato, M and Webster, T J (2004) 'Nonobiotechnology: implications for the future of nanotechnology in orthopedic applications' *Expert Rev Med Dev,* **1**, 105–114.

Schaefer, D W and Justice, R S (2007) 'How nano are nano-composites?' *Macromolecules,* **40**, 8501–8517.

Scaglione, S, Guarino, V, Sandri, M, Tampieri, A, Ambrosio, L and Quarto, R (2012) '*In vivo* lamellar bone formation in fibre coated MgCHA-PCL-composite scaffolds' *J Mater Sci Mater Med,* **23**, 117–128.

Schaefer, S, Detsch, R, Uhl, F, Deisinger, U and Ziegler, G (2011) 'How degradation of calcium phosphate bone substitute materials is influenced by phase composition and porosity' *Adv Eng Mater,* **13**, 342–350.

Shen, X, Chen, L, Cai, X, Tong, T, Tong, H and Hu, J (2011) 'A novel method for the fabrication of homogeneous hydroxyapatite/collagen nanocomposite and nanocomposite scaffold with hierarchical porosity' *J Mater Sci: Mater Med,* **22**, 299–305.

Shin, T J, Park, S Y, Kim, H J, Lee, H J and Youk, J H (2010) 'Development of 3-D poly(trimethylenecarbonate-co-epsilon-caprolactone)-block-poly(p-dioxanone) scaffold for bone regeneration with high porosity using a wet electrospinning method' *Biotechnol Lett,* **32**, 877–882.

Sultana, N and Wang, M (2008) 'Fabrication of HA/PHBV composite scaffolds through the emulsion freezing/freeze-drying process and characterisation of the scaffolds' *J Mater Sci: Mater Med,* **19**, 2555–2561.

Sultana, N and Khan, T H (2012) '*In vitro* degradation of PHBV scaffolds and nHA/PHBV composite scaffolds containing hydroxyapatite nanoparticles for bone tissue engineering' *J Nanomater,* **2012**, 12.

Sun, F, Zhou, H and Lee, J (2011) 'Various preparation methods of highly porous hydroxyapatite/polymer nanoscale biocomposites for bone regeneration' *Acta Biomater,* **7**, 3813–3828.

Takahashi, Y and Tabata, Y (2004) 'Effect of the fiber diameter and porosity of non-woven PET fabrics on the osteogenic differentiation of mesenchymal stem cells' *J Biomater Sci Polym Ed,* **15**, 41–57.

Vallet-Regi, M (2001) 'Ceramics for medical applications' *J Chem Soc Dalton Trans,* 97–108.

Vaquette, C and Cooper-White, J (2013) 'A simple method for fabricating 3-D multilayered composite scaffolds' *Acta Biomater,* **9**, 4599–4608.

Verma, D, Katti, K and Katti, D (2006) 'Bioactivity in *in situ* hydroxyapatite-polycaprolactone composites' *J Biomed Mater Res,* **78A**, 772–780.

Wang, H, Zuo, Y, Zou, Q, Cheng, L, Huang, D, Wang, L and Li, Y (2009) 'Nano-hydroxyapatite/polyamide66 composite tissue-engineering scaffolds with anisotropy in morphology and mechanical behaviors' *J Polym Sci Part A: Polym Chem,* **47**, 658–669.

Wang, M and Bonfield, W (2001) 'Chemically coupled hydroxyapatite–polyethylene composites: structure and properties' *Biomaterials,* **22**, 1311–1320.

Wang, X, Song, G and Lou, T (2010a) 'Fabrication and characterization of nano-composite scaffold of PLLA/silane modified hydroxyapatite' *Med Eng Phys,* **32**, 391–397.

Wang, Y, Dai, J, Zhang, Q, Xiao, Y and Lang, M (2010b) 'Improved mechanical properties of hydroxyapatite/poly(e-caprolactone) scaffolds by surface modification of hydroxyapatite' *Appl Surf Sci,* **256**, 6107–6112.

Wang, Y, Xiao, Y, Huang, X and Lang, M (2011a) 'Preparation of poly(methyl methacrylate) grafted hydroxyapatite nanoparticles via reverse ATRP' *J Colloid Interface Sci,* **360**, 415–421.

Wang, Y, Zhang, X, Yan, J, Xiao, Y and Lang, M (2011b) 'Surface modification of hydroxyapatite with poly(methyl methacrylate) via surface-initiated ATRP' *Appl Surf Sci,* **257**, 6233–6238.

Wang, M (2006) 'Composite scaffolds for bone tissue engineering' *Am J Biochem Biotech,* **2**, 80–84.

Wang, Y-W, Wu, Q, Chen, J and Chen, G-Q (2005) 'Evaluation of three-dimensional scaffolds made of blends of hydroxyapatite and poly(3-hydroxybutyrate-co-3-hydroxyhexanoate) for bone reconstruction' *Biomaterials,* **26**, 899–904.

Wang, Y-W, Wu, Q, Chen, J and Chen, G-Q (2004) 'Attachment proliferation and differentiation of osteoblasts on random biopolyester poly(3-hydroxybuterate-co-3-hydroxyvalerate) scaffolds' *Biomaterials,* **25**, 669–675.

Wang, M (2003) 'Developing bioactive composite materials for tissue replacement' *Biomaterials,* **24**, 2133–2151.

Wang, M, Berry, C, Braden, M and Bonfield, W (1998) 'Young's and shear moduli of ceramic particle filled polyethylene' *J Mater Sci, Mater Med,* **9**, 621–624.

Wang, M and Bonfield, W (1996) *Polymer Processing Towards AD2000*, Singapore, 203–204.

Wang, M, Porter, D and Bonfield, W (1994) 'Processing, characterization, and evaluation of hydroxyapatite reinforced polyethylene composites' *Brit Ceramic Transac,* **93**, 91–94.

Wei, G and Ma, P X (2004) 'Structure and properties of nano-hydroxyapatite/polymer composite scaffolds for bone tissue engineering' *Biomaterials,* **25**, 4749–4757.

Wu, C K A, Grøndahl, L, Jack, K S, Foo, M X, Trau, M, Hume, D A and Cassady, A I (2006) 'Reduction of the *in vitro* pro-inflammatory response by macrophages to poly(3-hydroxybutyrate-co-3-hydroxyvalerate)' *Biomaterials,* **27**, 4715–4725.

Wu, C, Ramaswamy, Y, Zhu, Y, Zheng, R, Appleyard, R, Howard, A and Zreiqat, H (2009) 'The effect of mesoporous bioactive glass on the physiochemical, biological and drug-release properties of poly(dl-lactide-co-glycolide) films' *Biomaterials,* **30**, 2199–2208.

Yang, S, Leong, K F, Du, Z and Chua, C K (2001) 'The design of scaffolds for use in tissue engineering. Part I. Traditional factors' *Tissue Eng,* **7**, 679–689.

Yunos, D M, Bretcanu, O and Baoccaccini, A R (2008) 'Polymer-bioceramic composites for tissue engineering scaffolds' *J Mater Sci,* **43**, 4433–4442.

Zeng, L, Wang, H, Fu, G, Jiang, J and Zhang, X (2010) 'A new approach for synthesis of the comb-shaped poly (ε-caprolactone) brushes on the surface of nano-hydroxyapatite by combination of ATRP and ROP' *J Colloid Interface Sci,* **352**, 36–42.

Zhang, P, Hong, Z, Yu, T, Chen, X and Jing, X (2009) '*In vivo* mineralization and osteogenesis of nanocomposite scaffold of poly(lactide-co-glycolide) and hydroxyapatite surface-grafted with poly(l-lactide)' *Biomaterials,* **30**, 58–70.

Zhou, S, Zhang, L, Wang, Y-Y, Zuo, Y, Gao, S-B and Li, Y-B (2012) 'A novel hydroxyapatite/ethylene-vinyl acetate/copolymer 66 composite for hard tissue regeneration' *J Macromol Sci B,* **51**, 1–11.

12

Marine organisms for bone repair and regeneration

S. A. CLARKE and P. WALSH, Queen's University Belfast, UK

DOI: 10.1533/9780857099037.3.294

Abstract: Bioprospecting has led to increased interest in potential applications for marine organisms and their by-products. As a rich source of mineralising porous organisms, our seas and oceans could provide new directions for bone tissue engineering, particularly in the supply of biomimetic templates that may enhance *in vivo* and *ex vivo* bone formation. In this chapter we examine the history of marine organism use in this field; exploring how these organisms could be utilised, given the problems of sustainability, and reviewing the current evidence to support their use for bone repair and regeneration.

Key words: mineralising marine organisms, calcium carbonate biomaterials, complex porous structures, bone substitute materials, bone tissue engineering.

12.1 Introduction

Bone is a composite consisting of an organic component (mainly collagens with some non-collagenous proteins) within an inorganic (calcium phosphate, CaP) matrix and this composition creates unique material properties which add up to more than the sum of the parts. This template is not unique in the animal kingdom and, in the marine environment, there are many organisms that exploit similar engineering to provide mineralised skeletons for support and function. Species richness and the relatively untapped potential of the world's oceans and seas has led to an explosion of interest in bioprospecting: exploiting these resources for applications in a wide range of areas from biofuels, electronics, food and agriculture to medicine, including for the support of bone repair.

Nowadays, there are a multitude of conventional and contemporary bone substitute materials on the market that are deemed capable of stimulating bone regeneration, however, their translation to the clinic remains stagnant. Poor or variable clinical outcomes make clinicians sceptical and uncertain about their use leading to the continued favour of autologous bone grafts. The complexity of the chemical, biological and mechanical interactions in tissue regeneration has set researchers a difficult task when designing bone substitute materials. Hence research continues for an ideal substitute material that emulates bone function. Marine derived and inspired bioceramics have shown good potential as substitute materials, offering the desirable physiochemical characteristics required for bone repair.

12.2 Why marine organisms?

There are a number of reasons why marine organisms could help in the quest to obtain reliable bone repair. Firstly, there is an array of organisms which produce some form of mineral within their anatomy ranging from the submicron silica frustules of diatoms to the aragonite skeletons of lobster and cuttlefish. The minerals found in marine organisms are generally silica-based or calcium carbonate in the form of calcite or aragonite (Table 12.1). Although not many utilise the CaP of mammalian bone, they could still be a useful resource, either by providing scaffolds directly to support bone growth for *ex vivo* or *in situ* tissue engineering or as study organisms to elucidate the mechanisms that allow mineral production at ambient temperatures in alkaline aqueous conditions. The mechanisms governing bone mineralisation, involving interactions between the supporting matrix and soluble factors in the surrounding milieu under the control of biomolecules and enzymes, are still poorly understood and are difficult to control or replicate *in vitro*. Some marine organisms, such as crayfish, which require rapid remineralisation during the moult cycle following shedding of their exoskeleton, use amorphous calcium carbonate as a precursor for mineralisation in ambient conditions (Akiva-Tal *et al.*, 2011). This strategy adopted by these animals is in contrast to the way we currently approach *in vivo* bone tissue engineering which is to provide a prefabricated, highly crystalline CaP scaffold rather than allowing mineral formation to occur *in situ*. Therefore, perhaps we can learn from these creatures by instead using poorly crystalline bone substitute materials (Huang *et al.*, 2010) or even using amorphous calcium carbonate precursors which have been shown to generate single crystal structures that retain micro- and nanometre features, thus allowing greater control and order (Li and Qi, 2008). This technology has the potential for use in a number of disciplines including the development of functional materials.

The second reason why marine organisms could be useful for bone repair is the porous nature of many of their structures. Early synthetic bone substitutes were often solid particles of CaP, usually hydroxyapatite (HA) which is closest to the mineral phase of bone but, although bioactive, these materials are difficult for the body to resorb and are still visible *in situ* many years after implantation. To improve bone integration into the synthetic material, Hulbert and Klawitter suggested the use of porous structures as early as

Table 12.1 Examples of mineralising marine organisms

Phylum	Class	Common name	Skeletal mineral
Heterokontophyta	Bacillariophyta	Diatoms	Silica
Haptophyta	Prymnesiophyceae	Coccolithophorids	$CaCO_3$
Radiolaria	Polycystina	Radiolarians	Silica
Foraminifera	Granuloreticulosa	Foraminiferans	$CaCO_3$
Rhodophyta	Florideophyceae	**Red algae**	$CaCO_3$/calcite
Ectoprocta	Stenolaemata	Bryozoans	$CaCO_3$
Porifera	Demospongia	**Sponges**	$CaCO_3$ and/or silica
Cnidarians	Anthozoa	**Corals**	$CaCO_3$/ aragonite
Mollusca	Gastropoda	**Snails**, limpets	$CaCO_3$/ aragonite
	Bivalvia	**Clams, mussels**	$CaCO_3$/ aragonite
	Cephalopods	Squid, **cuttlefish**	$CaCO_3$/ aragonite
Arthropoda	Crustacea	**Lobster, crab,** shrimp	$CaCO_3$/ aragonite
Brachiopoda	e.g., Lingulata	Lampshells	$CaCO_3$ or $CaPO_4$
Echinodermata	Asteroidea	**Starfish**	$CaCO_3$/calcite
	Echinoidea	**Sea urchins**	$CaCO_3$/calcite
	Ophiuroidea	Brittle stars	$CaCO_3$/calcite
	Holothuroidea	Sea cucumbers	$CaCO_3$/calcite

Note: Those that have already been investigated for bone tissue engineering are highlighted in bold.
Source: Reprinted from Clarke *et al.* (2011) with permission from Elsevier.

1970 (Hulbert *et al.*, 1970, 1973). The introduction of porosity not only created a dichotomy between mechanical strength and percentage porosity, but also sparked huge debate over optimum pore features: pore size, connectivity of the pores (i.e., whether they are open channels connecting to others or are closed cul-de-sacs), their tortuosity (the 'difficulty' of the route through the material) and the overall porosity (some materials have >90% porosity with very little actual ceramic providing a much larger surface area). Decades on, there has been little consensus on the optimum pore size either for resorption or for bone ingrowth (Bohner *et al.*, 2011), with suggestions ranging from mean pore sizes of 100 μm to as large as 500 μm diameter (Karageorgiou and Kaplan, 2005). More recently, the addition of microporosity (<10 μm) has been shown to enhance bone repair (Hing *et al.*, 2005; Lan Levengood *et al.*, 2010) perhaps by improving fluid flow and promoting neovascularisation (Cunningham *et al.*, 2010).

In truth, there is probably no single optimum pore size and a range of pore diameters and pore features may allow the material to function on a

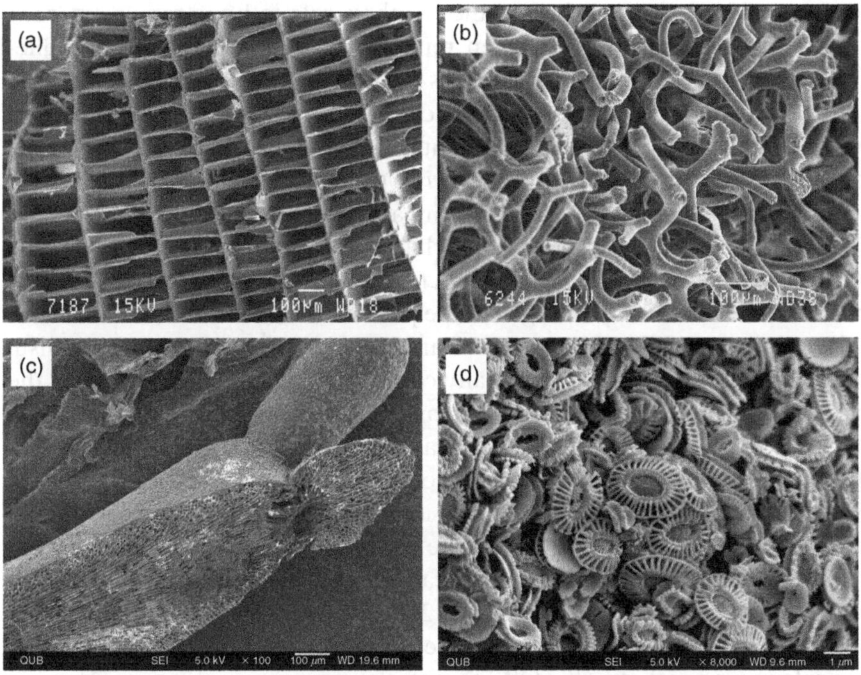

12.1 Scanning electron micrographs of (a) cuttlefish (*Sepia officinalis*) (b) sponge (*Spongia agaricina*) (c) red algae (*Corallina officinalis*) and (d) coccolithophores (*Emiliania huxleyi*) demonstrating a range of macro and microporous structures. (*Source*: *E. huxleyi* micrograph courtesy of Katherine Fee. Reprinted from Clarke *et al.* (2011) with permission from Elsevier.)

number of different levels. Marine organisms provide a natural reservoir of porous materials across the spectrum from submicron to millimetre-sized pores (Fig. 12.1) allowing them to be tailored for specific applications, for example to enhance bone ingrowth, material resorption or neovascularisation depending on the specific clinical requirements. Furthermore, the naturally occurring interconnectivity, pore size distribution and tortuosity provide unique templates for material design that cannot be replicated synthetically (Sharma and Elisseeff, 2004), allowing marine organisms to be used either as templates for material design (see Section 12.4), or directly as bone substitute materials (see Section 12.3).

Another area of research activity in improving CaP bone substitute materials has been in the addition of other ions to the ceramic base in what have been termed 'substituted CaPs'. The basis of this approach is that the mineral phase of bone is non-stoichiometric HA that contains a number of minor ions such as carbonate (CO_3^{2-}), magnesium (Mg^{2+}) and silicate (SiO_4^{4-}). Secondary ions, which were once classified as impurities, have now been recognised for their importance as their role in bone repair emerges, for example, magnesium has been reported to enhance osteoblast adhesion, angiogenesis in porous structures (Holzapfel *et al.*, 2012) and increase bioresorption (Fellah *et al.*, 2008; Dorozhkin, 2012). Those with the largest evidence base are carbonated CaPs and silicon-substituted CaPs. Both of these materials have been shown *in vivo* to enhance bone formation over pure CaP alone (Landi *et al.*, 2003; Mangano *et al.*, 2004; Spence *et al.*, 2008; Coathup *et al.*, 2011), however the mechanism is unknown and there is some controversy over whether it is a direct effect of the substituted ion or an indirect effect caused by changes to the physical characteristics of the material (Bohner, 2009; Habibovic and Barralet, 2011). Whichever mechanism is active, ion substitution is generally considered a positive addition to the base ceramic material and marine organisms are a rich source of these ions. For direct use of the organisms as scaffolds, they could provide access to alternative biogenic ions in the form of their base mineral. In other cases, marine organisms are a rich source of ions which can become incorporated into a biomaterial when converting the base mineral into CaP. For example, the aragonite skeleton of cuttlefish can be converted to carbonated HA (Ivankovic *et al.*, 2010; Kasioptas *et al.*, 2010; Rocha *et al.*, 2005) and two species of sea urchins rich in Mg^{2+} ions were found to produce Mg-substituted β-tricalcium phosphate (β-TCP) during the conversion process (Vecchio *et al.*, 2007b). By simply altering the biogenic source of calcite or aragonite, a range of ion substitutions could be possible, producing an array of materials with tailored properties using relatively simple manufacturing techniques compared to some manufacturing processes currently required to produce ion-substituted ceramics (Kannan *et al.*, 2007).

One major disadvantage of porous ceramic materials as bone substitutes is their relative lack of mechanical strength and their limited use in load bearing applications (Zhang *et al.*, 2007). A possible strategy for solving this problem is to mimic the complex structures of marine organisms. Perhaps the most studied material in this regard is nacre, the lining layer of many seashells, which has inspired several modifications incorporating the best that nature and synthetic chemistry has to offer (Ortiz and Boyce, 2008). Other potential structures providing inspiration include the exoskeletons of arthropods and brachiopods which not only consist of a composite of biomineral and chitin, reminiscent of the collagen and HA of bone (Giraud-Guille *et al.*, 2004), but they also have a complex 3D organisation. The arthropod shell has been described as 'twisted plywood' or a Bouligand structure which provides additional mechanical strength (Raabe *et al.*, 2005; Chen *et al.*, 2008) and the micro- and nano-hardness of calcitic shells of brachiopods is increased by the addition of chitin (Schmahl *et al.*, 2008). Attempting to recreate some of these complex structures is a necessary evolution of biomimetics which began many decades ago with the use of synthetic HA and the simple principle of replacing like with like. Indeed, concerns about sustainability of some of these creatures which have shown promise as osteoconductive matrices such as the seastar, *Pisaster giganteus*, (Martina *et al.*, 2005) may restrict the use of these organisms to that of study organisms rather than using them directly as bone substitutes.

12.3 Marine organisms used directly as biomaterials

Stony corals of the phylum Cnidaria have the longest history of use as a biomaterial for bone repair: its use was first proposed by White in the early 1970s (White *et al.*, 1972) and has been used clinically since the 1980s (Vago, 2008). Its most widespread use has been following conversion to a CaP bone substitute (reviewed in Section 12.4) but it has also been used in its natural form. Similar to CaPs, the calcium carbonate structure of coral is osteoconductive but has little inherent osteoinductive capacity. *In vitro*, corals have been shown to support human primary osteoblast and osteoclast growth (Ehrlich *et al.*, 2006) and proliferation and differentiation of a mouse-derived mesenchymal stem cell (MSC) line (Abramovitch-Gottlib *et al.*, 2006; Gravel *et al.*, 2006) using the species *Porites lutea* (Abramovitch-Gottlib *et al.*, 2006; Gravel *et al.*, 2006), *Millepora dichotoma* (Abramovitch-Gottlib *et al.*, 2006) and Anthozoa: Gorgonacea (Ehrlich *et al.*, 2006) either as a demineralised matrix (Ehrlich *et al.*, 2006), combined with chitosan in a composite (Gravel *et al.*, 2006) or in its natural state (Abramovitch-Gottlib *et al.*, 2006). Yet in studies investigating ectopic bone formation in non-skeletal sites, it was only with the addition of bone marrow derived stromal cells that new bone formation within the coral scaffold was seen (Ohgushi, 1997; Chen *et al.*, 2007).

In contrast to CaPs, calcium carbonate is much more soluble and is more quickly resorbed *in vivo*. Ohgushi (1997) noted more osteoclast formation and faster resorption of the natural calcium carbonate coral scaffold when compared to one that had been converted to HA (coralline HA) when implanted in a non-skeletal site but this phenomenon is also seen following skeletal implantation. After just 6–8 weeks implantation in either a cortical or a cancellous defect site, two different species of coral had been completely resorbed in a canine model (Guillemin *et al.*, 1987) and the fast resorption rate of coral was associated with non-healing of a sheep metatarsal defect that necessitated replacement of the scaffold with a coralline HA to provide longer term support of new bone growth (Bensaïd *et al.*, 2005). This resorption is affected by the size of the original implant and can be even further enhanced by the addition of growth factors (Geiger *et al.*, 2007) or whole bone marrow (Petite *et al.*, 2000) designed to augment bone repair but which were instead found to remove the scaffold too fast for new bone formation to occur. Therefore, use of coral clinically requires careful consideration of the size of the defect, coral particles and construct and could benefit from

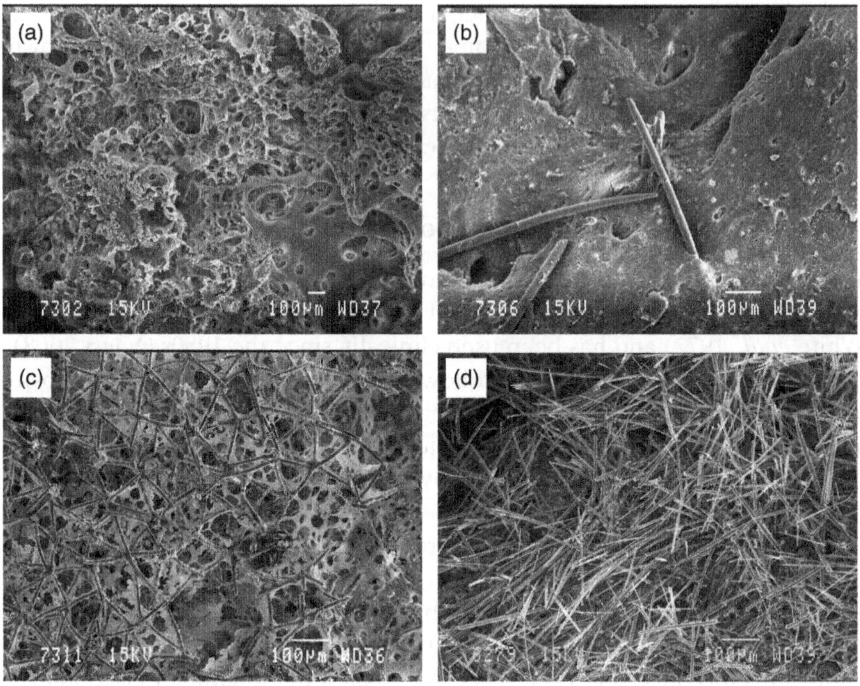

12.2 Scanning electron micrographs of marine demosponges showing varying spicule size, spicule shape and composition. (a): *Dysidea fragilis* (b): *Pachymatisma johnstonia* (c): *Haliclona simulans* (d): *Suberites ficus.*

the addition of autologous osteogenic cells which seem to speed up the formation of new bone to something approaching that of the resorption rate of coral to allow at least partial healing of the bone defect (Louisia *et al.*, 1999; Petite *et al.*, 2000; Cui *et al.*, 2007; Viateau *et al.*, 2007).

Few other organisms have been investigated for their ability to support bone growth in their natural state but those that have include seastars (phylum Echinodermata) (Martina *et al.*, 2005), cuttlefish (phylum Mollusca) (Okumuş and Yildirim, 2005) and sponges (phylum Porifera) (Green *et al.*, 2003; Abramovitch-Gottlib *et al.*, 2006; Ehrlich *et al.*, 2006; Zheng *et al.*, 2007; Schmahl *et al.*, 2008; Vago, 2008; Lin *et al.*, 2011). Each has their own unique chemical and physical properties which illustrate the diversity of materials that could be used to support bone growth. Coral, cuttlefish 'bone' and seastars all contain calcium carbonate: in coral and cuttlefish this takes the form of aragonite whereas the seastar mineral is magnesium rich-calcite (Martina *et al.*, 2005). Sponges are phylogenetically diverse: some have a spongin skeleton (a protein analogous to vertebrate collagen) with varying degrees of additional support provided by silicaceous spicules, some have calcite or aragonite spicules and some are spicule-free (Fig. 12.2). Of those used experimentally, *Spongia officinalis* consists of spongin fibres with only occasional, broken, foreign, silica spicules (Green *et al.*, 2003), the 'unidentified' sponges studied by Zheng *et al.* (2007) seem from the micrographs to be spicule-free and the *Callyspongiidae* species in (Lin *et al.*, 2011) has occasional, randomly orientated silica spicules. Recently, the discovery of a silica-aragonite-chitin biocomposite structure in a species of demosponge (Ehrlich *et al.*, 2010) has further increased the breadth of chemistries and mineralisation mechanisms available for exploitation in this invertebrate alone.

The four organisms used in these studies also provided a range of physical structures particularly in the diversity of pore sizes. *Pisaster giganteus*, the seastar, has pores less than 20 μm (Martina *et al.*, 2005) while coral and cuttlefish bone have pores in the range of 100–500 μm (Kasioptas *et al.*, 2010; Okumuş and Yildirim, 2005; Ben-Nissan, 2003), although the regular and uniform arrangement of cuttlefish pores ensures very different architecture to that of the randomly aligned and interconnected pores of the stony corals. Sponges have the largest porosity and pore size distribution of all with some species exhibiting a range from 100 to greater than 1000 μm diameter (Green *et al.*, 2003; Cunningham *et al.*, 2010).

Given this range of physiochemical properties in the potential scaffolds, it would be useful to compare their relative promise as bone graft materials; however this is difficult as each study employs different methodology, most have only been tested *in vitro* and not all *in vivo* studies use a positive control, such as autograft, as a universal comparator. Only cuttlefish and coral have been implanted in a bone defect site. Cuttlefish bone performed well in a metaphyseal defect in rabbits when compared to a bovine graft and both performed better than tricalcium phosphate (TCP) and demineralised bone matrix (Okumuş and Yildirim, 2005) but with only a 30-day implantation time and no direct measure of implant resorption it is difficult to know from this single study if this calcium carbonate-based material suffers from the same fast resorption issues seen with coral *in vivo*.

Seastar ossicles and several marine sponges (a total of seven species) have only been tested *in vitro* and all were capable of supporting bone cell attachment and differentiation using primary cell sources; human bone derived osteoblasts (Martina *et al.*, 2005); human bone marrow stromal cells (Green *et al.*, 2003; Martina *et al.*, 2005); and primary calvarial murine osteoblasts (Zheng *et al.*, 2007; Lin *et al.*, 2011). Some of these were qualitatively assessed and, without a standard control material, it is difficult to know how these results compare to other bone substitute materials and how they would translate to *in vivo* performance. For the sponges, in particular, there is a large range of chemical and physical properties within the same species and it is unlikely that all sponges will be suitable as bone scaffolds. Our own work suggests that some of the more spiculated sponges provide a hostile environment to cells (Fig. 12.2) and that cells preferentially attach to those sponges with a high spongin content (unpublished data).

12.4 Marine organisms used indirectly as biomaterials

Calcified marine organisms have evolved over millions of years for multipurpose functions. This has resulted in a diverse range of organic and inorganic precursors, with intricate structures and chemistries that are of interest in bone repair. Three indirect approaches, based on using marine organisms as a template to fabricate bone substitutes, have been investigated (Fan *et al.*, 2009): (1) chemical conversion of the inorganic marine matrix into a biomorphic substitute (Roy and Linnehan, 1974), (2) synthetic precursors cast using biomorphic moulds (White *et al.*, 1972, Cunningham *et al.*, 2010) and, more recently, (3) self-assembly of nanoparticles via chemical and molecular interactions, for example, electrostatic interactions (Espinosa *et al.*, 2009; Pouget *et al.*, 2009).

The first approach is the most commonly used method in the fabrication of marine materials for bone substitutes. It facilitates ionic substitution without destroying skeletal morphology of the marine organism. This is achieved in one of two ways, either through hydrothermal synthesis or by hydrothermal hot pressing. Hydrothermal synthesis involves heating the marine skeletons under alkaline conditions at a specific temperature-pressure in either a reaction vessel (Walsh *et al.*, 2008) or an autoclave (Roy and Linnehan, 1974; Kasioptas *et al.*, 2010). Processing conditions, such as temperature and pressure will significantly influence the material attributes of the resultant product (e.g., degree of crystallinity, grain size, specific surface area), as highlighted by the patent application of Jordanova-Spassova and co-workers (Jordanova-Spassova, 2002). This group demonstrated that it is possible to manipulate crystallite size (50–150 nm) during synthesis through temperature regulation (200–250°C). Hydrothermal hot pressing involves the solidification of synthetic HA powder, typically at 100–300°C (30 MPa, 2 h) (Hosoi *et al.*, 1996). This process was used in the fabrication of Pro-Osteon™ a commercially available bone substitute consisting of a coral skeleton (aragonite) with a synthetic HA coating (Chiroff *et al.*, 1975). The thickness of the coating, crystallinity, grain size and stoichiometry will all influence the bioresorption profile of these materials. Other coating techniques, such as electrophoretic deposition and thermal spraying, could also potentially be used to manipulate these properties (Dorozhkin, 2010), but as yet, have not been applied to marine skeletons.

Both hydrothermal synthesis and hydrothermal hot pressing require pretreatment to remove the organic phase of the marine skeleton. Pyrolysis, between 350 and 700°C, is the preferred choice as it also causes partial breakdown of the aragonite/calcite structure (aragonite: 300–450°C, calcite: > 600°C), which can improve the degree of conversion to HA. The synthesis of HA, or HA derivatives from synthetic or natural calcium precursors is relatively straightforward chemistry. It is formed by Ca^{2+} ions, surrounded by PO_4^{3-} and OH^- ions which fuse together in a hexagonal $P6_3/m$ (i.e., six-fold symmetry axis with three-fold helix and mirror plane) space group with specific cell dimensions (McLean and Nelson, 1982). In addition to various species of coral (Roy and Linnehan 1974; Sivakumar *et al.*, 1996; Hu *et al.*, 2001; Jinawath *et al.*, 2002; Murugan and Ramakrishna, 2004), variations of this technique have been used to convert the skeletons of clams (Vecchio *et al.*, 2007a), cuttlefish (Ivankovic *et al.*, 2010; Kasioptas *et al.*, 2010), red algae (Felicio-Fernandes and Laranjeira, 2000; Walsh *et al.*, 2008) and sea urchins (Vecchio *et al.*, 2007b; Marchegiani *et al.*, 2009) resulting in structures with various chemistries (Clarke *et al.*, 2011).

Other fabrication techniques have been reported that use marine skeletons (specifically of the sea snail *Pomacea lineata*) to successfully produce HA, such as microwave synthesis and precipitation (dePaula *et al.*, 2010). Techniques, such as sol-gel fabrication have been coupled with hydrothermal synthesis to improve the mechanical properties of these bioceramics and enable their use in load bearing applications (Ben-Nissan *et al.*, 2004).

The second conversion approach involves a casting technique to utilise the skeletal features of the marine organisms. This technique was first reported by White and co-workers in 1972, when they used coral as a template to produce bone substitute materials fabricated from TiO_2, Al_2O_3 and HA (White *et al.*, 1972). More recently, marine sponges have been used to produce templates whereby the structures of Dalmata Fina (*Spongia officinalis Linnaeus*, Adriatic Sea), Fina Silk (*Spongia zimocca*, Mediterranean) and Elephant Ear (*Spongia agaricina*, Caribbean) were successfully replicated with a HA slurry and slip cast technique (Cunningham *et al.*, 2010). The residual sponge material is burnt off by sintering at high temperatures.

The third approach is still very much in its developmental phase and uses a combination of the previous two approaches with novel chemistries to produce 'smart' bone substitutes, for example, pH- or temperature-sensitive (Espinosa *et al.*, 2009; Pouget *et al.*, 2009). Although, all of the approaches are relatively straightforward in terms of manufacturing, the influence of these variable material attributes on cell fate is unknown (Place *et al.*, 2009).

12.5 Components of marine organisms as biomaterial adjuncts

Most bone substitutes are osteoconductive but not osteoinductive and do not perform as well as autograft in a clinical environment. Because of this there has been much research to try to improve the performance of these substitutes. To this end, a number of components with a marine origin have been tested ranging from particles of shell linings (nacre), through the macromolecules of chitin and collagen, to smaller bioactives such as fucoidan and ulvan. Furthermore, there is a great deal of interest in the potential isolation of bioactive compounds with therapeutic, antimicrobial and antibacterial activity from marine organisms, particularly sponges. These could potentially be combined with bone substitute materials to provide long-term protection from post-surgical infections or may even have some inherent osteogenic capacity themselves (Lee *et al.*, 2011). Further discussion of these particular bioactive compounds, however, is beyond the scope of this article and the reader is directed elsewhere (D'Orazio *et al.*, 2012; Javed *et al.*, 2011; Lordan *et al.*, 2011).

Much of the work involving nacre has come from a single species of mollusc, *Pinctada maxima* and involves particles of crushed shell lining. Following early reports of an osteogenic effect of nacre particles in close proximity to osteoblasts *in vitro* (Silve *et al.*, 1992) and tight bony apposition to nacre particles with no inflammatory reaction *in vivo* (Lamghari *et al.*, 1999), nacre powder was used in periodontal surgery for the reconstruction of maxillary alveolar bone in eight patients (Atlan *et al.*, 1997). The authors report good results from biopsies at six months with bone ingrowth throughout the region, direct bone apposition to the nacre surface and gradual dissolution of the implant. The authors suggest that part of the success of this material is due to the release of soluble osteogenic factors from the nacre powder and subsequent publications from this group have demonstrated that water soluble fractions from nacre powder can increase alkaline phosphatase expression of MRC5 fibroblasts (Pereira Mouries *et al.*, 2002) and mineralisation and expression of some osteogenic genes in the murine osteoblast-like cell line MC3T3-E1 (Rousseau *et al.*, 2008). Not all results with this material have been positive however. Bahar *et al.* (2003) found that an inflammatory reaction did occur *in vivo* around nacre particles implanted in a subcutaneous, non-skeletal site unless the particles had been pretreated with calcium hydroxide – a treatment which presumably formed a CaP layer on the particle surface before implantation. Another limitation of the nacre powder could be its limited resorbability as it seems to be only partially resorbed by osteoclasts compared to native bone (Duplat *et al.*, 2007). Its removal *in vivo*, however, could involve other mechanisms and long-term studies are needed to investigate this.

The macromolecules chitin and collagen, which augment the mechanical properties of the mineral phase and provide nucleation sites for mineralisation in marine organisms, have been widely investigated for biomedical applications ranging from wound healing to tissue engineering. Chitin, and its chemical derivative chitosan, have been used experimentally on their own to fill bone defects (Muzzarelli *et al.*, 1993) but more often they form part of a composite material, combined with materials such as coral (Gravel *et al.*, 2006), HA (Maruyama and Ito, 1996; Malafaya and Reis, 2009), nano-HA (Kashiwazaki *et al.*, 2009; Thein-Han and Misra, 2009), β-TCP (Ito *et al.*, 1994) and poly(propylene carbonate) (Zhao *et al.*, 2002) or as part of a tri-composite with collagen and a ceramic filler (Zhang *et al.*, 2010; Lee *et al.*, 2012). Chitin and chitosan have been investigated for use as a bone substitute material but also as a biomaterial to support articular cartilage repair (Mattioli-Belmonte, *et al.*, 1999; Malafaya and Reis, 2009) and as a component of an injectible gel (Wang *et al.*, 2010) or CaP cement (Weir and Xu, 2010). This material shows huge potential as a biomaterial adjunct or a drug delivery system (Mattioli-Belmonte *et al.*, 1999) and is the focus of Chapter 10.

Collagen has been isolated from a number of marine sources such as jellyfish *Stomolophus nomurai meleagris* (Song *et al.*, 2006) and sponge, *Chondrosia reniformis* (Heinemann *et al.*, 2007). Collagen has been extensively investigated as a biomaterial for many decades. Mostly bovine derived, there are concerns about immunogenicity and zoonosis, in particular of the transmissible spongiform encephalopathies. Therefore collagen from marine sources has been proposed as a safer alternative. Hydrogels produced with collagen extracted from the jellyfish and sponge species above have been shown to produce porous structures which support cell attachment (Song *et al.*, 2006; Heinemann *et al.*, 2007) but more work is required to confirm the anticipated low immunogenicity before they could be available clinically.

Red algae, in particular mineralising species, have been investigated as potential bone graft materials (see Sections 12.4 and 12.7), but both green and brown algae may also have potential applications for bone repair in the form of the polysaccharides ulvan (derived from green seaweed) and fucoidan (derived from brown species). Ulvan has been shown to be multifunctional and could have anti-inflammatory, anticoagulant and antioxidant properties (Lahaye and Robic, 2007). It has also recently been combined with poly-D,L-lactic acid in a bone graft composite, where it has shown promise as a drug delivery system (Alves *et al.*, 2012) but these investigations are in the very early stages and no *in vivo* testing has been performed yet. Fucoidan is a sulphated polysaccharide with a similar breadth of therapeutic properties (Cho *et al.*, 2009) and potential drug delivery applications (Huang and Liu, 2012). In addition, this compound has the ability to sequester and augment the actions of fibroblast growth factors which are involved in tissue repair and angiogenesis (Matou *et al.*, 2002; Zemani *et al.*, 2005) and has shown osteogenic potential in *in vitro* studies with an osteosarcoma cell line, MG63s (Cho *et al.*, 2009), and human adipose-derived MSCs (Park *et al.*, 2012).

12.6 Commercially available marine-based products

Work on the use of corals in the 1970s resulted in the commercialisation of two marine derived products: Pro-Osteon™ (Biomet, USA) and Biocoral®

(Inoteb, Saint-Gonnery, France). These products are derived from different genera of stony corals (White and Shors 1989; Demers *et al.*, 2002), which have a similar skeletal morphology to human cancellous bone. Biocoral® is used in its pure mineral form, calcium carbonate, with limited processing that removes contaminants while retaining the original morphology and chemistry (Demers *et al.*, 2002). Pro-Osteon™ differs as it combines the natural morphology of coral with synthetic chemistry, to produce a substitute material that mimics the natural mineral phase of bone; however, the conversion is incomplete and the resultant biphasic material has an outer HA shell with an inner calcium carbonate core (White and Shors, 1989).

Although these products have been available for many years, their use in the clinic remains low. This is not surprising when several studies have raised concerns over the bioresorbability of coral substitute materials. In 1986, Holmes *et al.* reported poor resorption of Pro-Osteon™ 500 (derived from *Goniopora* coral and cited under Interpore 500, its former trade name) after implanting it into the proximal tibial metaphysis of dogs (Holmes *et al.*, 1986). Even after 1 year, only 66.6% resorption had occurred and the authors suggested that this may have limited the regeneration of new tissue. Others reported much faster resorption using Pro-Osteon™ in the femur of dogs (between 24% and 63% after 4 months) (Pollick *et al.*, 1995) and rabbits (25% resorption and 47% ingrowth after 3 months) (Leupold *et al.*, 2006). In 1998, an *in vivo* study using rats by Vuola *et al.* concluded that the resorption rate of Pro-Osteon™ 200 (Interpore 200) was unpredictable, varying from a few months to more than 1 year (Vuola *et al.*, 1998). Some have attributed the unpredictable resorption rates and unsuitability of coral derivatives as implant materials to the biphasic nature (surface HA: interior $CaCO_3$) of the material (Ben-Nissan, 2003) but this theory was not supported by a study performed in 2003, where modified Pro-Osteon™ was implanted in a bilateral metaphyseal-diaphyseal rabbit defect. A solid-state topotactic conversion (characterised by internal atomic displacements) was applied to the Pro-Osteon™ to achieve different layer thicknesses of HA but this failed to affect the resorption rate or volume of bone ingrowth in this model (Walsh *et al.*, 2003).

When comparing these studies, the lack of standardised models is apparent making it difficult to assess the true performance and capabilities of this material. However, in clinical trials the results were also found to be variable. Some authors reported no inflammatory response, with good osteoconductivity and osteointegration with no fibrous encapsulation using Pro-Osteon™ (Nunes *et al.*, 1997) although highlighting that the application of the material should be restricted to areas of low load bearing as a result of its inherent mechanical weakness and poor resorbability (Mahan and Carey, 1999). Yet others reported poor clinical results: Vuola *et al.* (2000) implanted Biocoral® into the iliac crest defects of ten patients. One implant became infected and was removed after 1.7 years. Out of the other nine implants, the authors only observed ingrowth in two patients and in 2 years none of the substitute material had resorbed fully. This variability in clinical performance may deter clinicians from using this material routinely.

The third commercially available product is derived from a species of mineralising red algae. AlgiPore® (also marketed as C GRAFT®, Algisorb® and AlgOss®) is a HA bone graft manufactured from *Corallina officinalis*. It is unlike its coral HA counterparts, as it is fabricated from a renewable feedstock, thus making it more commercially attractive. It was pioneered by Kasperk *et al.* (2005), following on from earlier work on the hydrothermal conversion of corals to bone graft substitutes (Roy and Linnehan, 1974). It gained FDA approval in 2003, after preclinical trials reported the long-term (up to 13 years) success rates of 95.6% in maxilla sinus grafting (Cottrell and Wolford, 1998; Ewers, 2005). This product differs from other bone graft substitute products on the market in that it contains a unique internal microporous (5–10 μm), interconnected morphology. A further advantage to this algal-derived HA is its high surface area, which can be beneficial when adding bioactive compounds or genes to enhance the therapeutic outcome; however, the high temperature and pressure synthesis methodologies employed in its manufacture produce a crystalline product that has similar concerns about resorbability as those found with coral-based products (Ewers, 2005).

12.7 Commercialisation concerns

It is currently estimated that more than 500 000 bone-grafting procedures are carried out each year in the United States alone (der Stok *et al.*, 2011). This figure is set to increase with a rapidly ageing population, heightening the need for more 'off-the-shelf' bone graft substitutes that are cost-effective and deliver on clinical performance. This creates a need to simplify material design and manufacture, thus reconciling clinical needs with financial imperatives (Place *et al.*, 2009). The unique morphologies and structures of marine materials, could offer simple solutions to complex problems; however, it would be misguided to assume that corals and other marine organisms are cheap and in plentiful supply. Although marine habitats are relatively unexploited resources, they can only be considered commercially viable if they are sustainable with minimal environmental impact. Products such as AlgiPore® that are fabricated from sustainable feedstocks, therefore hold significant potential in the rapidly growing field of bone repair. Although it is currently the only algal-derived HA on the market, other studies have demonstrated the potential of mineralised algae derivatives in bone repair (Turhani *et al.*, 2005; Walsh *et al.*, 2008).

Porites and *Goniopora* corals that are used in Pro-Osteon™ and Biocoral®, are from the scleractinian coral species which are harvested mainly from the Pacific and Indian Oceans (Shors, 1999). Although these genera are not classified as endangered (Shamberger *et al.*, 2011), coral reefs are under threat from climate change (Tambutté *et al.*, 2011), environmental stresses (Lough and Cooper, 2011) and overexploitation (Harriott, 2003). Recent studies on coral reefs have indicated a rapid decrease (Tambutté *et al.*, 2011) and, in some instances, cessation (Carricart-Ganivet, 2011) in calcification, which has been attributed to ocean warming and acidification (Tambutté *et al.*, 2011). A few studies, however, have reported opposite trends in calcification, suggesting that elevated seawater temperatures are increasing the rate of calcification (Harriott, 2003). Under- and over-estimation in calcification rates could be attributed to spatial environmental gradients, such as depth or, more plausibly, by anomalies in the measurement of skeletal extension rates in corals (Lough and Cooper, 2011). Regardless, scleractinian coral is essentially a non-renewable resource that should be protected, as recognised by the Convention on International Trade in Endangered Species (CITES) treaty (Harriott, 2003). This makes the future of all coral-derived products uncertain. In addition, harvesting of corals, or any marine organism can not only have direct effects on ecosystems by, for example, reducing native populations so that they are replaced by other taxa (Cole *et al.*, 2011), but also indirect effects, such as depleting fish feedstocks (Brown and Taylor, 1999). Therefore, marine preservation should always be taken into consideration when selecting marine organisms for commercial applications.

12.8 Marine organisms as inspiration

In their simplest terms, mineralised biological tissues consist of an organic matrix reinforced by an inorganic phase. However, years of evolution have produced a hierarchical structure that give much greater mechanical properties than any that have been achieved by man synthetically (Chen *et al.*, 2008). It achieves a trade-off between strength, toughness, porosity, stiffness and elasticity. Collagen molecules form linear arrays producing fibre strands with minerals impregnated between the interstitial pockets of the molecules. The fibre strands entwine, forming a cross-linked matrix giving it toughness and elasticity. These properties are enhanced further by the mineral phase providing rigidity and strength for load-bearing functions (Vallet-Regi and Gonzalez-Calbet, 2004). Thousands of studies have been performed in an attempt to find the 'ideal' bone substitute material, based on adaptations on this matrix. A wide variety of materials and processing methodologies have been explored including a range of biodegradable polymers, bioactive ceramics, conventional scaffold fabrication techniques such as solvent casting, fibre knitting, freeze-drying, gas foaming and rapid-prototyping techniques, for example, fused deposition modelling/electrospinning (Leong *et al.*, 2003; Liu *et al.*, 2007). Yet, all of these composite attempts have dismally failed to replicate mineralised biological tissue. Problems associated with these technologies include ease and cost of manufacture, poor material properties and the impractically of implanting precision blocks of materials to irregularly sized defects. For example, rapid-prototyping has proven successful in facial reconstruction of complex fractures, whereby the defect area is mapped out using CT/MRI scans and reproduced using computer-aided design/modelling packages coupled with rapid-prototyping (Bagaria *et al.*, 2011), however, the cost of using this technology for routine surgeries is prohibitive.

More recently, sophisticated techniques such as self-assembly and 'intelligent' biomaterials, which supposedly function in tune with the innate repair mechanisms of the body (Kopeček, 2007; Pérez *et al.*, 2012) have shown good promise (Burns, 2009; Place *et al.*, 2009). Biomimetic approaches have tended to focus on material attributes, such as pore features, with little or no attention given to their unique physiochemical attributes (Mayer, 2005) and the interplay with other biological factors, for example, enzymes. New approaches must address physical and chemical material attributes as a system approach and their effect on cell fate to develop the materials of tomorrow.

12.9 Conclusions

Several strategies could be and have been employed in the application of marine organisms to the field of bone repair, ranging from the direct use of porous materials to support bone growth, through chemical and physical conversion of the mineralised structures to bone-mimicking ceramics, to extraction of individual elements as biomaterial adjuncts. Yet, although many marine organisms have been investigated for use as bone substitute materials or for their ability to provide scaffolds for bone tissue engineering, this area of research remains limited and the oceans are still an untapped resource to a large extent.

With a better understanding of the mechanisms of biomineralisation and the potential of manipulating intrinsic material chemistries and geometries to control biological outcomes, the range of materials found in the marine environment could prove highly adaptable and accelerate this research field towards its ultimate goal which is the development of 'smart biomaterials' that are application, site and patient specific. With this the potential of marine material in bone tissue repair may finally be unlocked.

12.10 Sources of further information

Ehrlich, H. (2010) *Biological Materials of Marine Origin,* Dordrecht, Springer. ISBN 978-94-007-5729-5

Handbook of Biomineralisation – 3 volumes, edited by Edmund Bäuerlein, Peter Behrens, Matthias Epple, Wiley. ISBN: 978-3-527-31641-0

12.11 References

Abramovitch-Gottlib, L., Geresh, S. and Vago, R. 2006. Biofabricated Marine Hydrozoan: A Bioactive Crystalline Material Promoting Ossification of Mesenchymal Stem Cells. *Tissue Engineering*, **12**, 729–739.

Akiva-Tal, A., Kababya, S., Balazs, Y. S., Glazer, L., Berman, A., Sagi, A. and Schmidt, A. 2011. In Situ Molecular NMR Picture of Bioavailable Calcium Stabilized As Amorphous Caco3 Biomineral in Crayfish Gastroliths. *Proceedings of tthe National Academy of Sciences of the United States of America*, **108**, 14763–14768.

Alves, A., Pinho, E. D., Neves, N. M., Mano, J. M., Sousa, R. A. and Reis, R. L. 2008. Ulvan: A New Natural Derived Biomaterial Obtained from Renewable Marine Resources. *Tissue Engineering Part A*, **14**, 793–793.

Atlan, G., Delattre, O., Berland, S., Lefaou, A., Nabias, G., Cot, D. and Lopez, E. 1999. Interface Between Bone and Nacre Implants in Sheep. *Biomaterials*, **20**, 1017–1022.

Bagaria, V., Deshpande, S., Rasalkar, D. D., Kuthe, A. and Paunipagar, B. K. 2011. Use of Rapid Prototyping and Three-Dimensional Reconstruction Modeling in the Management of Complex Fractures. *European Journal of Radiology*, **80**, 814–820.

Bahar, H., Yaffe, A. and Binderman, I. 2003. The Influence of Nacre Surface and Its Modification on Bone Apposition: A Bone Development Model on Rats. *Journal of Periodontology*, **74**, 366–371.

Ben-Nissan, B. 2003. Natural Bioceramics: From Coral to Bone and Beyond. *Current Opinion in Solid State and Materials Science*, **7**, 283–288.

Ben-Nissan, B., Milev, A. and Vago, R. 2004. Morphology of Sol-Gel Derived Nano-Coated Coralline Hydroxyapatite. *Biomaterials*, **25**, 4971–4975.

Bensaid, W., Oudina, K., Viateau, V., Potier, E., Bousson, V., Blanchat, C., Sedel, L., Guillemin, G. and Petite, H. 2005. De Novo Reconstruction of Functional Bone by Tissue Engineering in the Metatarsal Sheep Model. *Tissue Engineering*, **11**, 814–824.

Bohner, M. 2009. Silicon-Substituted Calcium Phosphates – A Critical View. *Biomaterials*, **30**, 6403–6406.

Bohner, M., Loosli, Y., Baroud, G. and Lacroix, D. 2011. Commentary: Deciphering the Link Between Architecture and Biological Response of a Bone Graft Substitute. *Acta Biomaterialia*, **7**, 478–484.

Brown, P. J. and Taylor, R. B. 1999. Effects of Trampling by Humans on Animals Inhabiting Coralline Algal Turf in the Rocky Intertidal. *Journal of Experimental Marine Biology and Ecology*, **235**, 45–53.

Burns, J. W. 2009. Biology Takes Centre Stage. *Nature Materials*, **8**, 441–443.

Carricart-Ganivet, J. P. 2011. Coral Skeletal Extension Rate: An Environmental Signal or a Subject to Inaccuracies? Journal of Experimental Marine Biology and Ecology, 405, 73–79.

Chen, F., Feng, X., Wu, W., Ouyang, H., Gao, Z., Cheng, X., Hou, R. and Mao, T. 2007. Segmental Bone Tissue Engineering by Seeding Osteoblast Precursor Cells into Titanium Mesh-Coral Composite Scaffolds. *International Journal of Oral and Maxillofacial Surgery*, **36**, 822–827.

Chen, P. Y., Lin, A. Y. M., Lin, Y. S., Seki, Y., Stokes, A. G., Peyras, J., Olevsky, E. A., Meyers, M. A. and Mckittrick, J. 2008a. Structure and Mechanical Properties of Selected Biological Materials. *Journal of the Mechanical Behavior of Biomedical Materials*, **1**, 208–226.

Chen, P. Y., Lin, A. Y. M., Mckittrick, J. and Meyers, M. A. 2008b. Structure and Mechanical Properties of Crab Exoskeletons. *Acta Biomaterialia*, **4**, 587–596.

Chiroff, R. T., White, E. W., Weber, J. N. and Roy, D. M. 1975. Tissue Ingrowth of Replamineform Implants. *Journal of Biomedical Materials Research*, **9**, 29–45.

Cho, Y. S., Jung, W. K., Kim, J. A., Choi, I. W. and Kim, S. K. 2009. Beneficial Effects of Fucoidan on Osteoblastic Mg-63 Cell Differentiation. *Food Chemistry*, **116**, 990–994.

Clarke, S. A., Walsh, P., Maggs, C. A. and Buchanan, F. 2011. Designs from the Deep: Marine Organisms for Bone Tissue Engineering. *Biotechnology Advances*, **29**, 610–617.

Cole, V. J., Mcquaid, C. D. and Nakin, M. D. V. 2011. Marine Protected Areas Export Larvae of Infauna, but Not of Bioengineering Mussels to Adjacent Areas. *Biological Conservation*, **144**, 2088–2096.

Cottrell, D. A. and Wolford, L. M. 1998. Long-Term Evaluation of the Use of Coralline Hydroxyapatite in Orthognathic Surgery. *Journal of Oral and Maxillofacial Surgery*, **56**, 935–941.

Cui, L., Liu, B., Liu, G., Zhang, W., Cen, L., Sun, J., Yin, S., Liu, W. and Cao, Y. 2007. Repair of Cranial Bone Defects with Adipose Derived Stem Cells and Coral Scaffold in a Canine Model. *Biomaterials*, **28**, 5477–5486.

Cunningham, E., Dunne, N., Walker, G., Maggs, C., Wilcox, R. and Buchanan, F. 2010. Hydroxyapatite Bone Substitutes Developed Via Replication of Natural Marine Sponges. *Journal of Materials Science-Materials in Medicine*, **21**, 2255–2261.

D'orazio, N., Gammone, M. A., Gemello, E., De Girolamo, M., Cusenza, S. and Riccioni, G. 2012. Marine Bioactives: Pharmacological Properties and Potential Applications Against Inflammatory Diseases. *Marine Drugs*, **10**, 812–833.

Demers, C., Hamdy, C. R., Corsi, K., Chellat, F., Tabrizian, M. and Yahia, L. 2002. Natural Coral Exoskeleton as a Bone Graft Substitute: A Review. *Bio-Medical Materials and Engineering*, **12**, 15–35.

Depaula, S. M., Huila, M. F. G., Araki, K. and Toma, H. E. 2010. Confocal Raman and Electronic Microscopy Studies on the Topotactic Conversion of Calcium Carbonate from Pomacea Lineate Shells into Hydroxyapatite Bioceramic Materials in Phosphate Media. *Micron*, **41**, 983–989.

Dorozhkin, S. V. 2010. Bioceramics of Calcium Orthophosphates. *Biomaterials*, **31**, 1465–1485.

Dorozhkin, S. V. 2012. Biphasic, Triphasic and Multiphasic Calcium Orthophosphates. *Acta Biomaterialia*, **8**, 963–977.

Duplat, D., Chabadel, A., Gallet, M., Berland, S., Bedouet, L., Rousseau, M., Kamel, S., Milet, C., Jurdic, P., Brazier, M. and Lopez, E. 2007. The *In Vitro* Osteoclastic Degradation of Nacre. *Biomaterials*, **28**, 2155–2162.

Ehrlich, H., Etnoyer, P., Litvinov, S. D., Olennikova, M. M., Domaschke, H., Hanke, T., Born, R., Meissner, H. and Worch, H. 2006. Biomaterial Structure in Deep-Sea Bamboo Coral (Anthozoa : Gorgonacea : Isididae): Perspectives for the Development of Bone Implants and Templates for Tissue Engineering. *Materialwissenschaft Und Werkstofftechnik*, **37**, 552–557.

Ehrlich, H., Simon, P., Carrillo-Cabrera, W., Bazhenov, V. V., Botting, J. P., Ilan, M., Ereskovsky, A. V., Muricy, G., Worch, H., Mensch, A., Born, R., Springer, A., Kummer, K., Vyalikh, D. V., Molodtsov, S. L., Kurek, D., Kammer, M., Paasch, S. and Brunner, E. 2010. Insights into Chemistry of Biological Materials: Newly Discovered Silica-Aragonite-Chitin Biocomposites in Demosponges. *Chemistry of Materials*, **22**, 1462–1471.

Espinosa, H. D., Rim, J. E., Barthelat, F. and Buehler, M. J. 2009. Merger of Structure and Material in Nacre and Bone – Perspectives on De Novo Biomimetic Materials. *Progress in Materials Science*, **54**, 1059–1100.

Ewers, R. 2005. Maxilla Sinus Grafting With Marine Algae Derived Bone Forming Material: A Clinical Report of Long-Term Results. *Journal of Oral and Maxillofacial Surgery*, **63**, 1712–1723.

Fan, T. X., Chow, S. K. and Di, Z. 2009. Biomorphic Mineralization: From Biology to Materials. *Progress in Materials Science*, **54**, 542–659.

Felicio-Fernandes, G. and Laranjeira, M. C. M. 2000. Calcium Phosphate Biomaterials from Marine Algae. *Hydrothermal Synthesis and Characterisation. Quimica Nova*, **23**, 441–446.

Fellah, B. H., Gauthier, O., Weiss, P., Chappard, D. and Layrolle, P. 2008. Osteogenicity of Biphasic Calcium Phosphate Ceramics and Bone Autograft in a Goat Model. *Biomaterials*, **29**, 1177–1188.

Geiger, F., Lorenz, H., Xu, W., Szalay, K., Lorenz, H., Claes, L., Augat, P. and Richter, W. 2007. VEGF Producing Bone Marrow Stromal Cells (BMSC) Enhance Vascularization and Resorption of a Natural Coral Bone Substitute. *Bone*, **41**, 516–522.

Giraud-Guille, M. M., Belamie, E. and Mosser, G. 2004. Organic and Mineral Networks in Carapaces, Bones and Biomimetic Materials. *Comptes Rendus Palevol*, **3**, 503–513.

Gravel, M., Vago, R. and Tabrizian, M. 2006. Use of Natural Coralline Biomaterials as Reinforcing and Gas-Forming Agent for Developing Novel Hybrid Biomatrices: Microarchitectural and Mechanical Studies. *Tissue Engineering*, **12**, 589–600.

Green, D., Howard, D., Yang, X., Kelly, M. and Oreffo, R. O. C. 2003. Natural Marine Sponge Fiber Skeleton: A Biomimetic Scaffold for Human Osteoprogenitor Cell Attachment, Growth, and Differentiation. *Tissue Engineering*, **9**, 1159–1166.

Guillemin, G., Patat, J. L., Fournie, J. and Chetail, M. 1987. the Use of Coral As A Bone-Graft Substitute. *Journal of Biomedical Materials Research*, **21**, 557–567.

Habibovic, P. and Barralet, J. E. 2011. Bioinorganics and Biomaterials: Bone Repair. *Acta Biomaterialia*, **7**, 3013–3026.

Harriott, V. J. 2003. Can Corals Be Harvested Sustainably? Ambio, 32, 130–133.

Heinemann, S., Ehrlich, H., Douglas, T., Heinemann, C., Worch, H., Schatton, W. and Hanke, T. 2007. Ultrastructura Studies on the Collagen of the Marine Sponge Chondrosia Reniformis Nardo. *Biomacromolecules*, **8**, 3452–3457.

Hing, K. A., Annaz, B., Saeed, S., Revell, P. A. and Buckland, T. 2005. Microporosity Enhances Bioactivity of Synthetic Bone Graft Substitutes. *Journal of Materials Science-Materials in Medicine*, **16**, 467–475.

Holmes, R. E., Bucholz, R. W. and Mooney, V. 1986. Porous Hydroxyapatite as a Bone-Graft Substitute in Metaphyseal Defects – A Histometric Study. *Journal of Bone and Joint Surgery-American Volume*, **68a**, 904–911.

Holzapfel, B. M., Reichert, J. C., Schantz, J. T., Gbureck, U., Rackwitz, L., Noth, U., Jakob, F., Rudert, M., Groll, J. and Hutmacher, D. W. 2013. How Smart Do Biomaterials Need To Be? A Translational Science and Clinical Point of View. *Advanced Drug Delivery Reviews*, **65**, 581–603.

Hosoi, K., Hashida, T., Takahashi, H., Yamasaki, N. and Korenaga, T. 1996. New Processing Technique for Hydroxyapatite Ceramics by the Hydrothermal Hot–Pressing Method. *Journal of the American Ceramic Society*, **79**, 2771–2774.

Hu, J., Russell, J. J., Ben-Nissan, B. and Vago, R. 2001. Production and Analysis of Hydroxyapatite from Australian Corals Via Hydrothermal Process. *Journal of Materials Science Letters*, **20**, 85–87.

Huang, M. S., Wu, H. D., Teng, N. C., Peng, B. Y., Wu, J. Y., Chang, W. J., Yang, J. C., Chen, C. C. and Lee, S. Y. 2010. In Vivo Evaluation of Poorly Crystalline Hydroxyapatite-Based Biphasic Calcium Phosphate Bone Substitutes for Treating Dental Bony Defects. *Journal of Dental Sciences*, **5**, 100–108.

Huang, Y. C. and Liu, T. J. 2012. Mobilization of Mesenchymal Stem Cells by Stromal Cell-Derived Factor-1 Released from Chitosan/Tripolyphosphate/Fucoidan Nanoparticles. *Acta Biomaterialia*, **8**, 1048–1056.

Hulbert, S. F., Cooke, F. W., Klawitte.Jj, Leonard, R. B., Sauer, B. W., Moyle, D. D. and Skinner, H. B. 1973. Attachment of Prostheses to Musculoskeletal System by Tissue Ingrowth and Mechanical Interlocking. *Journal of Biomedical Materials Research*, **7**, 1–23.

Hulbert, S. F., Young, F. A., Mathews, R. S., Klawitter, J. J., Talbert, C. D. and Stelling, F. H. 1970. Potential of Ceramic Materials as Permanently Implantable Skeletal Prostheses. *Journal of Biomedical Materials Research*, **4**, 433–56.

Ito, M., Yamagishi, T., Yagasaki, H., Hashem, A. and Oshida, Y. 1994. Experimental Development of A Chitosan Bonded Beta-Tricalcium Phosphate Bone Filling Paste. *Journal of Dental Research*, **73**, 138–138.

Ivankovic, H., Tkalcec, E., Orlic, S., Ferrer, G. G. and Schauperl, Z. 2010. Hydroxyapatite Formation from Cuttlefish Bones: Kinetics. *Journal of Materials Science-Materials in Medicine*, **21**, 2711–2722.

Javed, F., Qadir, M. I., Janbaz, K. H. and Ali, M. 2011. Novel Drugs from Marine Microorganisms. *Critical Reviews in Microbiology*, **37**, 245–249.

Jinawath, S., Pongkao, D. and Yoshimura, M. 2002. Hydrothermal Synthesis of Hydroxyapatite from Natural Source. *Journal of Materials Science-Materials in Medicine*, **13**, 491–494.

Jordanova-Spassova, M. 2002. Hydroxyapatite Material Containing Tricalcium Phosphate With Microporous Structure. US Patent No. 0114755 A1.

Kannan, S., Rocha, J. H. G., Agathopoulos, S. and Ferreira, J. M. F. 2007. Fluorine-Substituted Hydroxyapatite Scaffolds Hydrothermally Grown from Aragonitic Cuttlefish Bones. *Acta Biomaterialia*, **3**, 243–249.

Karageorgiou, V. and Kaplan, D. 2005. Porosity of 3d Biomaterial Scaffolds and Osteogenesis. *Biomaterials*, **26**, 5474–5491.

Kashiwazaki, H., Kishiya, Y., Matsuda, A., Yamaguchi, K., Iizuka, T., Tanaka, J. and Inoue, N. 2009. Fabrication of Porous Chitosan/Hydroxyapatite Nanocomposites: Their Mechanical and Biological Properties. *Bio-Medical Materials and Engineering*, **19**, 133–140.

Kasioptas, A., Geisler, T., Putnis, C. V., Perdikouri, C. and Putnis, A. 2010. Crystal Growth of Apatite by Replacement of an Aragonite Precursor. *Journal of Crystal Growth*, **312**, 2431–2440.

Kasperk, C., Ewers, R., Simons, B. and Kasperk, R. 1988. Algae-Derived (Phycogene) Hydroxylapatite – A Comparative Histological Study. *International Journal of Oral and Maxillofacial Surgery*, **17**, 319–324.

Kopecek, J. 2007. Hydrogel Biomaterials: A Smart Future? Biomaterials, 28, 5185–5192.

Lahaye, M. and Robic, A. 2007. Structure and Functional Properties of Ulvan, a Polysaccharide from Green Seaweeds. *Biomacromolecules*, **8**, 1765–1774.

Lamghari, M., Almeida, M. J., Berland, S., Huet, H., Laurent, A., Milet, C. and Lopez, E. 1999. Stimulation of Bone Marrow Cells and Bone Formation by Nacre: *In Vivo* and *In Vitro* Studies. *Bone*, **25**, 91s–94s.

Lee, S. B., Kwon, J. S., Lee, Y. K., Kim, K. M. and Kim, K. N. 2012. Bioactivity and Mechanical Properties of Collagen Composite Membranes Reinforced by Chitosan and Beta-Tricalcium Phosphate. *Journal of Biomedical Materials Research Part B-Applied Biomaterials*, **100b**, 1935–1942.

Lee, S. U., Kwak, H. B., Pi, S. H., You, H. K., Byeon, S. R., Ying, Y. C., Luesch, H., Hong, J. Y. and Kim, S. H. 2011. *In Vitro* and *In Vivo* Osteogenic Activity of Largazole. *Acs Medicinal Chemistry Letters*, **2**, 248–251.

Leong, K. F., Cheah, C. M. and Chua, C. K. 2003. Solid Freeform Fabrication of Three-Dimensional Scaffolds for Engineering Replacement Tissues and Organs. *Biomaterials*, **24**, 2363–2378.

Leupold, J. A., Barfield, W. R., An, Y. H. and Hartsock, L. A. 2006. A Comparison of Pro-Osteon, Dbx, and Collagraft in a Rabbit Model. *Journal of Biomedical Materials Research Part B-Applied Biomaterials*, **79b**, 292–297.

Levengood, S. K. L., Polak, S. J., Wheeler, M. B., Maki, A. J., Clark, S. G., Jamison, R. D. and Johnson, A. J. W. 2010. Multiscale Osteointegration as a New Paradigm for the Design of Calcium Phosphate Scaffolds for Bone Regeneration. *Biomaterials*, **31**, 3552–3563.

Li, C. and Qi, L. M. 2008. Bioinspired Fabrication of 3d Ordered Macroporous Single Crystals of Calcite from a Transient Amorphous Phase. *Angewandte Chemie-International Edition*, **47**, 2388–2393.

Lin, Z., Solomon, K. L., Zhang, X. L., Pavlos, N. J., Abel, T., Willers, C., Dai, K. R., Xu, J. K., Zheng, Q. J. and Zheng, M. H. 2011. *In Vitro* Evaluation of Natural Marine Sponge Collagen as a Scaffold for Bone Tissue Engineering. *International Journal of Biological Sciences*, **7**, 968–977.

Liu, C., Xia, Z. and Czernuszka, J. T. 2007. Design and Development of Three-Dimensional Scaffolds for Tissue Engineering. *Chemical Engineering Research and Design*, **85**, 1051–1064.

Lordan, S., Ross, R. P. and Stanton, C. 2011. Marine Bioactives as Functional Food Ingredients: Potential to Reduce the Incidence of Chronic Diseases. *Marine Drugs*, **9**, 1056–1100.

Lough, J. M. and Cooper, T. F. 2011. New Insights from Coral Growth Band Studies in an Era of Rapid Environmental Change. *Earth-Science Reviews*, **108**, 170–184.

Louisia, S., Stromboni, M., Meunier, A., Sedel, L. and Petite, H. 1999. Coral Grafting Supplemented with Bone Marrow. *Journal of Bone and Joint Surgery-British Volume*, **81b**, 719–724.

Mahan, K. T. and Carey, M. J. 1999. Hydroxyapatite as a Bone Substitute. *Journal of the American Podiatric Medical Association*, **89**, 392–397.

Malafaya, P. B. and Reis, R. L. 2009. Bilayered Chitosan-Based Scaffolds for Osteochondral Tissue Engineering: Influence of Hydroxyapatite on *In Vitro* Cytotoxicity and Dynamic Bioactivity Studies in a Specific Double-Chamber Bioreactor. *Acta Biomaterialia*, **5**, 644–660.

Marchegiani, F., Cibej, E., Vergni, P., Tosi, G., Fermani, S. and Falini, G. 2009. Hydroxyapatite Synthesis from Biogenic Calcite Single Crystals into Phosphate Solutions at Ambient Conditions. *Journal of Crystal Growth*, **311**, 4219–4225.

Martina, M., Subramanyam, G., Weaver, J. C., Hutmacher, D. W., Morse, D. E. and Valiyaveettil, S. 2005. Developing Macroporous Bicontinuous Materials as Scaffolds for Tissue Engineering. *Biomaterials*, **26**, 5609–5616.

Maruyama, M. and Ito, M. 1996. *In Vitro* Properties of a Chitosan-Bonded Self-Hardening Paste with Hydroxyapatite Granules. *Journal of Biomedical Materials Research*, **32**, 527–532.

Matou, S., Helley, D., Chabut, D., Bros, A. and Fischer, A. M. 2002. Effect of Fucoidan on Fibroblast Growth Factor-2-Induced Angiogenesis *In Vitro*. *Thrombosis Research*, **106**, 213–221.

Mattioli-Belmonte, M., Gigante, A., Muzzarelli, R. A. A., Politano, R., De Benedittis, A., Specchia, N., Buffa, A., Biagini, G. and Greco, F. 1999. N,N-Dicarboxymethyl Chitosan as Delivery Agent for Bone Morphogenetic Protein in the Repair of Articular Cartilage. *Medical and Biological Engineering and Computing*, **37**, 130–134.

Mayer, G. 2005. Rigid Biological Systems as Models for Synthetic Composites. *Science*, **310**, 1144–1147.

Mclean, J. D. and Nelson, D. G. A. 1982. High-Resolution N-Beam Lattice Images of Hydroxyapatite. *Micron*, **13**, 409–413.

Mouries, L. P., Almeida, M. J., Milet, C., Berland, S. and Lopez, E. 2002. Bioactivity of Nacre Water-Soluble Organic Matrix from the Bivalve Mollusk Pinctada Maxima in Three Mammalian Cell Types: Fibroblasts, Bone Marrow Stromal Cells and Osteoblasts. *Comparative Biochemistry and Physiology B-Biochemistry and Molecular Biology*, **132**, 217–229.

Murugan, R. and Ramakrishna, S. 2004. Coupling of Therapeutic Molecules Onto Surface Modified Coralline Hydroxyapatite. *Biomaterials*, **25**, 3073–3080.

Muzzarelli, R. A. A., Zucchini, C., Ilari, P., Pugnaloni, A., Belmonte, M. M., Biagini, G. and Castaldini, C. 1993. Osteoconductive Properties of Methylpyrrolidinone Chitosan in an Animal-Model. *Biomaterials*, **14**, 925–929.

Nunes, C. R., Simske, S. J., Sachdeva, R. and Wolford, L. M. 1997. Long-Term Ingrowth and Apposition of Porous Hydroxylapatite Implants. *Journal of Biomedical Materials Research*, **36**, 560–563.

Ohgushi, H. 1997. Coral Derived Porous Framework Having Different Chemical Compositions as a Scaffold for Osteoblastic Differentiation. Porous Materials for Tissue Engineering.

Okumus, Z. and Yildirim, O. S. 2005. the Cuttlefish Backbone: A New Bone Xenograft Material? *Turkish Journal of Veterinary and Animal Sciences*, **29**, 1177–1184.

Ortiz, C. and Boyce, M. C. 2008. Materials Science – Bioinspired Structural Materials. *Science*, **319**, 1053–1054.

Park, S. J., Lee, K. W., Lim, D. S. and Lee, S. 2012. the Sulfated Polysaccharide Fucoidan Stimulates Osteogenic Differentiation of Human Adipose-Derived Stem Cells. *Stem Cells and Development*, **21**, 2204–2211.

Perez, R. A., Won, J. E., Knowles, J. C. and Kim, H. W. 2013. Naturally and Synthetic Smart Composite Biomaterials for Tissue Regeneration. *Advanced Drug Delivery Reviews*, **65**, 471–496.

Petite, H., Viateau, V., Bensaid, W., Meunier, A., De Pollak, C., Bourguignon, M., Oudina, K., Sedel, L. and Guillemin, G. 2000. Tissue-Engineered Bone Regeneration. *Nature Biotechnology*, **18**, 959–963.

Place, E. S., Evans, N. D. and Stevens, M. M. 2009. Complexity in Biomaterials for Tissue Engineering. *Nature Materials*, **8**, 457–470.

Pollick, S., Shors, E. C., Holmes, R. E. and Kraut, R. A. 1995. Bone-Formation and Implant Degradation of Coralline Porous Ceramics Placed in Bone and Ectopic Sites. *Journal of Oral and Maxillofacial Surgery*, **53**, 915–922.

Pouget, E. M., Bomans, P. H. H., Goos, J. A. C. M., Frederik, P. M., De With, G. and Sommerdijk, N. A. J. M. 2009. The Initial Stages of Template-Controlled Caco3 Formation Revealed By Cryo-Tem. *Science*, **323**, 1455–1458.

Raabe, D., Sachs, C. and Romano, P. 2005. the Crustacean Exoskeleton as an Example of a Structurally and Mechanically Graded Biological Nanocomposite Material. *Acta Materialia*, **53**, 4281–4292.

Rocha, J. H. G., Lemos, A. F., Kannan, S., Agathopoulos, S. and Ferreira, J. M. F. 2005. Hydroxyapatite Scaffolds Hydrothermally Grown from Aragonitic Cuttlefish Bones. *Journal of Materials Chemistry*, **15**, 5007–5011.

Rousseau, M., Boulzaguet, H., Biagianti, J., Duplat, D., Milet, C., Lopez, E. and Bedouet, L. 2008. Low Molecular Weight Molecules of Oyster Nacre Induce Mineralization of the Mc3t3-E1 Cells. *Journal of Biomedical Materials Research Part A*, **85a**, 487–497.

Roy, D. M. and Linnehan, S. K. 1974. Hydroxyapatite Formed from Coral Skeletal Carbonate by Hydrothermal Exchange. *Nature*, **247**, 220–222.

Schmahl, W. W., Griesshaber, E., Merkel, C., Kelm, K., Deuschle, J., Neuser, R. D., Goetz, A. J., Sehrbrock, A. and Mader, W. 2008. Hierarchical Fibre Composite Structure and Micromechanical Properties of Phosphatic and Calcitic Brachiopod Shell Biomaterials – An Overview. *Mineralogical Magazine*, **72**, 541–562.

Shamberger, K. E. F., Feely, R. A., Sabine, C. L., Atkinson, M. J., Decarlo, E. H., Mackenzie, F. T., Drupp, P. S. and Butterfield, D. A. 2011. Calcification and Organic Production on a Hawaiian Coral Reef. *Marine Chemistry*, **127**, 64–75.

Sharma, B. and Elisseeff, J. H. 2004. Engineering Structurally Organized Cartilage and Bone Tissues. *Annals of Biomedical Engineering*, **32**, 148–159.

Shors, E. C. 1999. Coralline Bone Graft Substitutes. *Orthopedic Clinics of North America*, **30**, 599–+.

Silve, C., Lopez, E., Vidal, B., Smith, D. C., Camprasse, S., Camprasse, G. and Couly, G. 1992. Nacre Initiates Biomineralization by Human Osteoblasts Maintained *In Vitro*. *Calcified Tissue International*, **51**, 363–369.

Sivakumar, M., Kumar, T. S. S., Shantha, K. L. and Rao, K. P. 1996. Development of Hydroxyapatite Derived from Indian Coral. *Biomaterials*, **17**, 1709–1714.

Song, E., Kim, S. Y., Chun, T., Byun, H. J. and Lee, Y. M. 2006. Collagen Scaffolds Derived from a Marine Source and Their Biocompatibility. *Biomaterials*, **27**, 2951–2961.

Tambutte, S., Holcomb, M., Ferrier-Pages, C., Reynaud, S., Tambutte, E., Zoccola, D. and Allemand, D. 2011. Coral Biomineralization: from the Gene to the Environment. *Journal of Experimental Marine Biology and Ecology*, **408**, 58–78.

Thein-Han, W. W. and Misra, R. D. K. 2009. Biomimetic Chitosan-Nanohydroxyapatite Composite Scaffolds for Bone Tissue Engineering. *Acta Biomaterialia*, **5**, 1182–1197.

Turhani, D., Watzinger, E., Weissenbock, M., Cvikl, B., Thurnber, D., Wittwer, G., Yerit, K. and Ewers, R. 2005. Analysis of Cell-Seeded 3-Dimensional Bone

Constructs Manufactured *In Vitro* with Hydroxyapatite Granules Obtained from Red Algae. *Journal of Oral and Maxillofacial Surgery*, **63**, 673–681.

Vago, R. 2008. Beyond the Skeleton: Cnidarian Biomaterials as Bioactive Extracellular Microenvironments for Tissue Engineering. *Organogenesis*, **4**, 18–22.

Vallet-Regi, M. and Gonzalez-Calbet, J. M. 2004. Calcium Phosphates as Substitution of Bone Tissues. *Progress in Solid State Chemistry*, **32**, 1–31.

Van Der Stok, J., Van Lieshout, E. M. M., El-Massoudi, Y., Van Kralingen, G. H. and Patka, P. 2011. Bone Substitutes in the Netherlands – A Systematic Literature Review. *Acta Biomaterialia*, **7**, 739–750.

Vecchio, K. S., Zhang, X., Massie, J. B., Wang, M. and Kim, C. W. 2007a. Conversion of Bulk Seashells to Biocompatible Hydroxyapatite for Bone Implants. *Acta Biomaterialia*, **3**, 910–918.

Vecchio, K. S., Zhang, X., Massie, J. B., Wang, M. and Kim, C. W. 2007b. Conversion of Sea Urchin Spines to Mg-Substituted Tricalcium Phosphate for Bone Implants. *Acta Biomaterialia*, **3**, 785–793.

Viateau, V., Guillemin, G., Bousson, V., Oudina, K., Hannouche, D., Sedel, L., Logeart-Avramoglou, D. and Petite, H. 2007. Long-Bone Critical-Size Defects Treated With Tissue-Engineered Grafts: A Study on Sheep. *Journal of Orthopaedic Research*, **25**, 741–749.

Vuola, J., Bohling, T., Kinnunen, J., Hirvensalo, E. and Asko-Seljavaara, S. 2000. Natural Coral as Bone-Defect-Filling Material. *Journal of Biomedical Materials Research*, **51**, 117–122.

Vuola, J., Taurio, R., Goransson, H. and Asko-Seljavaara, S. 1998. Compressive Strength of Calcium Carbonate and Hydroxyapatite Implants after Bone-Marrow-Induced Osteogenesis. *Biomaterials*, **19**, 223–227.

Walsh, P. J., Buchanan, F. J., Dring, M., Maggs, C., Bell, S. and Walker, G. M. 2008. Low-Pressure Synthesis and Characterisation of Hydroxyapatite Derived from Mineralise Red Algae. *Chemical Engineering Journal*, **137**, 173–179.

Walsh, W. R., Chapman-Sheath, P. J., Cain, S., Debes, J., Bruce, W. J. M., Svehla, M. J. and Gillies, R. M. 2003. A Resorbable Porous Ceramic Composite Bone Graft Substitute in a Rabbit Metaphyseal Defect Model. *Journal of Orthopaedic Research*, **21**, 655–661.

Wang, W., Itoh, S., Aizawa, T., Okawa, A., Sakai, K., Ohkuma, T. and Demura, M. 2010. Development of an Injectable Chitosan/Marine Collagen Composite Gel. *Biomedical Materials*, 5.

Weir, M. D. and Xu, H. H. K. 2010. Osteoblastic Induction on Calcium Phosphate Cement-Chitosan Constructs for Bone Tissue Engineering. *Journal of Biomedical Materials Research Part A*, **94a**, 223–233.

White, E. and Shors, E. C. 1989. Coated Biomaterials and Methods for Making Same. US Patent Application.

White, R. A., White, E. W. and Weber, J. N. 1972. Replamineform – New Process for Preparing Porous Ceramic, Metal, and Polymer Prosthetic Materials. *Science*, **176**, 922–and.

Zemani, F., Benisvy, D., Galy-Fauroux, I., Lokajczyk, A., Colliec-Jouault, S., Uzan, G., Fischer, A. M. and Boisson-Vidal, C. 2005. Low-Molecular-Weight Fucoidan Enhances the Proangiogenic Phenotype of Endothelial Progenitor Cells. *Biochemical Pharmacology*, **70**, 1167–1175.

Zhang, F., Chang, J., Lu, J., Lin, K. and Ning, C. 2007. Bioinspired Structure of Bioceramics for Bone Regeneration in Load-Bearing Sites. *Acta Biomaterialia*, **3**, 896–904.

Zhang, Y. Z., Reddy, V. J., Wong, S. Y., Li, X., Su, B., Ramakrishna, S. and Lim, C. T. 2010. Enhanced Biomineralization in Osteoblasts on a Novel Electrospun Biocomposite Nanofibrous Substrate of Hydroxyapatite/Collagen/Chitosan. *Tissue Engineering Part A*, **16**, 1949–1960.

Zhao, F., Yin, Y. J., Lu, W. W., Leong, J. C., Zhang, W. J., Zhang, J. Y., Zhang, M. F. and Yao, K. D. 2002. Preparation and Histological Evaluation of Biomimetic Three-Dimensional Hydroxyapatite/Chitosan-Gelatin Network Composite Scaffolds. *Biomaterials*, **23**, 3227–3234.

Zheng, M. H., Hinterketiser, K., Solomon, K., Kunert, V., Pavlos, N. J. and Xu, J. 2007. Collagen-Derived Biomaterials in Bone and Cartilage Repair. *Macromolecular Symposia*, **253**, 179–185.

Index

CPSIA information can be obtained at www.ICGtesting.com
Printed in the USA
LVOW02*0721150914

404065LV00001B/6/P

9 780857 094971